T0243077

LONDON MATHEMATICAL SOCIETY LECTURE NOTE SERIES

Managing Editor: Professor N.J. Hitchin, Mathematical Institute, University of Oxford, 24–29 St Giles, Oxford OX1 3LB, United Kingdom

The titles below are available from booksellers, or, in case of difficulty, from Cambridge University Press.

London Mathematical Society Lecture Note Series. 291

Tits Buildings and the Model Theory of Groups

Edited by

Katrin Tent
Universität Würzburg

CAMBRIDGE UNIVERSITY PRESS
Cambridge, New York, Melbourne, Madrid, Cape Town, Singapore,
São Paulo, Delhi, Dubai, Tokyo, Mexico City

Cambridge University Press
The Edinburgh Building, Cambridge CB2 8RU, UK

Published in the United States of America by Cambridge University Press, New York

www.cambridge.org
Information on this title: www.cambridge.org/9780521010634

First published 2002

A catalogue record for this publication is available from the British Library

Library of Congress Cataloguing in Publication Data

Tits buildings and the model theory of groups / edited by Katrin Tent.
    p. cm. – (London Mathematical Society lecture note series; 291)
Includes bibliographical references and index.
ISBN 0 521 01063 2
1. Buildings (Group theory). 2. Group theory.
I. Tent, Katrin, 1963– II. Series.
QA174.2.T5 2001
512'.2–dc21   2001043521

ISBN 978-0-521-01063-4 Paperback

Contents

Preface

One off-spring of the happy marriage of algebra and geometry is the close liasion between groups and Tits buildings. Recently, this connection has acquired a logic angle by injecting a dose of model theory. Groups, on the other hand, have always been a central topic in model theory, and so the connection to geometry and Tits buildings is not entirely surprising. The workshop on Tits buildings and the model theory groups held in Würzburg, Germany, in September 2000, brought together for the first time a number of specialists from both sides, incidence geometry and model theory, with the interest to learn from each other. Hence, speakers were encouraged to give introductory talks to their area accdessible also to 'the other side'. The conference started with an introduction to Tits buildings and continued with introduction to special cases of these buildings, generalized polygons, twin buildings and twin trees. To supply examples, one session explained the terminology in the context of the classical groups. These talks were hoped to be helpful for logicians working on the model theory of groups.

On the other hand, many of the geometric concepts used turned out to be in the range of model theory of first order structures, and so, model theoretic constructions have produced useful examples and counterexamples in geometry. A number of talks were thus concerned with different variations of Hrushovski-like constructions. In this volume, we have tried to collect those expositions which have a survey character and might be a useful reference both to model theorists and geometers.

In the context of the Cherlin-Zil'ber Conjecture, ideas from finite group theory as well as geometric ideas come into play, so we were happy to also include a talk by Gernot Stroth, who gave a survey about the state of the classification of the finite simple groups. This classification plays a crucial role in the approach to the conjecture presented in Tuna Altinel's paper.

Unfortunately, due to lack of time, several of the speakers could not contribute their exposition to this volume, among them Gernot Stroth, Elisabeth Bouscaren and Françoise Delon, who had planned to write a survey on groups definable in different classes of fields, and Richard Weiss, who explained the classification of the Moufang polygons (see the article by Van Maldeghem, this volume).

Dugald Macpherson contributed a paper whose results are due to his student Keith Johnson. It presents in some detail a particular application of the Hrushovski construction to Jordan permutation groups.

We are confident that the articles collected here will serve as a useful reference for logicians as well as geometers.

For financial support we are indebted to the *Deutsche Forschungsgemein-schaft* and the State of Bavaria. Without their support, the conference would not have been possible.

Thanks are due to Linus Kramer for his work in coordinating and editing the geometry papers and to Oliver Bletz and Nils Roeshr for their efforts in formatting the files into a proper manuscript. Finally, I would also like to thank Roger Astley from Cambridge University Press for his quick and uncomplicated handling of all questions concerning the publication.

Katrin Tent
Würzburg, June 2001

List of speakers and talks

Tuna Altinel (Lyon)
Simple groups of finite Morley rank and even type

John Baldwin (Chicago)
Model theoretic constructions of algebraic objects

Andreas Baudisch (Berlin)
CM-Trivial stable groups

Elisabeth Bouscaren (Paris)
Groups definable in separably closed fields I

Steven Buechler (Notre Dame)
Stability theory for bilinear forms

Françoise Delon (Paris)
Groups definable in separably closed fields II

Theo Grundhöfer (Würzburg)
Basics on Buildings

Linus Kramer (Würzburg)
Classical groups and theor buildings

Dugald Macpherson (Leeds)
Definable sets in algebraically closed valued fields

Bernhard Mühlherr (Dortmund)
Twin Buildings

François Point (Mons)
Separably closed fields viewed as modules over a non commutative skew polynomial ring

Bruno Poizat (Lyon)
Les amalgames de Hrushovski: tentative de calssement; puissance et limites de la methode

Mark Ronan (Chicago)
Twin Trees

Gernot Stroth (Halle)
The classification of the finite sample groups – a characteristic p approach

Katrin Tent (Würzburg)
BN-pairs and the Cherlin-Zil'ber Conjecture

Hendrik Van Maldeghem (Gent)
Introduction the generalized polygons

Richard Weiss (Tufts)
The classification of Moufang Polygons

Martin Ziegler (Freiburg)
G-compact theories

Basics on buildings

Theo Grundhöfer (Würzburg)

This is an introductory survey on buildings for non-specialists. Buildings are natural (and far-reaching) generalizations of projective spaces, so we first look at projective planes and projective spaces from this point of view. We give a general definition of buildings as sets with a kind of distance function which takes its values in a Coxeter group. Then we state the fundamental and definitive classification results obtained by Jacques Tits. The Moufang condition is a strong geometric homogeneity condition for buildings, which arises in connection with Tits' classification. We also consider BN-pairs in groups, which correspond to a weaker homogeneity condition for buildings. In an appendix, we indicate various other possibilities to define buildings.

Historically, buildings have been introduced by Tits in order to give geometric interpretations of all simple Lie groups, including the exceptional types. It turned out that buildings are relevant also for the classification of simple algebraic groups and in other areas of mathematics; here we just mention Mostow's rigidity theorem [29], the classification of isoparametric submanifolds by Thorbergsson [47], and the proof of a conjecture of Margulis (about the asymptotic geometry of symmetric spaces of non-compact type) by Kleiner-Leeb [26]; see also [4, 9, 13, 21, 28, 35, 37, 42, 52]. For applications of buildings in model theory see Tent [46].

1 Projective planes and projective spaces

Incidence structures are triples $(P, \mathcal{L}, \Delta)$ of sets with $\Delta \subseteq P \times \mathcal{L}$; the elements of P are called points, the elements of $\mathcal{L}$ are called lines, and the relation Δ describes the incidence: $(p, L) \in \Delta$ means that the point p is on the line L (and then (p, L) is called a flag). Such an incidence structure $(P, \mathcal{L}, \Delta)$ is a *projective plane* if the usual axioms hold:

(i) any two distinct points are on a unique line;

(ii) dually, any two distinct lines have a unique point in common;

(iii) every point is on at least three lines, and every line has at least three points.

In any projective plane $(P, \mathcal{L}, \Delta)$, there are precisely six possibilities for the relative (combinatorial) position of two flags $(p, L), (p', L') \in \Delta$, and we want to assign distances to these six possibilities. That is, we define a distance function δ on Δ^2 (with values in some group to be revealed a little later), as follows: if $p = p'$ and $L = L'$, then $\delta((p, L), (p, L)) = 1$, as we write our group multiplicatively. If $p \neq p'$ and $L = L'$, then $\delta((p, L), (p', L')) = s_1$ (for situation number 1), and dually, for $p = p'$ and $L \neq L'$ we define $\delta((p, L), (p', L')) = s_2$. For the remaining three cases, we have $p \neq p'$ and $L \neq L'$, and we define $\delta((p, L), (p', L')) = s_1 s_2$ if p' is on L, furthermore $\delta((p, L), (p', L')) = s_2 s_1$ if p is on L', and finally $\delta((p, L), (p', L')) = s_1 s_2 s_1$ for two flags in general position (which means that p is not on L' and p' is not on L).

We entertain the idea that products like $s_1 s_2$ or $s_1 s_2 s_1$ should correspond to sequences of flags where each flag has distance s_i from its successor. Sequences of flags like $(p, L), (p', L), (p', L')$ for p' on L are examples of galleries, as defined in Section 3. Vaguely speaking, distance s_i means that component number i of the flag has changed. This motivates the relations $s_i^2 = 1$, and also $s_1 s_2 s_1 = s_2 s_1 s_2$, hence $(s_1 s_2)^3 = 1$. Thus we arrive at the non-abelian group $\langle s_1, s_2 \mid s_1^2 = s_2^2 = (s_1 s_2)^3 = 1 \rangle \cong S_3 \cong \mathrm{Dih}_6$, the symmetric group of degree 3, which is also the dihedral group of order 6.

Let us now consider projective spaces of finite dimension n (we define projective spaces axiomatically, by replacing axiom (ii) with the Veblen-Young axiom, see e.g. [27] 1.5 or [41] 4.1.3 or [16] 2.1). Since we have already covered projective planes, we assume that $n \geq 3$. By the fundamental theorem of projective geometry, such a projective space is always the lattice $PG(n, K)$ of all subspaces of the vector space K^{n+1} over some skew field K (the existence of non-desarguesian projective planes means that this is not true for $n = 2$). We use the vector space structure to define (in the lemma below) a distance function δ on the set Δ of all maximal flags of $PG(n, K)$; a maximal flag is a chain of subspaces of maximal lenght, i.e.

$$\Delta = \{(U_0, U_1, \dots, U_{n+1}) \mid U_i < U_{i+1}, \dim U_i = i \text{ for all } i\}.$$

Note that each ordered basis of K^{n+1} yields an element of Δ: just define U_i to be the subspace generated by the first i basis vectors.

Lemma. *For any two maximal flags $U = (U_0, \dots, U_{n+1})$ and $V = (V_0, \dots, V_{n+1})$ of $PG(n, K)$, there exists a basis $b_1, \dots, b_{n+1}$ of K^{n+1} and a permuta-*

tion $\pi \in S_{n+1}$ such that

$$
\begin{aligned}
U_i &= \langle b_1, \dots, b_i \rangle \text{ and} \\
V_i &= \langle b_{\pi(1)}, \dots, b_{\pi(i)} \rangle \text{ for } 0 \le i \le n+1.
\end{aligned}
$$

Furthermore, π does not depend on the basis, hence the "distance map" δ : $\Delta^2 \to S_{n+1} : (U, V) \mapsto \pi$ is well-defined.

The proof of this lemma is an elementary exercise in linear algebra. One can proceed by induction on n to show the existence of a basis B such that suitable subsets of B generate the subspaces U_i, V_i (consider the subspaces $V_i \cap U_n$ for $0 \le i \le n+1$). This also gives the existence of π. The fact that π does not depend on the basis follows from the equation

$$
\pi(i) = \min\{j \mid V_i \subseteq V_{i-1} + U_j\},
$$

which describes π as a kind of Jordan-Hölder permutation (the maximal flags are the composition series of the vector space K^{n+1}; compare Brown [9] p. 84, Abels [1, 2] for more about this point of view).

We remark that the symmetric group S_{n+1} is generated by the n special transpositions $s_i = (i, i+1)$ with $1 \le i \le n$, and that

$$
\delta(U, V) = s_i \iff U_i \ne V_i \text{ and } U_j = V_j \text{ for all } j \ne i,
$$

that is, if U and V differ only at position i. This fits for $n = 2$ with our definition of distance between flags of projective planes.

We say that the pair (Δ, δ) as above is the *flag system* of the corresponding projective space. It is a feature of the theory of buildings that flags are the basic objects, whereas points, lines and other subspaces appear as derived objects (as special residues, see Section 7); in fact, the flag systems in this section are buildings of type A_n (as defined in Section 3).

2 Coxeter groups

In order to generalize projective spaces (to buildings), we first have to generalize the symmetric groups (to Coxeter groups).

A *Coxeter group* (of finite rank n) is a group W with a presentation by generators and relations of the form

$$
W = \langle s_1, \dots, s_n \mid (s_i s_j)^{m_{ij}} = 1 \text{ for all } i, j \rangle,
$$

where $m_{ij} \in \mathbb{N} \cup \{\infty\}, m_{ii} = 1$ (hence $s_i^2 = 1$), and $m_{ij} = m_{ji} \ge 2$ for $i \ne j$; if $m_{ij} = \infty$, then we omit the relation $(s_i s_j)^{m_{ij}} = 1$. Each Coxeter group has, by definition, a distinguished set $S = \{s_1, \dots, s_n\}$ of generators. The pair (W, S) is called a *Coxeter system*. One can show that the order of $s_i s_j$ is precisely m_{ij} (and not some divisor of m_{ij}), hence the $n \times n$-matrix $(m_{ij})_{i,j}$ is determined by (W, S).

The matrix $(m_{ij})_{i,j}$ is described very concisely by the corresponding *(Coxeter) diagram*. This is the graph with vertices $1, 2, \dots, n$ which has $m_{ij} - 2$ bonds between i and j. So we draw one edge o——o if $m_{ij} = 3$, two edges o==o if $m_{ij} = 4$, and for $m_{ij} = m > 4$ we draw o——$\overset{m}{}$——o .

As examples, we mention that o is the Coxeter diagram for the Coxeter group $W \cong S_2$ of rank 1, and that o——o describes the Coxeter group $W \cong S_3 \cong \mathrm{Dih}_6$. The diagram o——o can be generalized in two ways: o——$\overset{m}{}$——o belongs to the group $W = \langle s_1, s_2 \mid s_1^2 = s_2^2 = (s_1 s_2)^m = 1 \rangle$, which is the dihedral group Dih_{2m} of order $2m$, and the diagram o——o——o----o——o——o with n vertices gives the Coxeter group $W = S_{n+1}$; here one has to prove that S_{n+1} has the presentation

$$\langle s_1, \dots, s_n \mid s_i^2 = 1 = (s_i s_{i+1})^3, s_i s_j = s_j s_i \text{ for } |i - j| > 1 \rangle,$$

see Robinson [34] 2.2.2.

These examples illustrate a main point of Section 2: Coxeter groups are a simultaneous generalization of symmetric groups and dihedral groups.

It is easy to see that the decomposition of the diagram into its connected components corresponds to a decomposition of the Coxeter system (W, S) as a direct product. If the diagram is connected, then (W, S) is said to be *irreducible*.

In general, neither S nor $|S|$ is determined by the isomorphism type of W; for counterexamples consider the diagrams o——$\overset{2m}{}$——o and o——$\overset{m}{}$——o o for odd m, and note that the dihedral group Dih_{4m} is isomorphic to the direct product of Dih_{2m} with the cyclic group of order 2; for irreducible counterexamples see Mühlherr [30], Brady et al. [8].

A Coxeter system (W, S) is called *spherical* if W is a finite group. Coxeter [18] has determined all spherical Coxeter systems; they are just direct products of the irreducible spherical Coxeter systems shown in Table 1.

In [17] and in [19] Chapter XI, Coxeter considers discrete groups W of isometries of a Euclidean affine space $\mathbb{R}^m$ such that W is generated by a finite set S of reflections. If W fixes a point of $\mathbb{R}^m$, say 0, then W is a finite group which acts on the unit sphere of $\mathbb{R}^m$, and this leads to the spherical Coxeter systems (W, S). Otherwise, i.e. if such a discrete group W is infinite, the Coxeter system (W, S) is said to be *affine* (or Euclidean).

One can prove that an irreducible Coxeter system (W, S) is affine if, and only if, the group W is infinite and has an abelian (normal) subgroup of finite

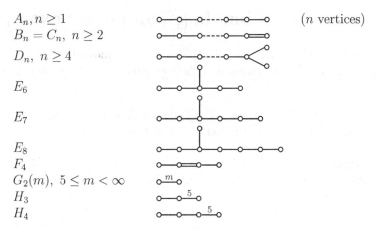

$A_n, n \geq 1$ (n vertices)

$B_n = C_n, \ n \geq 2$

$D_n, \ n \geq 4$

E_6

E_7

E_8

F_4

$G_2(m), \ 5 \leq m < \infty$

H_3

H_4

Table 1: Diagrams for irreducible spherical Coxeter systems

index. The simplest example of an affine Coxeter group is provided by the infinite dihedral group $\mathrm{Dih}_\infty = \langle x \mapsto \pm x + a \mid a \in \mathbb{Z} \rangle$, which is a semidirect product of $\mathbb{Z}$ with a cyclic group of order 2; this is the Coxeter group of type $\widetilde{A}_1$ ∘—∞—∘ . Every irreducible affine Coxeter group W of rank n is a semidirect product of a free abelian group $\mathbb{Z}^{n-1}$ with a finite Coxeter group W_0, which is generated by $n-1$ reflections and has the "crystallographic" property that it leaves a lattice of $\mathbb{R}^{n-1}$ invariant. These groups W_0 are called Weyl groups (the types H_3, H_4 and $G_2(m)$ with $m \neq 6$ from the list above are not Weyl groups). Coxeter's list of all irreducible affine Coxeter systems ([17]) can also be found in Bourbaki [7] VI, 4.3, p. 199, or in [36] Chapter 9. Here we just mention that for $n > 1$, the affine Coxeter diagram of type $\widetilde{A}_n$ is a circuit with $n + 1$ vertices.

There are also hyperbolic Coxeter groups, which arise as discrete groups generated by reflections in hyperbolic spaces, compare Humphreys [24] 6.8, 6.9 and the references given there.

For more information on Coxeter groups see the surveys by Cohen [15], de la Harpe [20], Taylor [44] Chapter 9, Scharlau [41] Section 2, and the books by Bourbaki [7], Humphreys [24], Hiller [23], Brown [9], Grove-Benson [22], Björner-Brenti [5] and Coxeter [19].

3   Definition of buildings, and a few examples

Following Tits [56], we introduce buildings as sets endowed with a "distance function" which takes its values in a Coxeter group. Let (W, S) be a Coxeter system.

Definition. *A* building *of type* (W, S) *is a set* Δ *with a "distance function"* $\delta : \Delta^2 \to W$ *such that the following two axioms hold:*

(i) *For each* $s \in S$, *the relation* $x \sim_s y : \iff \delta(x, y) \in \{1, s\}$ *is an equivalence relation on* Δ, *and each equivalence class of* $\sim_s$ *has at least two elements.*

(ii) *Let* $w = r_1 r_2 \ldots r_k$ *be a shortest representation of* $w \in W$ *(with* $r_i \in S$*), and let* $x, y \in \Delta$. *Then* $\delta(x, y) = w$ *if, and only if, there exists a sequence* $x_0 = x, x_1, x_2, \ldots, x_k = y$ *in* Δ *with* $x_{i-1} \neq x_i$ *and* $\delta(x_{i-1}, x_i) = r_i$ *for* $1 \leq i \leq k$.

The rank *of the building* Δ, *and of the Coxeter system* (W, S), *is the integer* $|S|$.

We offer some remarks, and more terminology. Δ is called *thick*, if all the equivalence classes in axiom (i) have at least three elements, and *thin*, if each of these equivalence classes has precisely two elements. A sequence $(x_0, \ldots, x_r)$ as in axiom (ii) is called a *gallery* of type $(r_1, r_2, \ldots, r_k)$ from x_0 to x_r. Axiom (ii) says that galleries in the building Δ are closely controlled by the shortest representations (hence by the relations) in the Coxeter group W. In fact, one can show that a gallery as in (ii) is uniquely determined by its extremities x, y and its type $(r_1, r_2, \ldots, r_k)$, see [54] 3.3, 3.7, [36] 3.1(v).

For variations and other possibilities to define buildings (e.g. as chamber systems or as simplicial complexes) see Section 7.

It is convenient to indicate the type (W, S) of a building Δ by the corresponding diagram. We say that Δ is irreducible, or spherical, or affine, if this holds for (W, S). In fact, it often suffices to consider only irreducible buildings, because each building is the direct product of irreducible buildings (see e.g. [36] 3.10).

Examples. (a) A building of rank 1 is just a set Δ containing at least two elements (since $W = \{1, s\}$, and δ is the obvious map, defined by $\delta(x, y) = 1 \iff x = y$).

(b) The definitions $\Delta := W$, $\delta(x, y) := x^{-1} y$ give a thin building Δ of type (W, S). Up to isomorphisms (i.e. isometries), this is the only thin building of type (W, S).

(c) The thick buildings of type A_n ○—○—○----○—○—○ are precisely the buildings obtained from projective spaces of dimension n, as described in Section 1; this is an early result of Tits, compare [49] 4.1, [50] p. 80, Ex. b), [51] 6.3, [54] 6.1.5, [41] 4.1.4. In particular, the thick buildings of type ○—○ are precisely the flag systems of projective planes, as described in Section 1. Therefore, projective geometry is contained as a special case in the theory of buildings.

(d) A thick building of type $G_2(m)$ ○—m—○ is the same thing as the (flag system of) a generalized m-gon, compare the contribution by Van Maldeghem [64] or his book [63]; here $3 \leq m < \infty$.

(e) A building of type $\widetilde{A}_1$ ○—∞—○ is the same thing as a tree without end points: let Δ consist of all edges of the tree, and use the fact that a tree is bipartite to define two equivalence classes on Δ (as in axiom (i) above). To phrase it differently: these trees, being bipartite, can be considered as incidence structures, and the corresponding flag systems (cp. Section 1) are the buildings of type $\widetilde{A}_1$.

(f) A building of type ○   ○ is the same as (the flag system of) a complete bipartite graph.

(g) By results of Tits [53] and Ronan-Tits [39], thick buildings belonging to a given diagram exist if, and only if, that diagram has no subdiagram H_3 ○—○—5—○ .

A building Δ of type (W, S) has plenty of sub-buildings: for every subset T of S and every $a \in \Delta$, the set $\{x \in \Delta \mid \delta(x, a) \in \langle T \rangle\}$ is called the T-*residue* of a, and this is a building of type $(\langle T \rangle, T)$; one can show that $(\langle T \rangle, T)$ is indeed a Coxeter system, its type is described by the subdiagram induced on T. The T-residues are precisely the equivalence classes of the T-*adjacency* $\sim_T$, which is the equivalence relation on Δ defined by $x \sim_T y \iff \delta(x, y) \in \langle T \rangle$. As examples, we mention that each subspace and each quotient space of a projective space Δ is a residue of Δ. These sub-buildings sometimes allow inductive proofs for properties of buildings. In fact, one can construct buildings of higher rank by some kind of amalgamation of buildings of rank 2, compare [48], [36] Chapter 7, [39], [41] 7.2.

More examples of buildings are described in Section 4. The examples above already show that buildings are a simultaneous generalization of projective spaces and generalized polygons.

For more information on buildings see Ronan [36], Brown [9, 10] and Scharlau [41]. For buildings of infinite rank compare Shult [43], and for twin buildings see Tits [57, 58], or [31] and [38] in this volume. Buildings are a special type of diagram geometries, see e.g. Taylor [44] Chapter 9 or the books by Pasini [32] and Buekenhout-Cohen [11].

4 Classification results

The concept of buildings acquires importance and power through the following classification results due to Jacques Tits. We state only vague versions of these classification results, and we describe more examples in the pertaining comments.

Theorem (Tits [51]). *All thick irreducible spherical buildings of rank at least 3 are "known".*

Here "known" means that these buildings can be described explicitly in algebraic terms. For example, the fundamental theorem of projective geometry says that each projective space of rank $n \geq 3$ is isomorphic to the lattice $PG(n, K)$ of subspaces of K^{n+1}, for some skew field K. Since the thick buildings of type A_n are just the flag systems of these projective spaces, we can consider the theorem above as a far-reaching generalization of the fundamental theorem of projective geometry.

Similarly, buildings of type C_n are flag systems of *polar spaces* (see Buekenhout-Shult [12], [44] p. 108, [16] Chapter 3, [41] 4.2.8, 4.2.5 or [27] 2.1 for an axiomatic definition of polar spaces as incidence structures). Let f be a non-degenerate form (quadratic or pseudo-quadratic or hermitian or symplectic) of Witt index n on some vector space (cp. [27] for more details). Then the set

$$\Delta = \{(U_0, U_1, \ldots, U_n) \mid U_i < U_{i+1}, f(U_i, U_i) = \{0\}\}$$

of all maximal totally isotropic flags (with respect to f) is a building of type C_n. This building Δ is thick unless f is a symmetric bilinear (or quadratic) form on a vector space of dimension $2n$ over a field of characteristic not 2; in that special case, a variation of the flag system above (the oriflamme system) leads to a thick building of type D_n, compare [51] Chapters 7 and 8, [41] 4.2 or [9] V.7. The classification result of Tits says that all thick buildings of type C_n or D_n with $n \geq 4$ arise in this fashion from a form f of Witt index n on some vector space (hence these buildings are embedded into the corresponding projective space; note that the vector space can have infinite dimension). For type C_3, there is one further class of examples, which can be described in terms of non-desarguesian Moufang planes, see Tits [51] 9.1. Type C_3 is the most difficult case in the proof of the above theorem. Compare

[41] 7.3 and [36] Chapter 8 for more detailed comments on the classification of spherical buildings.

We point out that the theorem above is false for rank 2, due to the existence of non-classical projective planes (and generalized polygons). The assumption that the building is irreducible of rank at least 3 can be replaced by the requirement that the building has no factor of rank 1 or 2.

Non-thick buildings can be described in terms of thick buildings (of possibly smaller rank), see Scharlau [40]; however, a full classification of all non-thick irreducible spherical buildings of rank at least 3 is impossible, because the associated thick buildings can have (non-classical) factors of rank 2. For non-thick generalized polygons see also [63] 1.6.

The analogous classification result for affine buildings reads

Theorem (Tits [55]). *All thick irreducible affine buildings of rank at least 4 are "known".*

Here "known" means that these buildings can be described by vector spaces over skew fields with a discrete valuation, compare Ronan [36] Chapters 9 and 10; for type $\widetilde{A}_n$ see also Brown [9] V.8.

The irreducible affine Coxeter systems of rank 3 are

$$\widetilde{A}_2 \quad \text{and} \quad \widetilde{B}_2 \quad \text{and} \quad \widetilde{G}_2 \quad .$$

The affine buildings of these types cannot be classified explicitly; descriptions of the types $\widetilde{A}_2$ and $\widetilde{B}_2$ have been given by Van Maldeghem, see [60, 61, 62] and [63] 9.7.

In [55], Tits introduces and classifies certain generalizations of affine buildings. These generalizations can be considered as metric spaces with "building-like" properties; as one-dimensional examples, we mention the $\mathbb{R}$-trees. Compare also Ronan [36] Appendix 3 on non-discrete buildings, Kleiner-Leeb [26], and Brown [9] VI.3 for metric properties of affine buildings.

5 Apartments, roots, and the Moufang condition

The Moufang condition is a very strong homogeneity condition for buildings; it requires the existence of many automorphisms of special type (with many fixed elements). Of course, by an *automorphism* of a building (Δ, δ) we mean an isometry with respect to δ, and we write Aut Δ for the group of all automorphisms of Δ.

We now need the concept of apartments in buildings. An *apartment* in a building Δ of type (W, S) is just an isometric copy in Δ of the thin building W. One can show that there are many apartments: any isometry of a subset of W into Δ extends to an isometry of W into Δ (see Tits [54] 3.7.4, [9] p. 90 or [36] 3.6). Hence any two elements of Δ are contained in an apartment of Δ, and more generally, any shortest gallery as in the Definition in Section 3 is contained in an apartment (cp. [36] 3.8). Due to this plenitude of apartments, many properties of buildings can be decided within an apartment, i.e. in the corresponding Coxeter group W.

For example, an apartment in a projective plane, considered as a building, is just a triangle (or rather, the flag system consisting of the 6 flags involved in a triangle, to be formally more correct). More generally, the apartments in a generalized m-gon Δ are the ordinary m-gons, and the apartments of a projective space $PG(n, K)$ are precisely the sets of $(n + 1)!$ maximal flags which can be obtained from a fixed unordered basis of K^{n+1}, as described in the lemma in Section 1.

We also need the notion of a root (i.e. a half-apartment), which can be defined by a folding, or via the geometric representation of a Coxeter group on a real vector space, using half-spaces, compare [51] 1.12, [9] p. 48, 55 and 66ff, [36] Section 2.2. Here we give a definition in terms of the Coxeter system (W, S) (cp. [7] p. 18). For each $s \in S$, let

$$P_s = \{w \in W \mid \text{ no shortest representation of } w \text{ starts with } s\}.$$

Then $W = P_s \cup sP_s$ is a partition of the thin building W (into the two "opposite roots" P_s and sP_s; for Weyl groups, the isometric images $aP_s, a \in W$ of these sets P_s do correspond to the roots of a root system). A *root* α in an arbitrary building Δ of type (W, S) can be defined as an isometric image α of P_s in Δ, for some $s \in S$. Each root of Δ is contained in an apartment of Δ, by the extension property mentioned above. The set

$$\overline{\alpha} := \{x \in \Delta \mid \text{there exist } y, z \in \alpha \text{ with } \delta(x, y) = \delta(y, z) \in S\}$$

is the union of all residues of rank 1 in Δ which contain two elements of α; one has $\alpha \subseteq \overline{\alpha}$, unless $|S| = 1$. (In the terminology of simplicial complexes, see Section 7, these residues are the panels of Δ containing a wall of α which does not belong to the boundary of α.)

Let (W, S) be a spherical Coxeter system without factors of type A_1 (i.e. without isolated vertices in its diagram), let Δ be a building of type (W, S), and let α be a root of Δ. Then the stabilizer

$$U_\alpha = \{g \in \operatorname{Aut} \Delta \mid g(x) = x \text{ for each } x \in \overline{\alpha}\}$$

is called a *root group* of Δ. We say that Δ satisfies the *Moufang condition* (or: Δ is a Moufang building), if U_α acts transitively on the set $\{A \mid A$ is an apartment in Δ with $\alpha \subset A\}$, for each root α of Δ. This action of U_α is then sharply transitive (because we have excluded factors of type A_1). Actually, it suffices to require the transitivity of U_α only for the roots α in a fixed apartment.

For a root α of a projective plane, one has $\overline{\alpha} = \{(x, L) \mid x$ is a point on $L\} \cup \{(p, X) \mid X$ is a line through $p\}$ for some flag (p, L), and U_α is then the group of all elations with center p and axis L. Thus the Moufang condition says in this special case that the projective plane admits all conceivable elations, and this is equivalent to the usual definition of a Moufang plane (see Pickert [33]). For the Moufang condition in generalized polygons compare the contribution by Van Maldeghem [64] Section 5, or [63], and for concrete descriptions of root groups in some buildings of higher rank see Kramer [27].

The importance of the Moufang condition was pointed out already in the Addenda (p. 274) of Tits [51]:

Theorem (Tits). *Each thick irreducible spherical building of rank at least 3 satisfies the Moufang condition.*

This result does not require the full classification carried out in [51], in fact it can be derived from Tits' strong extension theorem [51] 4.16 (compare [51] p. 274, [53] 3.5, [36] 6.6, 6.7 or [41] 5.3.5). By the fundamental theorem of projective geometry, each projective space of dimension at least 3 admits many automorphisms (they are induced by linear or semi-linear maps) and does indeed satisfy the Moufang condition (compare [36] p. 67). The theorem above extends this observation from projective spaces to spherical buildings.

That theorem is false for buildings of rank 2, there are many generalized polygons (projective planes and others) which violate the Moufang condition. It is false also for non-thick irreducible spherical buildings, compare the comments on non-thick buildings in Section 4. However, Tits and Weiss have recently achieved a complete classification of all generalized polygons which satisfy the Moufang condition, see [59] or the information provided in [63] Chapter 5 and Appendix C.

For extensions and variations of the Moufang condition compare Ronan [36] p. 73f, Van Steen [66], Van Maldeghem-Van Steen [65].

6 BN-pairs

A BN-pair in a group G is a pair B, N of subgroups such that axioms (i) - (iv) below are satisfied; by the proposition below, these axioms say that G acts strongly transitively on a building, in the sense of the following definition.

Definition. Let (Δ, δ) be a building of type (W, S), and let G be a group which acts on Δ as a group of automorphisms (i.e. isometries) of (Δ, δ). We say that G is *strongly transitive*, if the following two conditions hold:

(1) For each $w \in W$, the group G acts transitively on $\delta^{-1}(w) = \{(x, y) \in \Delta^2 \mid \delta(x, y) = w\}$.

(2) For some apartment A, the set-wise stabilizer $G_{\{A\}}$ acts transitively on A.

Condition (2) means that $G_{\{A\}}$ induces on A precisely the Coxeter group W. If Δ is spherical, then this definition can be simplified, compare [36] p. 56; in fact, for spherical buildings the transitivity of G on $\delta^{-1}(w)$ where w is the unique longest element of W (i.e. the transitivity of G on "opposite" pairs) implies conditions (1) and (2) above.

With the notation of the definition, let $B = G_a$ for some $a \in A$, and $N = G_{\{A\}}$. Then B and N enjoy the following properties (see [36] 5.2, [15] 7.13, [41] 4.3.11):

(i) $G = \langle B, N \rangle$

(ii) $H := B \cap N$ is a normal subgroup of N, and the factor group N/H can be identified with W

(iii) $BsBwB \subseteq BwB \cup BswB$ for $s \in S, w \in W$

(iv) $sBs \neq B$ for $s \in S$.

Note that the double cosets like BwB appearing in (iii) are well-defined, because $w \in W = N/H$ is a coset of H, and $H \leq B$.

Properties (i) - (iv) can be taken as defining axioms for a BN-*pair* in an (a priori) arbitrary group G; we only have to modify (ii), as follows

(ii') $H := B \cap N$ is a normal subgroup of N, and the factor group $W := N/H$ is a Coxeter group with distinguished generating set S,

in order to introduce the Coxeter system. Such a quadruple (G, B, N, S) is also called a *Tits system*. In fact, the assumption that W is a Coxeter group can be omitted, and one can derive a stronger version of axiom (iii), see Brown [9] V.2, Theorem. (For a geometric interpretation of axiom (iii) see Carter [14] p. 301 or Taylor [44] p. 86.)

These axioms lead to the *Bruhat decomposition*, which says that G is the disjoint union

$$G = \bigcup_{w \in W} BwB,$$

compare [36] 5.1. For the general linear group $G = \mathrm{GL}_n F$ over a field F, acting on the corresponding projective space, the stabilizer B of a suitable maximal flag consists of all upper triangular matrices in G; the Bruhat decomposition says in this case that every invertible matrix can be written in the form $b\pi b'$, where π is a permutation matrix (which is uniquely determined) and b, b' are upper triangular matrices.

The Bruhat decomposition allows to go from a BN-pair to a building, as follows (cp. [36] 5.3, [41] 4.3.10):

Proposition. *Let (G, B, N, S) be a Tits system, as defined above. Define $\Delta = G/B = \{gB \mid g \in G\}$ and $\delta : \Delta^2 \to W$ by*

$$\delta(gB, hB) = w \iff g^{-1}h \in BwB, \text{ for } g, h \in G.$$

Then (Δ, δ) is a thick building of type (W, S), and G is strongly transitive on Δ.

Hence, a BN-pair (in a group G) is the same thing as a building with a strongly transitive action (of G).

The cosets gB are the elements of the building Δ defined in the proposition above. We can replace gB by its stabilizer gBg^{-1} in G, thus identifying Δ with the conjugacy class of B. The conjugates of subgroups between B and G are usually called *parabolic subgroups*; they are the stabilizers of (arbitrary, non-maximal) flags of Δ, considered as an incidence geometry (compare Section 7 and [36] 5.4).

Simple (or semi-simple) algebraic groups and Lie groups contain natural BN-pairs. In order to be a little bit more explicit, let G be a (semi-) simple algebraic group defined over a field k, and let T be a maximal k-split torus in G. We obtain a BN-pair (with a finite Coxeter group which is in fact a Weyl group) in the group $G(k)$ of all k-rational points, if we take for B a minimal k-parabolic subgroup with $B \supset T$, and for N the normalizer of T. If k is algebraically closed, then B is a Borel subgroup, i.e. a maximal connected solvable subgroup. We mention that such a BN-pair can be trivial, in the sense that $B = N = G$ (then the associated building is just a singleton and has rank 0); this happens precisely if G is anisotropic over k, i.e. if $T = \{1\}$. The spherical buildings associated with these BN-pairs can be used for the classification of all (semi-) simple algebraic groups up to their "anisotropic parts"; these anisotropic parts depend strongly on the field k. For more details compare Scharlau [41] 4.4 and the references given there, in particular Borel [6], or the brief account in Brown [9] p. 203f. For Chevalley groups see also Carter [14] 8.2.1; in fact, for a Chevalley group G over a finite field of characteristic p, one can take for B the normalizer of a Sylow p-subgroup of G ([14] 8.6).

Similarly, every (semi-) simple real Lie group G has a natural BN-pair; here one takes for B a maximal element among all closed subgroups which are extensions of a compact group by a connected solvable group, compare Warner [67] 1.2.3. The corresponding building has rank 0 if, and only if, the connected component G^1 of G is compact. See also Borel [6] §24C and Mitchell [28].

The Moufang condition for a building Δ implies that $G = \operatorname{Aut}\Delta$ is strongly transitive, hence G contains a BN-pair. See Tits [51] 13.36 for the additional conditions which are satisfied by these special BN-pairs. In fact, strong transitivity is a homogeneity condition for buildings which is much weaker than the Moufang condition; this can be seen from the constructions of Tits [53], Kegel-Schleiermacher [25] or Tent [46, 45]. The classification of all BN-pairs in groups of a given category is usually a difficult problem, or impossible.

Special cases: rank 1 or 2. A building Δ of rank 1 is just a set containing at least two elements, and a group G is strongly transitive on Δ precisely if G acts doubly transitively on the set Δ. A group G is strongly transitive on a generalized n-gon Γ if, and only if, G acts transitively on the set of all ordered ordinary n-gons of Γ. In particular, for projective planes the strong transitivity is just the transitivity on ordered triangles.

For more examples of BN-pairs see Kramer [27] in this volume, or Brown [9], Taylor [44], Borel [6] §23C.

7 Appendix

In this appendix we mention several other possibilities to define buildings, and we explain some connections between these different approaches to buildings.

In Section 3, we have defined a building of type (W, S) as a set Δ endowed with a map $\delta : \Delta^2 \to W$ satisfying two axioms. According to Tits [57] (cp. also [10]) one can replace these two axioms by the following requirements (where $x, y, z \in \Delta, w \in W, s \in S$)

$$\delta(x, y) = 1 \iff x = y;$$

If $\delta(x,y) = w$ and $\delta(y,z) = s$, then $\delta(x,z) \in \{w, ws\}$, and if no shortest representation of w ends with s, then in fact $\delta(x,z) = ws$;

Given $x, y \in \Delta$ and $s \in S$, there exists $z \in \Delta$ such that $\delta(x,z) = \delta(x,y)s$.

These requirements for δ are closer to the axioms for BN-pairs (Section 6), and they lend themselves to the definition of twin buildings (see Tits [57, 58] or [31]). Abels [3] considers an arbitrary group W with a generating set S, and he infers from the requirements above that (W, S) is a Coxeter system, provided that (Δ, δ) is thick (in the sense of Section 3).

Axiom (ii) from Section 3 allows to define the distance function δ in terms of the equivalence relations $\sim_s$ with $s \in S$. This means that we can eliminate δ, which leads to another definition of buildings as sets Δ with equivalence relations $\sim_s$, where $s \in S$. A *chamber system* is just a set (whose elements are called chambers) endowed with a collection of equivalence relations. Hence one can consider buildings as special chamber systems, as in Tits [54, 56], Ronan [36], Scharlau [41] 1.2, 3.3. In this spirit, Cohen [15] 7.4 gives the following definition: a building is a connected chamber system such that every simple closed path with minimal type is trivial (i.e. consists of a single chamber). For example, the set Δ of all maximal flags of a projective space (as in Section 1) is a chamber system with the equivalence relations which express that two elements of Δ differ at most at some fixed position i.

Conversely, if a building is given as a chamber system, then one can introduce galleries (and their types) as in Section 3, and the axioms for that chamber system say that axiom (ii) from Section 3 gives a well-defined mapping $\delta : \Delta^2 \to W$.

In order to define buildings as special chamber systems, one can also employ the "gate-property", which is equivalent to the existence of certain projection maps on the chamber system, see Scharlau [41] 5.1 and his notes on p. 589, Shult [43].

A *simplicial complex* with vertex set V is a collection X of finite subsets of V (called simplices) such that every singleton $\{v\}$ with $v \in V$ is a simplex, and every subset of a simplex A is a simplex (called a face of A); X is partially ordered by inclusion. Often a building is defined as a simplicial complex X which is endowed with a large family of subcomplexes (called apartments) which are Coxeter complexes (to be described below), compare Tits [51], [9], [41] 3.2.1. A maximal simplex is now called a chamber, and a panel is defined to be a second maximal simplex. Let Δ be the set of all chambers of X. The axioms for the building X include the requirement that any two chambers are contained in an apartment. This allows to introduce a well-defined distance function $\delta : \Delta^2 \to W$ (and equivalence relations $\sim_s$) on Δ, so we arrive at a building in the sense of Section 3 (or at a chamber system).

The earliest definitions introduced buildings as incidence geometries, with somewhat more complicated axioms, compare Tits [48, 49, 50]. Here we just indicate the general framework of incidence geometry, and we suppress all axioms. A *geometry* $\Gamma = (V_1, V_2, ..., V_n, *)$ *of rank* n consists of n sets $V_1, ..., V_n$ which are mutually disjoint and not empty, and of a symmetric relation $*$ on $V := \bigcup_{i=1}^n V_i$ with the property that $*$ restricted to any V_i is the identity of V_i (i.e. $\forall x, y \in V_i : x * y \iff x = y$). A flag of Γ is defined as a subset $F \subseteq V$ with $x * y$ for all $x, y \in F$.

Examples can be obtained very naturally from projective geometry: let K be any skew field, let V_i consist of all i-dimensional subspaces of the vector space K^{n+1}, and define $x * y \iff x \subseteq y$ or $y \subseteq x$. Furthermore, every incidence structure $(P, \mathcal{L}, \Delta)$ as in Section 1 yields a geometry $(P, \mathcal{L}, *)$ of rank 2, simply by making Δ symmetric.

Let Γ be a geometry of rank n as above. Then the set X of all flags (now called simplices) of Γ is a simplicial complex with vertex set $V = \bigcup_{i=1}^n V_i$; this simplicial complex X is called the flag complex of Γ. Thickness means that every flag of size $n - 1$ (now called a panel) is contained in at least 3 flags of size n (now called chambers).

Conversely, let (Δ, δ) be a building of rank n as defined in Section 3, and let (W, S) be the corresponding Coxeter system, with $S = \{s_1, s_2, ..., s_n\}$. Let V_i be the set of all $(S \setminus \{s_i\})$-residues in Δ (in other words: V_i is the quotient of Δ with respect to $(S \setminus \{s_i\})$-adjacency, see Section 3). Define $x * y \iff x \cap y \neq \emptyset$, for $x \in V_i, y \in V_j$. Then $\Gamma = (V_1, ..., V_n, *)$ is a geometry of rank n, and $V = \bigcup_{i=1}^n V_i$ consists of all residues of rank $n - 1$ in Δ. The residues of rank r in Δ correspond bijectively to the flags of size $n - r$ of Γ (map a residue R to the set of all residues of rank $n - 1$ which contain R).

As an example, we describe the *Coxeter complex* Σ associated to a Coxeter system (W, S), starting from the thin building W with distance map $(x, y) \mapsto x^{-1}y$. The simplices of Σ are the residues of W, i.e. all cosets $w\langle T \rangle$ with $w \in W, T \subseteq S$, and the ordering is reverse inclusion. Thus the chambers of Σ "are" the elements of W, and the panels of Σ are the (co)sets $\{w, wt\}$ with $w \in W, t \in S$.

References

[1] H. Abels, The gallery distance of flags, Order 8 (1991) 77 - 92

[2] H. Abels, The geometry of the chamber system of a semimodular lattice, Order 8 (1991) 143 - 158

[3] H. Abels, The group of values of a W-distance is a Coxeter group, in: Group theory from a geometrical viewpoint, Proc. Trieste 1990, pp. 296 - 301, World Sci. Publishing 1991

[4] P. Abramenko, Twin buildings and applications to S-arithmetic groups, Lecture Notes in Math. 1641, Springer 1996

[5] A. Björner, F. Brenti, Combinatorics of Coxeter groups, to appear (Springer)

[6] A. Borel, Linear algebraic groups, Springer 1991 (Second enlarged edition)

[7] N. Bourbaki, Groupes et algèbres de Lie, Chapitres 4 - 6, Hermann Paris 1968, Masson Paris 1981

[8] N. Brady, J.P. McCammond, B. Mühlherr, W.D. Neumann, Rigidity of Coxeter groups and Artin groups, preprint (2000)

[9] K.S. Brown, Buildings, Springer 1989, 1998

[10] K.S. Brown, Five lectures on buildings, in: Group theory from a geometrical viewpoint, Proc. Trieste 1990, pp. 254 - 295, World Sci. Publ. 1991

[11] F. Buekenhout, A.M. Cohen, Diagram geometries, to appear

[12] F. Buekenhout, E. Shult, On the foundations of polar geometry, Geom. Dedicata 3 (1974) 155 - 170

[13] K. Burns, R. Spatzier, Manifolds of nonpositive curvature and their buildings, Inst. Hautes Études Sci. Publ. Math. 65 (1987) 35 - 59

[14] R.W. Carter, Simple Groups of Lie Type, Wiley 1989

[15] A.M. Cohen, Coxeter groups and three Related Topics, in: Generators and Relations in Groups and Geometries, Proc. Castelvecchio Pascoli 1990, A. Barlotti et al. (eds), pp. 235 - 278, Kluwer 1991

[16] A.M. Cohen, Point-line spaces related to buildings, in: F. Buekenhout (ed.), Handbook of incidence geometry, pp. 647 - 737, North-Holland 1995

[17] H.S.M. Coxeter, Discrete groups generated by reflections, Ann. of Math. (2) 35 (1934) 588-621

[18] H.S.M. Coxeter, The complete enumeration of finite groups of the form $R_i^2 = (R_i R_j)^{k_{ij}} = 1$, J. London Math. Soc. 10 (1935) 21-25

[19] H.S.M. Coxeter, Regular polytopes, Dover Publ. 1973 (third edition)

[20] P. de la Harpe, An invitation to Coxeter groups, in: Group theory from a geometrical viewpoint, Proc. Trieste 1990, pp. 193 - 253, World Sci. Publishing 1991

[21] A. Dress, W. Terhalle, The tree of life and other affine buildings, Proc. Int. Congress Math. Berlin 1998, Vol. III, pp. 565 - 574, Doc. Math. 1998, extra Vol. III

[22] L.C. Grove, C.T. Benson, Finite reflection groups, Springer 1985 (second edition)

[23] H. Hiller, Geometry of Coxeter Groups, Pitman 1982

[24] J. E. Humphreys, Reflection Groups and Coxeter Groups, Cambridge University Press 1990

[25] O. Kegel, A. Schleiermacher, Amalgams and embeddings of projective planes, Geom. Dedicata 2 (1973) 379 - 395

[26] B. Kleiner, B. Leeb, Rigidity of quasi-isometries for symmetric spaces and Euclidean buildings, Inst. Hautes Études Sci. Publ. Math. 86 (1997) 115 - 197

[27] L. Kramer, Buildings and classical groups, this volume

[28] S. Mitchell, Quillen's theorem on buildings and the loops on a symmetric space, Enseign. Math. (2) 34 (1988) 123 - 166

[29] G.D. Mostow, Strong rigidity of locally symmetric spaces, Princeton University Press 1973

[30] B. Mühlherr, On isomorphisms between Coxeter groups, Designs, Codes and Cryptography 21 (2000) 189

[31] B. Mühlherr, Twin buildings, this volume

[32] A. Pasini, Diagram geometries, Oxford University Press 1994

[33] G. Pickert, Projektive Ebenen, Springer 1975 (second edition)

[34] D. Robinson, A course in the theory of groups, Springer 1982

[35] J. Rohlfs, T.A. Springer, Applications of buildings, in: F. Buekenhout (ed.), Handbook of incidence geometry, pp. 1085 - 1114, North-Holland 1995

[36] M. Ronan, Lectures on buildings, Academic Press 1989

[37] M. Ronan, Buildings: main ideas and applications, I. Main ideas, II. Arithmetic groups, buildings and symmetric spaces, Bull. London Math. Soc. 24 (1992) 1 - 51, 97 - 126

[38] M. Ronan, Twin trees and twin buildings, this volume

[39] M. Ronan, J. Tits, Building buildings, Math. Ann. 278 (1987) 291 - 306

[40] R. Scharlau, A structure theorem for weak buildings of spherical type, Geom. Dedicata 24 (1987) 77 - 84

[41] R. Scharlau, Buildings, in: F. Buekenhout (ed.), Handbook of incidence geometry, pp. 477 - 645, North-Holland 1995

[42] P. Schneider, Gebäude in der Darstellungstheorie über lokalen Zahlkörpern, Jahresber. Deutsch. Math.-Verein. 98 (1996) 135 - 145

[43] E. Shult, Aspects of buildings, in: Groups and Geometries, Proc. Siena 1996, pp. 177 - 188, Birkhäuser 1998

[44] D.E. Taylor, The geometry of the classical groups, Heldermann Verlag, Berlin 1992

[45] K. Tent, Very homogeneous generalized n-gons of finite Morley rank, J. London Math. Soc. (2) 62 (2000) 1 - 5

[46] K. Tent, Model theory of groups and BN-pairs, this volume

[47] G. Thorbergsson, Isoparametric foliations and their buildings, Ann. of Math. (2) 133 (1991) 429 - 446

[48] J. Tits, Les groupes de Lie exceptionnels et leur interprétation géométrique, Bull. Soc. Math. Belg. 8 (1956) 48 - 81

[49] J. Tits, Groupes algébriques semi-simples et géométries associées, in: Algebraical and Topological Foundations of Geometry, Proc. Utrecht 1959, H. Freudenthal (ed.), pp. 175 - 192, Pergamon Oxford 1962

[50] J. Tits, Géométries polyédriques et groupes simples, in: Atti della II Riunione del Groupement de Mathematiciens d'Expression Latine, Firenze-Bologna 1961, pp. 66-88, Edizioni Cremonese, Rome 1963

[51] J. Tits, Buildings of spherical type and finite BN-pairs, Lecture Notes in Math. 386, Springer 1974

[52] J. Tits, On buildings and their applications, Proc. Int. Congress Math. Vancouver 1974, Vol. 1, pp. 209 - 220, Canad. Math. Congress, Montreal, 1975

[53] J. Tits, Endliche Spiegelungsgruppen, die als Weylgruppen auftreten, Invent. Math. 43 (1977) 283 - 295

[54] J. Tits, A local approach to buildings, in: Ch. Davis et al. (eds), The geometric vein, pp. 519 - 547, Springer 1981

[55] J. Tits, Immeubles de type affine, in: L. Rosati (ed.), Buildings and the geometry of diagrams, Proc. Como 1984, pp. 159 - 190, Lecture Notes in Math. 1181, Springer 1986

[56] J. Tits, Buildings and group amalgamations, in: Proc. of Groups - St Andrews 1985, E. F. Robertson and C. M. Campbell (eds), pp. 110 - 127, Cambridge University Press 1986

[57] J. Tits, Résumé de cours, 1988/89, Annuaire du Collège de France, Paris

[58] J. Tits, Twin buildings and groups of Kac-Moody type, in: Groups, combinatorics & geometry, Proc. Durham 1990, pp. 249 - 286, Cambridge University Press 1992

[59] J. Tits, R. Weiss, The classification of Moufang polygons, to appear

[60] H. Van Maldeghem, Nonclassical triangle buildings, Geom. Dedicata 24 (1987) 123 - 206

[61] H. Van Maldeghem, Quadratic quaternary rings with valuation and affine buildings of type $\widetilde{C}_2$, Mitt. Math. Sem. Giessen 189 (1989) 1 - 159

[62] H. Van Maldeghem, An algebraic characterization of affine buildings of type $\widetilde{C}_2$, Mitt. Math. Sem. Giessen 198 (1990) 1 - 42

[63] H. Van Maldeghem, Generalized polygons, Birkhäuser 1998

[64] H. Van Maldeghem, An introduction to generalized polygons, this volume

[65] H. Van Maldeghem, K. Van Steen, Characterizations by automorphism groups of some rank 3 buildings. IV. Hyperbolic p-adic Moufang buildings of rank 3, Geom. Dedicata 75 (1999) 115 - 122

[66] K. Van Steen, Characterizations by automorphism groups of some rank 3 buildings. III. Moufang-like conditions, Geom. Dedicata 74 (1999) 225 - 240

[67] G. Warner, Harmonic analysis on semi-simple Lie groups I, Springer 1972

An Introduction to Generalized Polygons

Hendrik Van Maldeghem
Department of Pure Mathematics and Computer Algebra
Ghent University
Galglaan 2, 9000 Gent
BELGIUM
hvm@cage.rug.ac.be

Abstract

We present a survey on generalized polygons, emphasizing the infinite case and mentioning some connection with Model theory. We try to be complementary to the monograph [60]. In an appendix, we review the classification of Moufang polygons.

1 Some definitions of the notion of a generalized polygon

We start this paper by presenting some definitions of a generalized polygon that can be found in the literature. Most definitions start with a triple $\Gamma = (\mathcal{P}, \mathcal{L}, \mathrm{I})$ (which we will call an *incidence system*), where $\mathcal{P}$ and $\mathcal{L}$ are two sets the elements of which are called points and lines, respectively, and where $\mathrm{I} \subseteq (\mathcal{P} \times \mathcal{L}) \cup (\mathcal{L} \times \mathcal{P})$ is a symmetric relation, called the incidence relation. We use common terminology such as *a point lies on a line, a line goes through a point, a line contains a point, a line passes through a point,* etc., to denote the incidence between a point and a line.

A *path* in Γ is a sequence $(x_0, x_1, \ldots, x_k)$ of points and lines such that $x_{i-1} \mathrm{I} x_i$, for all $i \in \{1, 2, \ldots, k\}$, and $x_{i-1} \neq x_{i+1}$, for all $i \in \{1, 2, \ldots, k-1\}$ (if the latter is not satisfied, then we call it a *sequence*). We say that the path (or sequence) *joins the elements* x_0 *and* x_k (and we do not distinguish between a path and its "inverse" here). The positive natural number k is called the *length* of the path. If $x_0 = x_k$, then the path is called *closed*. Note that a closed path of Γ always has even length. A closed path $(x_0, x_1, \ldots, x_{2n} = x_0)$ of length $2n > 2$ is called an *ordinary n-gon* if $x_1 \neq x_{2n-1}$. The *girth* of Γ is the length of a closed path of minimal length, if such a path exists. If not, then the girth is said to be ∞. The *distance* $\delta(x, y)$ between two elements x, y of Γ is the length of a path of minimal length joining x and y, if such a path exists. If not, then the distance between x and y is by definition ∞. The *diameter* of Γ is the maximal value that can occur for the distance between two elements of Γ (where ∞ is by definition bigger than any natural number). Two elements at distance the diameter of Γ are called *opposite*. A *flag* of Γ is a pair $\{x, L\}$ consisting of a point x and a line L which are incident with each other. We say that a flag $\{x, L\}$ is contained in a path γ if x and L occur in γ at adjacent places. The set of flags of Γ will be denoted by $\mathcal{F}$.

Finally we call Γ *thick* (*firm*) if each element is incident with at least three (two) elements.

In the following, let $n \geq 3$ be a positive integer, and let $\Gamma = (\mathcal{P}, \mathcal{L}, \mathrm{I})$ be an incidence system.

(I) The first definition is the original one by Tits [43]. We call Γ a *generalized n-gon of type* (I) if

   (Ia) every two elements of $\mathcal{P} \cup \mathcal{L}$ can be joined by at most one path of length $< n$,

   (Ib) every two elements of $\mathcal{P} \cup \mathcal{L}$ can be joined by at least one sequence of length $\leq n$.

(II) The second definition is also due to Tits [47]. We call Γ a *generalized n-gon of type* (II) if

   (IIa) the diameter of Γ is equal to n,

   (IIb) the girth of Γ is equal to $2n$.

(III) The third definition is also due to Tits and emerges from his more general definition of "buildings". We call Γ a *generalized n-gon of type* (III) if

   (IIIa) every two elements of $\mathcal{P} \cup \mathcal{L} \cup \mathcal{F}$ are contained in a common ordinary n-gon of Γ,

   (IIIb) if $x, y \in \mathcal{P} \cup \mathcal{L} \cup \mathcal{F}$ are both contained in two ordinary n-gons $(x_0, x_1, \ldots, x_{2n})$ and $(y_0, y_1, \ldots, y_{2n})$, where $x_0, y_0 \in \mathcal{P}$, then there exists an integer k such that the map $\theta : x_i \mapsto y_{i+2k}$ (reading subscripts modulo $2n$) has the property $\theta(x) = x$ and $\theta(y) = y$ (it is clear how to define $\theta(x)$ if $x \in \mathcal{F}$).

(IV) The fourth definition was first explicitly mentioned by Schroth [31], who attributes it to folklore. We call Γ a *generalized n-gon of type* (IV) if

(IVa) there are no ordinary k-gons in Γ, with $k < n$,

(IVb) every two elements of $\mathcal{P} \cup \mathcal{L}$ of Γ are contained in an ordinary n-gon.

(V) The last definition is due to Abramenko and Van Maldeghem [2]. We call Γ a *generalized n-gon of type* (V) if

(Va) the diameter of Γ is equal to n,

(Vb) for every element $x \in \mathcal{P} \cup \mathcal{L}$ and every element $y \in \mathcal{P} \cup \mathcal{L}$ at distance $n - 1$ from x, there exists a unique element $x' \mathbf{I} x$ at distance $n - 2$ from y.

The most general definitions of this lot are undoubtedly the first one and the last one. For instance, if $\mathcal{P} = \{x, y, z\}$, $\mathcal{L} = \{K, L, M, N\}$ and $\mathbf{I}$ is defined by $K \mathbf{I} x \mathbf{I} L \mathbf{I} y \mathbf{I} M \mathbf{I} z \mathbf{I} N$, then Γ is a generalized 6-gon of both type (I) and (V), and also a generalized k-gon of type (I) for all $k \geq 6$. On the other hand, Γ is not a generalized n-gon of any type (II), (III), (IV) for any value of n. Notice that generalized polygons of type (II), (III) and (IV) are automatically firm. For types (III) and (IV) this is trivial, and for type (II) this follows from Lemma 1.5.10 of [60]. Also, the results in Chapter 1 of [60] and in Section 4 of [2] show the following theorem.

Theorem 1.1 *If* Γ *is firm, then the definitions* (I), (II), (III), (IV), (V) *are all equivalent.*

If Γ is firm, then we call a generalized polygon of any of the types above a *weak generalized polygon*. If Γ is moreover thick, then we call it a generalized polygon. This is the terminology used in [60], which conforms to the more general standard terminology of (weak) buildings.

For $n = 2$, one finds that a weak generalized 2-gon (digon) is a rather trivial geometry in which every point is incident with every line. For $n = 3$, a generalized 3-gon is nothing other than a projective plane. For higher values, one sometimes uses the terminology generalized *quadrangle* (4-gon), *pentagon* (5-gon), *hexagon* (6-gon), *heptagon* (7-gon) and *octagon* (8-gon). Grammatically, this is not very consistent, but it is borrowed from the usual terminology of regular n-gons in the Euclidean plane.

Let us end this section by explaining briefly why Tits defined generalized polygons in [43] originally as the ones of type (I) in that generality.

The reason is that in [43], Tits classifies the trialities of the triality quadrics (the buildings of type D_4) having at least one absolute point. The possible configurations of absolute points and lines precisely form a generalized hexagon (6-gon) of type (I), and some cases are not firm, just like some polarities in projective 3-space over a field of characteristic 2 have as set of absolute lines a plane line pencil. Moreover, the diameter of this absolute geometry can be strictly smaller than 6, when there are no closed paths. All these cases are covered by definition (I). Of course, the non-trivial examples are the firm ones, and people have restricted their attention only to those. On top of that, we will see below that also the firm non-thick ones can be reduced to thick ones; therefore usually people restrict themselves to the thick case.

Finally, note that, if $\Gamma = (\mathcal{P}, \mathcal{L}, \mathtt{I})$ is a (weak) generalized polygon, then so is $\Gamma^D = (\mathcal{L}, \mathcal{P}, \mathtt{I})$, which is called the *dual* of Γ.

2   Isomorphisms

Let $\Gamma_1 = (\mathcal{P}_1, \mathcal{L}_1, \mathtt{I}_1)$ and $\Gamma_2 = (\mathcal{P}_2, \mathcal{L}_2, \mathtt{I}_2)$ be two weak generalized polygons. A *morphism* φ from Γ_1 to Γ_2 is a pair of mappings $\varphi_1 : \mathcal{P}_1 \to \mathcal{P}_2$ and $\varphi_2 : \mathcal{L}_1 \to \mathcal{L}_2$ such that $\varphi_1(p)\mathtt{I}_2\varphi_2(L)$, for all $p \in \mathcal{P}_1$ and all $L \in \mathcal{L}_1$, whenever $p\mathtt{I}_1 L$. If the diameters of Γ_1 and Γ_2 are different, then there is some theory and there are interesting examples; we direct the reader to [16, 17]. From now on, we assume that Γ_1 and Γ_2 are both generalized n-gons, for some $n \geq 2$. If both φ_1 and φ_2 are bijections, and if also the inverses φ_1^{-1} and φ_2^{-1} define a morphism, then we say that φ is an *isomorphism*, and we call Γ_1 and Γ_2 isomorphic.

We have the following characterization of isomorphisms.

Theorem 2.1 *With the above notation, a morphism φ is an isomorphism whenever $n \geq 3$ and one of φ_1, φ_2 is bijective.*

Proof. Without loss of generality we may assume that φ_1 is bijective. First we show that φ_2 is surjective.

Let M be any line of Γ_2, and let q, q' be two arbitrary but distinct points of Γ_2 both incident with M. Let $p, p' \in \mathcal{P}_1$ be such that $\varphi_1(p) = q$ and $\varphi_1(p') = q'$. Let $(p = x_0, x_1, \ldots, x_k = p')$ be a path of length $k \leq n$ (see definition (I)). If the sequence $\gamma = (M, \varphi_1(x_0), \varphi_2(x_1), \varphi_1(x_2), \ldots, \varphi_2(x_{k-1}), \varphi_1(x_k), M)$ is a closed path, then $k + 2 \geq 2n$, implying $n + 2 \geq 2n$, hence $n = 2$, a contradiction. Consequently γ is not a path. We show by induction on $k \geq 2$ that all lines of γ coincide with M. For $k = 2$, this follows from the assumption $n \neq 2$. So suppose now $k > 2$. Since γ is not a path, and since we may assume that $\varphi_2(x_1) \neq M \neq \varphi_2(x_{k-1})$, the bijectivity of φ_1 implies that there exists $j \in \{2, 4, \ldots, k-4, k-2\}$ such that $\varphi_2(x_{j-1}) = \varphi_2(x_{j+1})$. So we may formally remove $\varphi_1(x_j)$ and $\varphi_2(x_{j+1})$ from γ, and obtain a sequence with the same properties as γ (except that it does not arise from a path in Γ_1, but that is not essential for the argument). The induction hypothesis implies $M = \varphi_2(x_1)$ after all. So φ_2 is surjective.

Next, we prove that φ_2 is injective. Suppose by way of contradiction that $\varphi_2(L) = \varphi_2(L')$, for two distinct lines $L, L' \in \mathcal{L}$. Let $(L = x_0, x_1, \ldots, x_k = L')$ be a path of length $k \leq n$. The image is a closed sequence, but cannot be a path, as before. Hence we again conclude that $\varphi_2(L) = \varphi_2(x_2)$. This means that we may assume that L and L' meet in a point p. Now consider any element x opposite p in Γ. By considering minimal paths joining x and L respectively L', we see that x, L, L' are contained in a common closed path γ of length $2n$ (joining p with itself). The image of γ is again a closed sequence, but can be shortened by removing $\varphi_1(p)$ in the beginning and a the end, leaving a closed sequence of length $2n - 2$. This cannot be a path, and a similar argument as before shows that all lines of that sequence coincide. We have shown that $\varphi_1(x) \mathrm{I}_2 \varphi_2(L)$, if n is even, and each point on x is mapped on $\varphi_2(L)$, if n is odd. In other words, every point p' at distance $n - 1$ or n from p is mapped onto $\varphi_2(L)$. Similarly, one shows that every point at distance $n - 1$ or n from such a point p' is mapped onto $\varphi_2(L)$. But it is easy to see that every point p'' lies at distance $n - 1$ or n from some such point p'. This of course contradicts the bijectivity of φ_1.

Hence φ_2 is bijective. Left to show is that the inverse is also a morphism. Suppose by way of contradiction that $p \in \mathcal{P}_1$ and $L \in \mathcal{L}_1$ are not incident, but $\varphi_1(p) \mathrm{I}_2 \varphi_2(L)$. The image of a minimal path joining p and L is a closed sequence which cannot be a path, contradicting the injectivity of both φ_1 and φ_2.

The theorem is proved.   □

There is another characterization of isomorphisms, which is proved in [15]. We denote the distance function in Γ_i by δ_i, $i = 1, 2$.

Theorem 2.2 *Let Γ_j, $j = 1, 2$ be a generalized n-gon as above, $n \geq 4$. Let $k \in \{1, 2, \ldots, n-1\}$ arbitrary.*

(i) *If k is even, then let $\varphi_1 : \mathcal{P}_1 \to \mathcal{P}_2$ be a bijection such that, for all $p, p' \in \mathcal{P}_1$, we have $\delta_1(p, p') = k$ if and only if $\delta(\varphi_1(p), \varphi_1(p')) = k$.*

(ii) *If k is odd, then let $\varphi_1 : \mathcal{P}_1 \to \mathcal{P}_2$ and $\varphi_2 : \mathcal{L}_1 \to \mathcal{L}_2$ be bijections such that, for all $p \in \mathcal{P}_1$ and $L \in \mathcal{L}_1$, we have $\delta_1(p, L) = k$ if and only if $\delta(\varphi_1(p), \varphi_2(L)) = k$.*

Then in case (i), *there is a unique bijection* $\varphi_2 : \mathcal{L}_1 \rightarrow \mathcal{L}_2$ *such that* (φ_1, φ_2) *is a morphism. In both cases* (i) *and* (ii), *the pair* (φ_1, φ_2) *is an isomorphism.*

For the case $k = n$, there do exist counterexamples, see [15] again. Also, a similar characterization using a map $\theta : \mathcal{F}_1 \rightarrow \mathcal{F}_2$ exists; see [1, 14].

The foregoing characterizations of isomorphisms are global characterizations. A beautiful local characterization is proved by Bödi and Kramer [4].

Theorem 2.3 *With the above notation, a morphism* φ *is an isomorphism whenever both* φ_1 *and* φ_2 *are surjective and for at least one element* $x \in \mathcal{P}_1 \cup \mathcal{L}_1$, *the restriction of* φ_i *to the set of elements of* Γ_1 *incident with* x *is bijective onto the set of elements of* Γ_2 *incident with* $\varphi_i(x)$, $i = 1$ *if* x *is a point,* $i = 2$ *if* x *is a line.*

3 The structure Theorem and subpolygons

As already mentioned before, the theory of weak generalized polygons can be reduced to the theory of (thick) generalized polygons. This is due to Tits [45], see also [70] and [60].

In the following, a *thick* element is a point or a line incident with at least three elements. The structure theorem roughly says that every weak n-gon arises from a generalized m-gon, where the dihedral group of order $2m$ is contained in the dihedral group of order $2n$. In this formulation, the result can be seen as a special case of the more general result for spherical buildings (the dihedral groups are the Weyl groups, see the paper of Grundhöfer in these proceedings). However, a more geometric formulation is the following.

Theorem 3.1 *Let* Γ *be a weak generalized n-gon. Then either there are no thick elements and* Γ *is basically an ordinary n-gon, or there are exactly two thick elements which are opposite, or the minimal distance* k *between two thick elements is not bigger than* $n/2$ *and the set of thick elements can be partitioned into two subsets* $\mathcal{P}'$ *and* $\mathcal{L}'$ *such that, if we define an incidence relation* I' *by* $p'\mathrm{I}'L'$ *if* $\delta(p', L') = k$, *then the incidence system* $\Gamma' = (\mathcal{P}', \mathcal{L}', \mathrm{I}')$ *is a (thick) generalized n/k-gon. In the latter case, every element of* $\Gamma_{ik}(x)$, *with* $x \in \mathcal{P}' \cup \mathcal{L}'$ *and* $1 \leq i \leq n/k$, *belongs to* $\mathcal{P}' \cup \mathcal{L}'$.

In the last case of the previous theorem we say that Γ is a *multiple* of Γ' (if $k = 2$, sometimes called the *double*). It is up to duality uniquely determined by Γ' and k.

Note that the elements p' and L' in Theorem 3.1, though regarded as a point and a line of Γ', respectively, are not necessarily a point and a line of Γ, respectively. Indeed, it can very well happen that all thick elements of Γ are points, for instance.

The above structure theorem is a very useful tool in the study of weak subpolygons of a given generalized polygon. Let us give an example. First we need some preliminaries about subpolygons.

A weak subpolygon $\Gamma' = (\mathcal{P}', \mathcal{L}', \mathbf{I}')$ of a weak generalized polygon $\Gamma = (\mathcal{P}, \mathcal{L}, \mathbf{I})$ consists of the sets $\mathcal{P}' \subseteq \mathcal{P}$ and $\mathcal{L}' \subseteq \mathcal{L}$ such that, with $\mathbf{I}'$ defined as the restriction of $\mathbf{I}$ to $(\mathcal{P}' \times \mathcal{L}') \cup (\mathcal{L}' \times \mathcal{P}')$, the incidence system $\Gamma' = (\mathcal{P}', \mathcal{L}', \mathbf{I})$ is a weak generalized polygon. Not much is known in case the diameters of Γ and Γ' are different, so from now on we assume that Γ and Γ' always have the same diameter.

The weak subpolygon Γ' of Γ is called a *full* (respectively *ideal*) weak subpolygon if, for every line L' (respectively point p') of Γ', every element of Γ incident with L' (respectively p') in Γ also belongs to Γ'.

For instance, Γ itself is a full and ideal weak subpolygon of itself. Conversely, it is not so hard to see that every full and ideal weak subpolygon of the weak generalized polygon Γ coincides with Γ (cp. Proposition 1.8.2 of [60]).

Another type of weak subpolygons Γ' of a generalized polygon Γ are the ones with the property that for every thick element x of Γ' all elements of Γ incident with x belong to Γ'. We call such a weak subpolygon containing at least one thick element a *solid* weak subpolygon.

An important class of sub-n-gons in the case of $n = 3$ (the projective planes) are the Baer subplanes. Let us recall the definition (also valid in the infinite case). A *Baer subplane* of a projective plane Γ is a subplane Γ' with the property that every element (point or line) of Γ is incident with at least one element of Γ'. Here is the generalization of this notion to higher n.

Let Γ be a weak generalized n-gon. A *Baer (weak) sub-n-gon* of Γ is a weak sub-$n$gon Γ' of Γ with the property that every element (point or line) of Γ is at distance $\leq n/2$ from at least one element of Γ'.

For odd n we find examples through the following theorem. For any generalized polygon Γ, any point or line x of Γ, and any positive integer i, we denote by $\Gamma_i(x)$, $\Gamma_{\leq i}(x)$ the set of elements of Γ at distance i, at distance at most i, respectively.

Theorem 3.2 *Let Γ be a generalized $(2m + 1)$-gon, and suppose that σ is an involution of Γ (an isomorphism of order 2 from Γ to itself). Let $\mathcal{P}'* *(respectively $\mathcal{L}'$) be the set of fixed points (respectively fixed lines) of σ. Denote by $\mathbf{I}'$ the induced incidence relation. Then either the incidence system $\Gamma' = (\mathcal{P}', \mathcal{L}', \mathbf{I}')$ is a Baer weak subpolygon of Γ, or there exist a unique point p and a unique line $L\mathbf{I}p$ of Γ such that $\mathcal{P}' = \Gamma_{\leq n}(p)$ and $\mathcal{L}' = \Gamma_{\leq n}(L)$.*

Proof. Suppose first that there is a closed path γ of length $4m + 2$ pointwise fixed by σ. Then clearly the girth of Γ' is equal to $4m + 2$, and the diameter is not smaller than $2m + 1$. But if x, y are two elements fixed by σ, and $\delta(x, y) \leq 2m$, then the unique path joining x and y is also fixed pointwise by σ, and so $\delta'(x, y) = \delta(x, y) \leq 2m$ (denoting the distance in Γ' with δ'). Suppose now that x and y are fixed by σ and are opposite each other. Applying the previous argument on x and an element of γ not opposite x, we see that there exists an element $x' \in \Gamma_1(x)$ fixed by σ. Now $\delta(x', y) = 2m$ and the previous argument shows $\delta'(x', y) = 2m$, implying $\delta'(x, y) = 2m + 1 = \delta(x, y)$. Hence the diameter of Γ' is equal to $2m + 1$ and by definition (II), Γ' is a generalized $(2m + 1)$-gon. Now let z be any element of γ. Then, since $2m + 1$ is odd, z^σ is not opposite z. Hence there is a unique path $(z = z_0, z_1, \ldots, z_{2k} = z^\sigma)$ of even length $2k \leq 2m$ joining z and z^σ. Clearly, this path is mapped onto $(z^\sigma = z_{2k}, z_{2k-1}, \ldots, z_1, z_0 = z)$. So we see that z_k is fixed by σ, with $\delta(z, z_k) = k \leq m$. We have shown that Γ' is a Baer weak subpolygon.

Hence we may assume that σ does not fix any closed path of length $4m + 2$ pointwise. Let $x \in \mathcal{P} \setminus \mathcal{P}'$ be arbitrary. There is a unique path $(x = x_0, x_1, \ldots, x_{2k} = x^\sigma)$ joining x and x^σ, with $\delta(x, x^\sigma) = 2k$. By choosing an arbitrary element $x_{-1} \in \Gamma_1(x) \setminus \{x_1\}$, we see that $\delta(x_{-1}, x^\sigma_{-1}) = 2k + 2$. Going on like this, we may assume that $\delta(x, x^\sigma) = 2m$, and the unique element z' of $\Gamma_m(x) \cap \Gamma_m(x^\sigma)$ is fixed by σ. Now consider an element y incident with x, and opposite x^σ. Then again $\delta(y, y^\sigma) = 2m$ and the unique element z'' of $\Gamma_m(y) \cap \Gamma_m(y^\sigma)$ is fixed by σ. It is easy to see that $z', x, y, z'', x^\sigma, y^\sigma$ belong to a common ordinary $(2m + 1)$-gon, and so z' is opposite z''. By varying y, we see that we obtain different elements z'' opposite z'. All these elements z'' are all either points or lines. Consequently, these elements are not mutually opposite, hence the unique paths joining them are also pointwise fixed by σ. In particular there is some element $u'\mathbf{I}z'$ fixed by σ. We conclude that there is a path $(z' = z'_0, z'_1, \ldots, z'_{2m+1} = z'')$ of length $2m + 1$ joining z' and z'' fixed pointwise by Γ. We now show that σ fixes $\Gamma_{\leq m}(z'_m) \cup \Gamma_{\leq m}(z'_{m+1})$ pointwise.

It suffices to show that, if $w' \in \Gamma_m(z'_m) \cap \Gamma_{m+1}(z'_{m+1})$, then w' is fixed by σ. Choose any element $w \in \Gamma_m(w') \cap \Gamma_{2m}(z'_m)$. Then clearly $\delta(w, z'_{m+1}) = 2m + 1$. As before, the middle element v' (for which holds $\delta(w, v') \leq m$) of the unique minimal path joining w and w^σ is fixed by σ. By the triangle inequality we have $\delta(z'_m, v') \geq m$ and $\delta(z'_{m+1}, v') \geq m + 1$. If $\delta(z'_{m+1}, v') > m + 1$, then some path of length $> 2m + 1$ joining v' and either z_0' or z_{2m+1}' is fixed pointwise by σ, implying that there is some ordinary $(2m+1)$-gon fixed by σ, a contradiction. Hence $\delta(z_{m+1}', v') = m + 1$ and so z_m' and v' are both points or lines, or one is a point and the other a line according whether m is even or odd. Hence, if $\delta(z_m', v') > m$, then $\delta(z_m', v') > m + 1$, and this leads similarly as above to a contradiction. All this implies $\delta(z_m', v') = m = \delta(w, v')$. But there is only one element at distance m from both z_m' and w, and it is w'.

The theorem is proved. □

So the previous theorem shows that Baer weak subpolygons arise naturally with automorphisms of order 2. We now show the following restriction in the non-thick case.

Theorem 3.3 *Let $\Gamma = (\mathcal{P}, \mathcal{L}, \mathbf{I})$ be a generalized $(2m + 1)$-gon, and suppose that $\Gamma' = (\mathcal{P}', \mathcal{L}', \mathbf{I}')$ is a Baer weak non-thick subpolygon. Then Γ' is solid.*

Proof. Since Γ' is non-thick, either it has no thick elements or there exists a unique maximal odd positive integer $k \neq 1$ dividing $2m + 1$ such that the distance between any two thick elements of Γ' is a multiple of k. So, if there exist thick elements in Γ', let $p \in \mathcal{P}'$ and $L \in \mathcal{L}'$ be two thick opposite elements, otherwise let them just be two opposite elements of Γ'.

Suppose by way of contradiction that Γ' is not solid. Hence we may assume without loss of generality that there is some point $a\mathbf{I}L$ of Γ that does not belong to Γ'. Let M be an arbitrary line of Γ' incident with p and let b be any point of Γ not belonging to Γ' and incident with M. The point b exists since M is not a thick line of Γ' (otherwise $k = 1$). Since both L, M belong to Γ', every element of the unique path of length $2n$ joining L and M belongs to Γ', hence $\delta(a, b)$ cannot be equal to $2m - 2$ and so $\delta(a, b) = 2m$. Consequently there is a unique element x of Γ at distance m from both a and b. Since Γ' is Baer weak subpolygon, there is an element z of Γ' at distance $\leq m$ from x. If $\delta(x, z) < m$, then all elements of the unique paths of length $\delta(z, L) < 2m + 1$ and $\delta(z, M) < 2m + 1$ joining z with L and M, respectively, belong to Γ', a contradiction since at least one of these paths contains x. So $\delta(x, z) = m$ and it follows that z is a point. Note that this argument also shows that $\delta(z, b) = \delta(z, a) = 2m$. If z were not opposite M, then x would belong to the unique path of length $2m - 1$ joining z and M, a contradiction. Hence z is opposite M, and, similarly, z is also opposite L.

Suppose first that Γ' contains thick elements, hence that k is well defined. Since k divides $2m + 1$, this implies that z is a thick point in Γ' (as its distance to L is a multiple of k). But for the same reason it now follows that also M is thick in Γ', and this is a contradiction.

So we may assume that Γ' is an ordinary $(2m + 1)$-gon. But in such a polygon, every element is opposite precisely one other element, again a contradiction since z is opposite both L and M.

The theorem is proved. □

It is actually not difficult to see that in the previous theorem k cannot be equal to $2m + 1$, if $m > 1$. Hence non-thick weak Baer subpolygons of generalized $(2m + 1)$-gons, with $m > 1$, are always multiples of other generalized polygons.

For generalized n-gons with $n = 2m$ even, the previous theorem is not valid. Indeed, for generalized digons, every ordinary sub-2-gon is a weak non thick Baer subpolygon. But one can do better. Let $\mathbb{K}$ be any field and let Γ be the generalized quadrangle arising from a nonsingular quadric in $\mathbf{PG}(4, \mathbb{K})$ of maximal Witt index (see below). Consider a hyperplane section that corresponds with a nonsingular quadric of maximal Witt index in that hyperplane (a ruled nonsingular quadric). This defines a weak non thick full subquadrangle Γ' of Γ. Now delete one line and all points incident with it from Γ' and denote the weak non thick subquadrangle that remains by Γ'''. It is clear that Γ' is a Baer weak subquadrangle which is not solid.

Actually, in the even case the definition of a Baer subpolygon is not optimal. One obtains better results by defining in this case a Baer (weak) subpolygon of a generalized $2m$-gon Γ as a weak subpolygon Γ' such that either every point or every line of Γ lies at distance $\leq m - 1$ from some element of Γ'. In that case it is easy to see that every weak non thick Baer subpolygon of a generalized $2m$-gon is either full or ideal, in particular solid. But also every thick Baer subpolygon must be full or ideal. Hence, in the even case, there is no completely satisfactory definition for what a Baer subpolygon should be.

4 Some examples — BN-pairs and the Tits property

In this section, we give some examples of generalized polygons. For a detailed list we once again refer to [60].

4.1 Some general constructions

First, let us show that we are not talking about the empty set in this paper. In fact, there exist generalized n-gons for any $n \geq 2$. These can be constructed by some free construction process introduced by Tits [47]. Briefly, this goes as follows. One starts with an incidence system with girth $\geq 2n$ and joins every pair of elements at distance $n+1$ by a path of length $n-1$ using new elements. This is repeatedly performed and in the limit one obtains a (usually thick, if the original incidence system is not too poor) generalized n-gon.

The generalized n-gons thus obtained have some symmetry. Indeed, at least the automorphisms of the original incidence system can be extended to automorphisms of the generalized n-gon, in the obvious way. This way it is possible to obtain involutions of generalized $(2m + 1)$-gons, for all $m \geq 1$, fixing Baer subpolygons. Also, by modifying the construction, one can construct for each n generalized n-gons Γ with an automorphism group G acting transitively on the set of pairs (f, Σ), where f is a flag and Σ is an ordinary sub-n-gon of Γ containing f. This is called the *Tits property* for (Γ, G) in [60]. These generalized polygons correspond precisely to groups with a (saturated) BN-pairs of rank 2. We will not give a precise definition of this popular notion. We content ourselves by mentioning that such a group is basically equivalent with an automorphism group G of a generalized polygon Γ such that the pair (Γ, G) has the Tits property. The notion "BN" refers to the fact that the stabilizer of a flag is denoted by B (and called the *Borel subgroup*) and the stabilizer of an ordinary n-gon containing the flag stabilized by B is denoted by N. Then, putting $T := B \cap N$, the quotient N/T is well defined and is the *Weyl group* of the BN-pair. For a generalized n-gon, it is isomorphic to the dihedral group of order $2n$.

Now we note that a weak generalized n-gon is thick precisely if it contains an ordinary $(n+1)$-gon as a subgeometry, see Lemma 1.3.2 of [60]. A natural question is whether there exist generalized n-gons Γ with an automorphism group acting transitively on the set of pairs (f, Ω) with f a flag and Ω an ordinary $(n + 1)$-gon containing f. Let us call this the *Joswig property* for (Γ, G), since it was Michael Joswig who first asked me if one could classify such pairs in the finite case (and this gave rise to the papers [42] and [58], where this classification is completed).

In the infinite case, Katrin Tent [32] constructs for each $n \geq 2$ infinitely many examples of pairs (Γ, G), with Γ a generalized n-gon, possessing the Joswig property. Her method uses model theory and we refer to the paper of Poizat in these proceedings. Hence a classification of all such pairs seems out of reach, unlike the restriction to the finite case. However, if one requires a regular action on the pairs (f, Ω) as above, then using involutions and the theory of weak Baer subplanes introduced above, one can show that in the odd case (by which we mean generalized n-gons with n odd) we necessarily have a Pappian projective plane i.e., a projective plane arising naturally from a vector space of dimension 3 over a commutative field, see below. In the even case this problem is still open, although we hope to finish the cases $n = 4, 6$

under the additional assumption of self-duality soon.

The above discussion shows that, whenever we have a group with a BN-pair of rank 2, then we have a generalized polygon with a lot of automorphisms. Standard examples of such BN-pairs are given by semi-simple and almost simple algebraic groups of relative rank 2, by certain classical groups and groups of mixed type, and by the Ree groups of characteristic 2, as defined by Tits [49] over arbitrary (not necessarily perfect) fields of characteristic 2 admitting a so-called *Tits endomorphism*, which is a square root of the Frobenius.

Most examples of the previous paragraph can also be constructed in a more geometric way. The so-called Desarguesian projective planes arise from 3-dimensional vector spaces over not necessarily commutative fields by considering the vector lines and vector planes, with natural incidence relation. Also, any polarity of projective 3-space having fixed lines defines a (weak) generalized quadrangle by considering the fixed lines and the fixed incident point-plane pairs. Similarly, any triality of the triality quadric having fixed lines gives rise to a (weak) generalized hexagon. In order to construct generalized octagons, one considers polarities of so-called metasymplectic spaces (these are equivalent to buildings of type F_4).

Even more geometrically, one can check that the geometry of points and lines of any quadric Q in any projective space (here necessarily over a commutative field, and of dimension at least 4 — in order to avoid the non thick case) is a generalized quadrangle, provided that Q contains lines, but no planes (we say that the Witt index is 2), and that no point on Q is collinear on Q with every other point (we say that Q is nonsingular).

Also, nonsingular Hermitian varieties containing lines but no planes define generalized quadrangles. In fact, if the characteristic of the underlying field is equal to 2, then these generalized quadrangles sometimes contain a lot of subquadrangles that are not themselves obtained from a Hermitian variety or a quadric. For instance, if Γ is the generalized quadrangle arising from a nonsingular quadric of Witt index 2 in projective 4-space over a field of characteristic 2, or equivalently (but dually), Γ arises from symplectic polarity in projective 3-space over the same field, then the subquadrangles of Γ are precisely the so-called *quadrangles of mixed type*, or briefly the *mixed quadrangles*, because they arise from groups of mixed type (they are called the "indifferent quadrangles" by Jacques Tits and Richard Weiss because of their indifferent behaviour with respect to the Steinberg relations corresponding to a root system).

The examples of the last two paragraphs are special cases of the classical quadrangles, which we review now briefly, but in detail.

4.2 The classical quadrangles

The definitions in this subsection are based on Chapter 10 of [7] and Chapter 8 of [44].

Let $\mathbb{K}$ be a skew field and σ an anti automorphism (that means $(ab)^\sigma = b^\sigma a^\sigma$, for all $a, b \in \mathbb{K}$) of order at most 2. Let V be a — not necessarily finite dimensional — right vector space over $\mathbb{K}$ and let $g : V \times V \to \mathbb{K}$ be a $(\sigma, 1)$-linear form, i.e., for all $v_1, v_2, w_1, w_2 \in V$ and all $a_1, a_2, b_1, b_2 \in \mathbb{K}$, we have

$$g(v_1 a_1 + v_2 a_2, w_1 b_1 + w_2 b_2) =$$
$$a_1^\sigma g(v_1, w_1) b_1 + a_1^\sigma g(v_1, w_2) b_2 + a_2^\sigma g(v_2, w_1) b_1 + a_2^\sigma g(v_2, w_2) b_2.$$

Denote $\mathbb{K}_\sigma := \{t^\sigma - t : t \in \mathbb{K}\}$. We define $q : V \to \mathbb{K}/\mathbb{K}_\sigma$ as

$$q(x) = g(x, x) + \mathbb{K}_\sigma,$$

for all $x \in V$. We call q a *pseudo quadratic*, or more precisely, a σ-*quadratic form (over $\mathbb{K}$)*. Let W be a subspace of V. We say that q is *anisotropic over W* if $q(w) = 0$ if and only if $w = 0$, for all $w \in W$ (where we have written the zero vector as 0, and the element $0 + \mathbb{K}_\sigma$ also as 0). It is *non degenerate* if it is anisotropic over the subspace $\{v \in V \mid g(v, w) + g(y, x)^\sigma = 0,$ for all $w \in V\}$. From now on we assume that q is non degenerate.

Noting that, if $q(v) = 0$, then $q(vk) = 0$, for all $k \in \mathbb{K}$, we can define the *Witt index* of q as the dimension of the maximal subspaces of V contained in $q^{-1}(0)$.

For a non degenerate σ-quadratic form q over $\mathbb{K}$ with Witt index 2, we define the following geometry $\Gamma = \mathsf{Q}(V, q)$. The points of Γ are the 1-spaces in $q^{-1}(0)$; the lines are the 2-spaces in $q^{-1}(0)$, and incidence is symmetrized inclusion.

One can now show that $\mathsf{Q}(V, q)$ is a weak generalized quadrangle; it is non thick if and only if the dimension of V is equal to 4 and σ is the identity (and consequently $\mathbb{K}$ is commutative). We call $\mathsf{Q}(V, q)$ and its dual *classical quadrangles*. If σ is the identity, then we say that the quadrangle is of *orthogonal type*. When the dimension of V is equal to 4, then we say that the classical quadrangle is of *reduced Hermitian type*.

The quadrics and Hermitian varieties that we mentioned above are classical quadrangles. For the quadrics, this is easy to see by taking σ the identity. For Hermitian varieties, see Chapter 2 of [60].

For V a 5-dimensional space and σ the identity, there is exactly one non degenerate pseudo quadratic form, up to isomorphism. The dual of the corresponding generalized quadrangle is called a *symplectic quadrangle* $\mathsf{W}(\mathbb{K})$, because $W(\mathbb{K})$ can be defined as the geometry of points and fixed lines of a 3-dimensional projective space over $\mathbb{K}$ with respect to a symplectic polarity.

By definition, we say that $W(\mathbb{K})$ is a *Pappian quadrangle*.

4.3 The split Cayley hexagons

We now define the Pappian hexagons, also called the split Cayley hexagons.

We define the generalized hexagon $\Gamma = H(\mathbb{K})$ as the geometry with as set of points the point set of the quadric $Q(6,\mathbb{K})$ in $\mathbf{PG}(6,\mathbb{K})$ with equation $X_0 X_4 + X_1 X_5 + X_2 X_6 = X_3^2$, and as set of lines the lines of $Q(6,\mathbb{K})$ the Grassmann coordinates of which satisfy the following six linear equations (where the *Grassmann coordinates* of a line containing the two points with coordinates $(x_0, \ldots, x_6)$ and $(y_0, \ldots, y_6)$ are defined, up to a common scalar multiple, as $p_{ij} = x_i y_j - y_i x_j$, $0 \leq i, j, \leq 6$):

$$p_{12} = p_{34}, \qquad p_{54} = p_{32}, \qquad p_{20} = p_{35},$$
$$p_{65} = p_{30}, \qquad p_{01} = p_{36}, \qquad p_{46} = p_{31}.$$

Incidence is natural. This description is due to Tits [43]. We call $H(\mathbb{K})$ a *split Cayley hexagon* and a *Pappian hexagon*. The split Cayley hexagons are precisely the generalized hexagons arising from "linear" trialities (i.e, where no field automorphism in the triality is involved).

All the preceding examples — apart from the free and model theoretic constructions — satisfy the so-called Moufang condition, see below and the appendix. There are also other constructions which give projective planes, generalized quadrangles and hexagons with a fairly large automorphism group, but not Moufang. To mention just one class of examples, consider the building at infinity of any rank 3 affine building. For other examples, the reader is directed to Chapter 3 of [60].

Let us end by constructing the smallest generalized hexagon $H(2)$ (i.e., $H(\mathbb{K})$ for $\mathbb{K}$ the field of two elements) in an alternative way. The construction is taken from [62]. It uses the smallest projective plane $\mathbf{PG}(2,2)$ where each element is incident with exactly three others. Let $\Gamma' = (\mathcal{P}', \mathcal{L}', \mathbf{I}')$ denote this projective plane. We define a new incidence system $\Gamma = (\mathcal{P}, \mathcal{L}, \mathbf{I})$ as follows. The set of points $\mathcal{P}$ is exactly $\mathcal{P}' \cup \mathcal{L}' \cup (\mathcal{P}' \times \mathcal{L}')$. The set of lines $\mathcal{L}$ is defined as follows. For every point p and every line $L\mathbf{I}'p$ of Γ', let x_1 and x_2 be the two points different from p and incident with L in Γ', and let M_1, M_2 be the two lines different from L incident with p in Γ'. Then the sets $\{p, L, (p, L)\}$ and $\{(p, L), (x_1, M_1), (x_2, M_2)\}$ are members of $\mathcal{L}$; the relation $\mathbf{I}$ is the natural one. We obtain a generalized hexagon where every element is incident with exactly three others. This hexagon and its dual are the only hexagons with that property, see [9].

4.4 Generalized polygons in special categories

We refer to [60] for a discussion of existence and uniqueness of small generalized polygons. We content ourselves here to mention that, if a generalized n-gon is finite, then Feit and Higman [11] proved that the number $s + 1$ of points incident with a line is a constant; likewise the number $t + 1$ of lines through a point is a constant; the number n is one of $2, 3, 4, 6, 8$, and if $n = 6$, then st is a perfect square, while if $n = 8$, the number $2st$ is a perfect square. Moreover, some inequalities must hold involving s, t, but we will not mention them explicitly.

The main result of the previous paragraph is that for finite generalized n-gons, the value n is highly restricted! This is a general phenomenon that occurs over and over again. For instance, disregarding the generalized digons, Moufang n-gons only exist for $n \in \{3, 4, 6, 8\}$ ([46, 48, 69]), compact connected n-gons only exist for $n \in \{3, 4, 6\}$ ([21, 22]), generalized n-gons with valuation only exist for $n \in \{3, 4, 6\}$ ([53]), generalized n-gons satisfying a certain geometric regularity condition (namely, for every i, $2 \leq i \leq n/2$, and every point p, the sets $\Gamma_i(p) \cap \Gamma_{n-i}(x)$, with $x \in \Gamma_n(p)$, are determined by any two of their elements) only exist for $n \in \{3, 4, 6\}$ ([57]), generalized n-gons arising from so-called "split" BN-pairs only exist for $n \in \{3, 4, 6, 8\}$ ([35, 37]). Hence it may be clear that the most important values for n are exactly $3, 4, 6$ and *sometimes* 8. It is very tempting to try to prove such restrictions in a model theoretic setting such as finite Morley rank. Here, it seems that whenever one can prove such a restriction, a full classification can be carried out. For instance, groups with a rank 2 BN-pair of finite Morley rank and with Weyl group of order $2n$ (hence corresponding to generalized n-gons), where the panels have Morley rank 1 (the strongly minimal case; a *panel* is the set of elements incident with a given element) have $n \in \{3, 4, 6\}$ and are precisely the Pappian polygons over algebraically closed fields, see [36]. These polygons share a number of interesting and interrelated properties (see throughout [60]).

Finally let us mention that every Moufang polygon together with its automorphism group is a pair that satisfies the Tits property, but not necessarily the Joswig property. It even is a nontrivial exercise to list all Moufang polygons with the Joswig property. So, the Joswig property is not a very natural property to study, in contrast to the Tits property, which translates naturally into the rank 2 BN-pair property for groups.

5   Characterization and classification results

In the previous section we already mentioned some classification results of
generalized n-gons satisfying certain conditions. In this section, we do this in
a more systematic way, and we emphasize the results that are waiting for an
analogy in the theory of geometries and groups of finite Morley rank.

5.1   The Moufang condition

A generalized n-gon $\Gamma = (\mathcal{P}, \mathcal{L}, \mathbf{I})$ satisfies the *Moufang condition*, or is said
to be *Moufang* if for every path $\gamma = (x_0, x_1, \ldots, x_n)$ of length n, the pointwise
stabilizer $G_{[\gamma]}$ of the set $\Gamma_1(x_1) \cup \Gamma_1(x_2) \cup \ldots \cup \Gamma_1(x_{n-1})$ acts transitively on
the set of ordinary n-gons containing γ.

It is not difficult to see that $G_{[\gamma]}$ in general actually acts semi regularly
on the set of ordinary n-gons containing γ. Hence, for a Moufang polygon, it
acts regularly, and it is called a *root group*. The group generated by all $G_{[\gamma]}$,
for γ ranging over the set of all paths of length n, is called the *little projective
group* of the Moufang polygon Γ. The groups $G_{[\gamma]}$ themselves are called the
root groups of Γ.

The Moufang condition is very important notion in the theory of generalized polygons, and of buildings in general. We will comment on the classification of Moufang polygons below, and in the appendix. Let us just say
now that this classification is one of the major achievements in the theory of
polygons, and even in the theory of buildings.

There is also a group free definition of a Moufang polygon. Since this
has a model theoretic consequence, we will discuss that definition now. It is
given by the following theorem, where, for two non opposite elements x, y,
the notation $\text{proj}_x y$ stands for the unique element of Γ incident with x and
at distance $\delta(x, y) - 1$ from y.

Theorem 5.1 *A generalized n-gon $\Gamma = (\mathcal{P}, \mathcal{L}, \mathbf{I})$ satisfies the Moufang condition if and only if for every path $\gamma = (x_1, x_2, \ldots, x_{n-1})$ of length $n-2$, for every closed path $(y_0, y_1, \ldots, y_{2n-1}, y_{2n} = y_0)$ of length $2n$ with $\delta(y_0, x_1) = n-1$,
and for every element $w \in \Gamma_1(\text{proj}_{x_1} y_0) \setminus \{x_1\}$, there exists a closed path
$(z_0, z_1, \ldots, z_{2n-1}, z_{2n} = z_0)$ of length $2n$ satisfying the following properties.*

(i) $\delta(z_0, x_1) = n - 1$ and $\delta(z_0, w) = n - 3$,

(ii) $\delta(x_i, y_j) = \delta(x_i, z_j)$, for all $i \in \{1, 2, \ldots, n-1\}$ and all $j \in \{1, 2, \ldots, 2n\}$,

(iii) whenever $\delta(x_i, y_j) \neq n$, for any $i \in \{1, 2, \ldots, n-1\}$ and any $j \in \{1, 2, \ldots, 2n\}$, then $\mathrm{proj}_{x_i} y_j = \mathrm{proj}_{x_i} z_j$.

For a proof of this theorem (for $n > 3$), see [41, 54]. Now, Linus Kramer (personal communication during the conference) remarked that a direct consequence of Theorem 5.1 is the following

Corollary 5.2 *The Moufang property is first order definable property (in the model theoretic sense).*

This answers affirmatively a question asked by Poizat on the conference.

We can now comment on the content of the paper [24]. One of the main results classifies all Moufang polygons of finite Morley rank. They are precisely the Pappian polygons (see above) over algebraically closed fields. The proof of this result heavily uses the classification of Moufang polygons as obtained by Tits and Weiss [51]. The main ingredient of the proof consists of proving that there is some interpretable field around. In the cases where the root groups are the additive groups of commutative fields, this is not so hard, but if this is not the case, various ad hoc arguments have to be found. This holds in particular for the mixed quadrangles where the root groups are vector spaces over fields of characteristic 2. Another example is provided by the class of Moufang quadrangles with non commutative root groups.

As an application, it is shown in [24] that any infinite simple group of finite Morley rank with an irreducible spherical BN-pair of rank ≥ 3 is interpretably isomorphic to a simple algebraic group over an algebraically closed field. This application, however, does not use the classification of Moufang polygons at all, although it uses the classification of irreducible spherical buildings of rank at least 3. Indeed, since all irreducible spherical buildings of rank at least 3 are classified, we have an explicit list of Moufang polygons occurring as a residue in such buildings, and that is all that is needed in order to prove the application. Of course, this is only a technical remark since the classification of Moufang polygons is — in my own opinion — more elementary and accessible than the classification of irreducible spherical buildings of rank at least 3.

The proofs of many characterization theorems consist of showing that the generalized polygon in question satisfies the Moufang condition.

5.2 Variations of the Moufang condition

There are a number of variations on the Moufang condition. Most of them can be found in [60], but we discuss them nevertheless.

Let $k \in \{2, 3, \ldots, n\}$. A generalized n-gon $\Gamma = (\mathcal{P}, \mathcal{L}, \mathrm{I})$ satisfies the k-*Moufang condition*, or is said to be k-*Moufang* if for every path $\gamma = (x_0, x_1, \ldots, x_k)$ of length k, the pointwise stabilizer $G_{[\gamma]}$ of the set $\Gamma_1(x_1) \cup \Gamma_1(x_2) \cup \ldots \cup \Gamma_1(x_{k-1})$ acts transitively on the set of ordinary n-gons containing γ.

A generalized n-gon $\Gamma = (\mathcal{P}, \mathcal{L}, \mathrm{I})$ satisfies the *half Moufang condition*, or is said to be *half Moufang* if for some $\mathcal{X} \in \{\mathcal{P}, \mathcal{L}\}$ and for every path $\gamma = (x_0, x_1, \ldots, x_n)$ of length n and with $x_1 \in \mathcal{X}$, the pointwise stabilizer $G_{[\gamma]}$ of the set $\Gamma_1(x_1) \cup \Gamma_1(x_2) \cup \ldots \cup \Gamma_1(x_{n-1})$ acts transitively on the set of ordinary n-gons containing γ.

We can combine these two definitions to obtain the following (new) notion.

Let k be an even integer with $2 \leq k \leq n$. A generalized n-gon $\Gamma = (\mathcal{P}, \mathcal{L}, \mathrm{I})$ satisfies the *half k-Moufang condition*, or is said to be *half k-Moufang* if for some $\mathcal{X} \in \{\mathcal{P}, \mathcal{L}\}$ and for every path $\gamma = (x_0, x_1, \ldots, x_k)$ of length k and with $x_1 \in \mathcal{X}$, the pointwise stabilizer $G_{[\gamma]}$ of the set $\Gamma_1(x_1) \cup \Gamma_1(x_2) \cup \ldots \cup \Gamma_1(x_{k-1})$ acts transitively on the set of ordinary n-gons containing γ.

Note that the preceding definition only makes sense if k is even, because otherwise every path of length $k - 2$ begins or ends with an element of $\mathcal{X}$, and so half-k-Moufang would automatically be equivalent with k-Moufang. Likewise, the half Moufang condition only makes sense for generalized n-gons with n even.

The following theorem is an easy exercise, see Section 6.8 of [60].

Theorem 5.3 *Every Moufang generalized n-gon is a half Moufang polygon, is a k-Moufang polygon for all $k \in \{2, 3, \ldots, n\}$, and is a half k-Moufang polygon for all even $k \in \{2, \ldots, n\}$.*

The ultimate conjecture is that every (half) k-Moufang generalized n-gon, $2 \leq k \leq n$, is a Moufang polygon. In some special cases, this has already been proved. For instance, for projective planes it has been shown in [64] that every half 2-Moufang projective plane is Moufang. For finite generalized polygons, the results in [59, 66] imply that every finite k-Moufang generalized n-gon, $2 \leq k \leq n$, is Moufang.

Hence it is reasonable to state the conjecture that every k-Moufang generalized n-gon of finite Morley rank, $2 \leq k \leq n$, is a Moufang polygon. An additional difficulty here is that we do not have automatically a restriction on n, unlike in the finite case.

The first interesting open case concerning half k-Moufang polygons are the half 2-Moufang generalized quadrangles. Even in the finite case, nothing is known without the classification of finite simple groups. Using the classification of finite simple groups one can show that every finite half $2k$-Moufang generalized $2m$-gon, $1 \leq k \leq m$, is a Moufang polygon. This follows basically from [8]. The case $k = m = 2$ can be handled without the classification of finite simple groups, see [40].

An interesting derived conjecture is that every half Moufang generalized polygon of finite Morley rank is Moufang.

We end this subsection by mentioning the following result for half Moufang polygons (see 6.4.9 of [60] and Proposition 5.1 of [35]). A *central elation* in a generalized $2m$-gon Γ is an automorphism fixing all elements of Γ at distance $\leq m$ from some fixed element c (which is then called the *centre* of the central elation).

Theorem 5.4 *If , for $m \geq 3$, a generalized $2m$-gon Γ is half Moufang and all elements of the corresponding root groups are central elations, then Γ is a Moufang hexagon.*

5.3 Split BN-pairs

The proof of the classification of Moufang polygons has a peculiar history. In the thirties, Ruth Moufang proved that Moufang projective planes are coordinatized by "alternative division rings", see [25]. These algebraic structures (the ones that are not skew fields) were classified by Bruck and Kleinfeld [6, 20], and proved to be a Cayley-Dickson division algebra over a field. Already in the late sixties, Tits proved (in some unpublished notes) that the root groups of any Moufang hexagon are coordinatized by a field and by a "quadratic Jordan algebra of degree 3" over that field. It is only during the past thirty years that these structures have been classified. Only in the late seventies, Tits [46, 48] proved that Moufang n-gons do not exist for $n \neq 3, 4, 6, 8$. In [69], Weiss used some ideas of Tits to compose an ingenious and short proof of the same restriction result. In the meantime a manuscript of Tits (entitled *Quadrangles de Moufang, I.*) circulated classifying the Moufang quadrangles satisfying additionally the geometric regularity condition for all elements mentioned above, and giving rise to all mixed quadrangles. Since the doubles of some of these quadrangles occur in the Moufang octagons, this can be seen as a preparation of the classification of Moufang octagons. Also, around the same time, Faulkner [10] classified certain classes of Moufang quadrangles and hexagons, starting from stronger assumptions, thus leaving out some cases. The classification of all Moufang octagons was published by Tits in 1983 [49], but Tits made no use of the previously mentioned preprint.

So the complete classification of Moufang quadrangles remained conjectural for several decades until Tits and Weiss started the joint project of writing the proof down in a book [51]. In early 1997, Richard Weiss discovered a new class of Moufang quadrangles, that was seemingly unrelated to the theory of semi-simple algebraic groups and groups of mixed type. However, shortly after Weiss' discovery, Mühlherr and Van Maldeghem [26] proved that these new quadrangles were of "exceptional type F_4" by constructing them alternatively inside buildings of type F_4, thus relating these quadrangles to groups of mixed type F_4.

Hence, the classification of all Moufang polygons (in particular, Moufang quadrangles) has been completed only recently. Some special cases, however, were already done many years ago. For instance, the compact connected case follows from the classification of simple Lie groups (for an explicit list see [18], see also Theorem 9.6.3 of [60]). The finite case follows from the classification by Fong and Seitz [12, 13] of the finite groups with an irreducible "split" BN-pair of rank 2. This means that, with previously mentioned notation, the subgroup B (which is the stabilizer of some flag f of the generalized polygon Γ) can be written as the (not necessarily semi-direct) product $B = UT$, where U is a normal nilpotent subgroup of B. If we take for U the group generated by all root groups related to paths containing the flag f, then the commutator (Steinberg) relations between the root groups (see Appendix A below) imply that U is nilpotent.

Consequently, in the finite case irreducible rank 2 split BN-pairs are essentially equivalent to Moufang polygons. In the infinite case, the natural BN-pair corresponding with a Moufang polygon is still split, and so one might wonder whether the converse is true in general. In other words, can we use the classification of Moufang polygons to classify the irreducible split BN-pairs of rank 2? If a geometric proof of this existed, this would at the same time serve as a revision of the proof of the result of Fong and Seitz mentioned above. Note that such a result also has to deal with generalized n-gons with $n \neq 3, 4, 6, 8$. This project has almost completely be carried out by Tent and Van Maldeghem [35, 37], resulting in the next theorem.

Theorem 5.5 *The generalized polygons corresponding with groups G with an irreducible rank 2 split BN-pair with Weyl group of order $2n \neq 16$ are Moufang, and G contains the little projective group of the Moufang polygon.*

Of course one does not expect that the case $n = 8$ is a true exception. It is still under consideration by Tent and Van Maldeghem.

Needless to say that the previous theorem is important for any kind of classification of classes of simple groups (as in the finite case), for instance the simple groups of finite Morley rank.

5.4 Other conditions and characterizations

Every Moufang generalized $2m$-gon contains central elations, as defined above. This follows from the proofs in [46, 48, 69] (for an explicit argument, see Corollary 5.4.7 of [60]). So, up to duality, every point of a Moufang $2m$-gon is the centre of a non trivial central elation. The converse of this observation for finite generalized hexagons and octagons has been proved by Walker [67, 68].

Theorem 5.6 *If every point of a finite generalized hexagon or octagon Γ is the centre of a non trivial central elation, then Γ satisfies the Moufang condition.*

Note that the proof of this result does not use the classification of finite simple groups. But the proof is hardly geometric and uses non trivial results from permutation group theory, such as Hering's trivial normalizer intersection theorem [19]. In terms of transitivity, the hypothesis of Theorem 5.6 is pretty weak. Indeed, no point or line transitivity can directly be derived. I am not aware of any similar result for other categories of generalized hexagons and octagons, such as compact connected hexagons, or, for instance, generalized $2m$-gons of finite Morley rank. In the latter case, also values for m different from 3 and 4 are conceivable.

Another result characterizes the finite Moufang polygons by means of some kind of relatives of the elements of the root groups, namely, the generalized homologies. Let x, y be two opposite elements of a generalized n-gon Γ. Then we call Γ (x, y)-*transitive* if for some element $z \mathrm{I} x$ the group H fixing $\Gamma_1(x) \cup \Gamma_1(y)$ pointwise acts transitively on the set $\Gamma_1(z) \setminus \{x, \mathrm{proj}_z y\}$. Note that this action is not necessarily regular (sharply transitive). We call Γ (x, y)-*quasi-transitive* if for some element $z \in \Gamma_2(x) \cap \Gamma_{n-2}(y)$ the group H fixing $\Gamma_1(x) \cup \Gamma_1(y)$ pointwise acts transitively on the set $\Gamma_1(z) \setminus \{\mathrm{proj}_z x, \mathrm{proj}_z y\}$. The elements of H are called *generalized homologies*. The following theorem joins results of different papers and authors [3, 38, 39, 55, 56].

Theorem 5.7 *If a finite generalized polygon is (x, y)-transitive for all pairs of opposite elements x, y, then it is a Moufang polygon (but not all finite Moufang polygons arise, in particular, all finite planes and hexagons do arise, but no octagon arises and the generalized quadrangles obtained from Hermitian varieties in 4-dimensional projective space do not arise either). This remains true for infinite generalized projective planes. If a finite generalized quadrangle is (x, y)-transitive for all pairs of opposite points x, y, then it is a Moufang quadrangle (and up to duality all finite Moufang quadrangles arise). Finally, if a finite generalized octagon is (x, y)-transitive for all pairs of opposite points*

x, y, and (L, M)-*quasi-transitive for all pairs of opposite lines L, M, then it is a Moufang octagon (and all finite Moufang octagons do occur).*

The hypotheses in the previous theorem imply a large transitivity of the automorphism group of the generalized polygon in question. However, for polygons of finite Morley rank this is not enough to classify, hence Theorem 5.7 has not yet an analogy for finite Morley rank polygons.

To end the group theoretic characterizations of (certain classes of) Moufang polygons, we refer to [33] for a characterization of certain Moufang polygons by means of the Tits property and a condition on the groups induced on $\Gamma_1(x)$, for x an element of the polygon.

Finally, we turn to some geometric and combinatorial characterizations. The motivation is here in part that these can be used to prove group theoretical ones as those above. This is in particular the case for generalized hexagons, as in this case the Moufang condition is really equivalent to the geometric condition of regularity (up to duality). This same regularity condition can be used to prove non existence of certain generalized n-gons with $n \neq 3, 4, 6$. For instance, this is how the proof of Theorem 5.4 works. Also, since in the finite Morley rank case the only Moufang polygons are the Pappian ones, we also give a characterization of these (and that characterization can be used as a definition of "Pappian").

We introduce some definitions. Let $\Gamma = (\mathcal{P}, \mathcal{L}, \mathbf{I})$ be a generalized n-gon, and let i be an integer with $2 \leq i \leq n/2$. An element $x \in \mathcal{P} \cup \mathcal{L}$ is called *distance-i-regular* if the incidence system $\Gamma_x^i = (\Gamma_i(x), \{\Gamma_i(x) \cap \Gamma_{n-i}(y) \mid \delta(x, y) = n\}, \overline{\in})$, where $\overline{\in}$ denotes inclusion made symmetrical, has girth ≥ 6. In other words, every line of Γ_x^i is determined by any two of its points. If an element $x \in \mathcal{P} \cup \mathcal{L}$ is distance-i-regular for all i, $2 \leq i \leq n/2$, then we say that x is *regular*. If an element x is distance-2-regular and for every pair of elements $y, z \in \Gamma_n(x)$, the intersection $\Gamma_2(x) \cap \Gamma_{n-2}(y) \cap \Gamma_{n-2}(z)$ is non empty, then x is called a *projective element*. The motivation for this name stems from the fact that in this case the incidence system

$$(\Gamma_2(x) \cup \{x\}, \{\Gamma_2(x) \cap \Gamma_{n-2}(y) \mid \delta(x, y) = n\} \cup \{\Gamma_1(z) \mid z\mathbf{I}x\}, \overline{\in})$$

is a projective plane, called the *derived plane (at x)* .

In the following theorem, we combine results of [29] and [57]. Note that regularity is void for projective planes and generalized digons; therefore we only consider generalized n-gons with $n \geq 4$ when dealing with regularity.

Theorem 5.8 *If a generalized n-gon has a distance-i-regular element, then either $i = n/2$ or $i \leq (n+2)/4$. If all points of a generalized n-gon are* *distance-i-regular, and if i is even, then either $i = n/2$ or $i \leq n/4$, and i divides n. Hence, if all points of a generalized n-gon, $4 \leq n$, are regular, then $n = 4, 6$. If all points of a generalized hexagon Γ are distance-2-regular, then they are also distance-3-regular, hence regular, and Γ is a Moufang hexagon (and, up to duality, every Moufang hexagon arises in this way).*

So one sees that (Moufang) hexagons behave particularly nice with respect to regularity. For Moufang quadrangles, the regularity of elements translates into the fact that root groups "at distance 2" commute. An important open question is whether we know all finite quadrangles all points of which are regular (the conjecture is we do, and then all such quadrangles would be Moufang). Also in other categories, the regularity of all points of a generalized quadrangle is not enough yet to pin it down. But to the best of my knowledge no generalized quadrangle is known all points of which are regular, but which is not Moufang. Also remark that there are Moufang quadrangles that have no regular elements at all (for instance, all exceptional ones including those of type F_4 mentioned before). Specialized to the case of finite Morley rank, an interesting open problem is whether Tent's [32] construction can be modified in such a way to yield generalized quadrangles of finite Morley rank all points of which are regular. If not, can one classify the generalized quadrangles of finite Morley rank all points of which are regular? More precisely, one could ask whether the symplectic quadrangles over algebraically closed fields are the only generalized quadrangles of finite Morley rank all points of which are regular and all panels of which have the same Morley rank. These are certainly very hard questions.

For generalized octagons, Theorem 5.8 above says that not all points can be distance-3-regular. All elements of the Moufang octagons, however, are distance-4-regular, but not a single element is distance-2-regular. The philosophy *"if a nice property holds for one example, then it should also hold for a standard example with a lot of symmetry"* lead me to prove the following (remarkable to me) result.

Theorem 5.9 *If Γ is a generalized octagon, then not all points, respectively lines, can be distance-2-regular.*

In fact, to the best of my knowledge, no generalized octagon with at least one distance-2-regular element is known.

Now we turn to generalized polygons with projective elements. Recall that a Pappian projective plane is one that arises from a commutative field as described above. The next result combines statements of Schroth [30], Ronan [29] and the author, see Theorems 6.2.1 and 6.3.1 in [60].

Theorem 5.10 *Let* Γ *be a generalized* n-*gon,* $n \in \{4, 6\}$, *and suppose that all points of* Γ *are (distance-2-)regular. If at least one point is projective, then all points are projective, all derived planes are Pappian and* Γ *is Pappian itself. All Pappian quadrangles and hexagons arise in this way.*

In fact, using the distance-2-regularity of a certain set of points of some generalized hexagon, one can define another derived structure at a point, which turns out to be a Pappian quadrangle if the hexagon is Pappian. This way, we obtain a short tower of Pappian polygons derived from each other in the above sense. See [65] for more details.

As an application, Theorem 5.10 has been used to classify the algebraic polygons in [23].

Let us also remark that the regularity can be used to prove several other geometric characterizations, such as for example the one with long hyperbolic lines, see [52].

We did not discuss classifications and characterizations of generalized polygons embedded in certain projective spaces. Here, too, the regularity conditions play an important role. Also, many questions are still open in that area, especially if one considers structures of finite Morley rank. The reason for omitting these results here is that they are gathered in a survey paper on embeddings of polygons that will appear around the same time of these proceedings, see [63].

Finally, another area I neglected in the present paper, although new results were found since [60] appeared, is the theory of "projectivities". Non opposition defines a bijection between the sets $\Gamma_1(x)$ and $\Gamma_1(y)$ for opposite elements x, y. Composing such bijection until we are back at x, we obtain a permutation group of $\Gamma_1(x)$, called the *group of projectivities of* x. It is, as a permutation group, independent of x, but depends on the type of x, i.e., whether x is a point of a line. Now there are some results recognizing certain classes of Moufang projective planes and quadrangles by just requiring some properties of the group(s) of projectivities. The new results in that direction can be found in [5] and in [61]. In the latter, the situation of sharply 2-transitive groups of projectivities is analyzed.

For more on the interaction between the theory of generalized polygons and model theory, we refer to the survey paper [34].

A Appendix: The classification of Moufang polygons

In this appendix, we give a short overview of the classification of the Moufang polygons, as obtained by Tits and Weiss [51]. The method of proceeding turned out to be completely different from the one outlined in my monograph [60]. In fact, this outline was suggested by Jacques Tits in his lectures at Collège de France; the co-authorship with Richard Weiss led to a different approach, although the basic ideas remain. For instance, in the quadrangle case, a fundamental idea is to first classify the *reduced* Moufang quadrangles, and then extend these (see below).

First, I will state the theorem, then describe some basic ideas of the proof, then I will consider some different cases in more detail. But I will give no proofs; these can be found in the forthcoming book [51]. Also, since [51] is not yet available, I will mention some alternative sources for certain results.

A.1 The classification result

The classification result can best be stated in the way that Tits wrote it down as a conjecture many years ago, see for instance [45]: all Moufang polygons arise from algebraic and classical groups *in one or the other way*. More precisely:

Theorem A.1 *All Moufang polygons arise in a canonical way from algebraic groups of relative rank 2, from classical groups of rank 2, from mixed groups of rank 2, from mixed groups of type F_4 and relative rank 2, and from the Ree groups of type $^2F_4$.*

In the course of the present proof of this result, it became clear that the quadrangles related to the mixed groups of type F_4 and relative rank 2 were missing in the detailed version of Tits' original conjecture. However, in 1997, the proof of the above theorem was completed by Tits and Weiss and these new Moufang quadrangles were proved to exist via their commutation relations. Soon afterwards, they were recognized by Mühlherr and Van Maldeghem [26] as quadrangles arising as the fixed point sets of certain involutions in certain mixed buildings of type F_4. Hence they can be seen as being of algebraic origin in the sense that they arise from a building associated to an algebraic group by considering first a subbuilding (the building of mixed type F_4), and then a subcomplex (the fixed point set under an appropriate involution). This is also the way in which the Moufang octagons arise, but one has to take a polarity rather than an involution as last step. Abstractly, this case is typical: most Moufang polygons arise as fixed point structure of an automorphism in some higher rank building, but in general this building is the building associated to a simple algebraic group and this automorphism is an algebraic one, unlike the cases of Moufang octagons and Moufang quadrangles of type F_4.

We now give a more detailed version of the list in Theorem A.1 above. We divide this list in two parts. The first part concerns the "classical" examples: these can be obtained from classical notions such as fields, skew fields, quadratic forms, and the corresponding groups are classical groups, or certain subgroups of classical groups. The second part consists of the "exceptional" examples; these are obtained from buildings of exceptional type, or from buildings of classical type in an "exceptional way" (for instance via a triality, see below). In fact, the first part of the list consists of those polygons that are obtained from buildings of classical type, if one would allow infinite dimensional phenomena. This does not occur in the exceptional cases because the rank of a building of exceptional type is bounded. Hence, in a certain sense, these behave much better.

I. The Moufang projective planes of classical type are those defined over a field (Pappian case) or a skew field (Desarguesian case). The Moufang quadrangles of classical type are those arising from a σ-quadratic form, and their subquadrangles (this includes the mixed quadrangles, see below). There are no Moufang hexagons nor octagons of classical type.

II. The Moufang projective planes of exceptional type are those related to an alternative field; they arise from buildings of type E_6. The Moufang quadrangles of exceptional type arise from buildings of type E_6, E_7, E_8 and F_4; in the latter case, the building of type F_4 is related to a mixed group of that type. The Moufang hexagons arise from buildings of type D_4 (via a triality; one example is the split Cayley hexagon related to the groups of type G_2), E_6 (2 cases) or E_8. Finally, all Moufang octagons arise from polarities of buildings of type F_4 (necessarily in characteristic 2)

A.2 About the proof

Concentrating on the list II above, one could guess that a good way of proceeding with the proof of the classification would be to somehow try to reconstruct the ambient building, or the corresponding algebraic group. In fact, this is approximately the way Tits looked at this problem. But the eventual proof is much more along the lines of the classification of Moufang octagons as it appeared in [49]. The idea is to translate the Moufang condition into a group theoretical condition, and then to work with groups instead of geometries. The condition involves certain subgroups — the root groups — and the conditions allow a *parametrization* of these. The properties of the root groups then translate into algebraic properties of the parametrization that allows one to recognize fields, skew fields, vector spaces, certain algebras, etc. As a last step, one has to classify the algebraic structures thus obtained (for instance, for projective planes, the algebraic structure arising is an alternative division rings, and those which are not skew fields are always certain 8-dimensional algebras — more exactly Cayley-Dickson division algebras — over a field).

Let us make the above more precise. Assume that Γ is a Moufang n-gon for some $n \in \{3,4,6,8\}$. Let Σ be any apartment of Γ and label the points and lines of Σ by the integers modulo $2n$ so that i is incident with $i+1$. For all i, Let U_i be the root group corresponding to the n-path $(i, i+1, \ldots, i+n)$ and denote by U_+ the group generated by the root groups $U_1, U_2, \ldots, U_n$. Note that U_+ is a subgroup of the little projective group G. Then Tits and Weiss [51] show that $(U_+; U_1, U_2, \ldots, U_n)$ is unique up to conjugation in G and up to the renumbering the elements of Σ, and — more importantly — that it determines Γ uniquely.

Now, the subgroups $U_+, U_1, U_2, \ldots, U_n$ have the property that every element $u \in U_+$ can be written in a unique way as a product $u_1 u_2 \ldots u_n$, with $u_i \in U_i$, $i \in \{1,2,\ldots,n\}$. Moreover, for $i,j \in \{1,2,\ldots,n\}$, the commutator $[U_i, U_j]$ is a subgroup of the product $U_{i+1}U_{i+2}\cdots U_{j-1}$ if $i < j-1$, and $[U_i, U_j]$ is trivial for $j = i+1$ (this implies in particular that the above product $U_{i+1}U_{i+2}\cdots U_{j-1}$ is a subgroup of U_+). Hence the $(n+1)$-tuple $(U_+; U_1, U_2, \ldots, U_n)$ is completely determined by the abstract groups U_i, $i \in \{1,2,\ldots,n\}$, and the commutation relations $[u_i, u_j]$, for $u_i \in U_i$, $u_j \in U_j$, $i,j \in \{1,2,\ldots,n\}$, $i < j$. For example, to see this, let $n = 4$, and suppose $u_1 u_2 u_3 u_4, u_1', u_2' u_3' u_4' \in U_1 U_2 U_3 U_4 = U_+$ (with self explaining notation). Suppose $[u_1, u_4^{-1}] = u_2'' u_3'' \in U_2 U_3$ (where $[a,b] - a^{-1}b^{-1}ab$). Then one calculates

$$(u_1 u_2 u_3 u_4)(u_1' u_2' u_3' u_4') = (u_1 u_1')(u_2 [u_1', u_3^{-1}] u_2'' u_2')(u_3 u_3'' [u_2' u_4^{-1}] u_3')(u_4 u_4').$$

Important remark. Of course not every $(n+1)$-tuple of groups $(U_+;$

$U_1, U_2, \ldots, U_n$) satisfying the properties of the previous paragraph gives rise to a Moufang n-gon. Indeed, we could take the direct product $U_+ := U_1 \times U_2 \times \cdots \times U_n$ of arbitrary groups $U_1, U_2, \ldots, U_n$, but these cannot produce any Moufang n-gon. There exists, however, a condition that provides a Moufang n-gon. That condition guarantees the existence of the groups U_0 and U_{n+1} through some requirement on the automorphism group of U_+. It is for instance stated explicitly in my monograph [60], Theorem 5.5.2.

Now the classification proceeds as follows. Geometrically, it is clear that the groups $U_1, U_3, \ldots$ are mutually isomorphic, and similarly for the groups $U_2, U_4, \ldots$. So one identifies in a certain, but precise, way the groups U_i, i odd, with a certain — additive — group A, and the groups U_j, j even, with a certain additive group B. If n is odd, then also A and B are identified. The commutation relations then define functions from $A \times B$, from $A \times A$, and/or from $B \times B$ to A and/or B. The task is to prove as many identities of these functions in order to be able to nail them down. For instance, sometimes one is able to prove that A has the structure of a field, and B is a vector space over that field. The functions define an algebra on B over A and this algebra should then be classified. We give explicit examples of this situation below.

We remark that the groups A and B have nilpotency class at most 2, but not both are of class 2. So at least one is commutative (hence the additive notation). This result, and many other restrictions on the commutation relations in the different cases $n = 3, 4, 6, 8$ can be found in Tits' paper [50]. In fact, from that paper one deduces that labels can be chosen such that A is abelian, and that B is also abelian for $n = 3, 6$.

Let us now treat the separate cases $n = 3, 4, 6, 8$.

A.3 Some different cases

In case $n = 3$ (a case that was already completely done by Moufang herself, see [25]), the geometry implies that the U_i are commutative (this is nothing else than saying that the translation group of a translation plane is abelian, a result that is by now classical), and the commutation relation $[u_1, u_3]$, $u_1 \in U_1$, $u_3 \in U_3$, defines a product in the additive group A. One then shows that the addition and the multiplication puts the structure of an alternative division ring on $A = B$. Later, Bruck and Kleinfeld [6] and Kleinfeld [20] showed that every alternative division ring that is not a skew field is a Cayley-Dickson algebra over a field (and that corresponds precisely to a form of E_6). That concludes the classification of Moufang projective planes.

The second class of Moufang polygons that is treated in the literature is the case $n = 8$ (by Tits [49]). This case is much more involved than $n = 3$.

But the result is in fact much simpler. The labels above can be chosen such that A is a field of characteristic 2, and $B = A \times A$ is the group defined by the addition $(a, b) + (a', b') = (a + a', b + b' + a'a^\sigma)$, where σ is a square root of the Frobenius (and called a *Tits endomorphism* in [60]). So Moufang octagons are in bijective correspondence with pairs (A, σ), where A is a field of characteristic 2 and σ is a Tits endomorphism.

The two cases $n = 3$ and $n = 8$ are at the same time typical and extreme cases. They have in common that the axioms of the algebraic system follow from the properties of the group U_+ induced by the geometry (and the commutation relations $[U_i, U_j]$ can be expressed in terms of the algebraic system), but they differ in that for $n = 8$ there is no point in trying to classify further the algebraic system obtained, while for $n = 3$ the non associative alternative division rings can be further classified. The cases $n = 3, 8$ are, however, not typical with respect to uniformity within one case of n. Indeed, for $n = 4$, the algebraic system is not described by one set of axioms, but a lot of cases occur. However, recent work of Tom De Medts (unpublished) shows that an alternative approach leads to one set of axioms from which all different cases can be derived.

Now let $n = 6$. Already in [50] it is proved that the group U_+ has a subgroup $U_+^\triangle$ generated by U_1, U_3, U_5 (up to the labels) such that the 4-tuple $(U_+^\triangle; U_1, U_3, U_5)$ defines a Moufang projective plane Γ'. The points and lines of that projective plane can, without loss of generality, be identified with certain points of the hexagon. The existence of the root groups U_0 and U_6 implies the existence of a duality in Γ' having a special commutation property with respect to the root groups U_3. This enables one to show that Γ' must be Pappian (see Proposition B.4 of [60] for more details). Hence A carries the structure of a field. The other commutation relations now imply that B is a vector space over A, endowed with a certain map $\# : B \to B$ satisfying a list of properties. The triple $(A, B, \#)$ is closely related to a certain Jordan algebra. These algebras have been studied by Albert, Jacobson and several of Jacobson's students. Finally, they were classified by Petersson and Racine [27, 28]. There are two main families appearing here. For the first family, A is a field of characteristic 3 and B is a subfield of A containing all third powers of A. As a vector space, B can be infinite dimensional over A. The roles of A and B can be interchanged. These examples are exactly the Moufang hexagons related to mixed groups of type G_2. For the second family, B has dimension $d = 1, 3, 9$ or 27 over A. The case $d = 1$ corresponds with the split Cayley hexagon over A, the case $d = 3$ with a separable cubic extension B of A (if this extension is a Galois extension, then we obtain the triality hexagons that we already mentioned before), in the case $d = 9$, the vector space B is either a skew field of dimension 9 over its centre A, or B is the

Jordan algebra of fixed elements of an involution σ of the second kind in a skew field of dimension 9 over its centre C, where C is a quadratic separable extension of A. Finally, if $d = 27$, then B is an exceptional Jordan algebra over A, and there are two main constructions to obtain these. See also [45] and [60], 5.5.13 for more details of this list.

Finally let $n = 4$. Here the rough strategy is to show first that every Moufang quadrangle has a subquadrangle such that, without loss of generality, $[U_1, U_3]$ is trivial (see [50]). Such a subquadrangle is called of reduced type. So the first step is to classify all Moufang quadrangles of reduced type. This splits further in two cases: when also $[U_2, U_4]$ is trivial, then we obtain the quadrangles related to mixed groups. This class was already classified by Tits in an unpublished preprint around 1976. A detailed description of these quadrangles can be found in [60], but also in [24]. When $[U_2, U_4]$ is nontrivial, then A carries the structure of a skew field, B of a vector space over A (or over the centre of A) and we obtain classical quadrangles. More exactly, the classical quadrangles of orthogonal type and of reduced Hermitian type occur here.

The second step is then the non reduced case. Hence one has to extend the quadrangles of reduced type. It turns out that the extensions of quadrangles of reduced Hermitian type give all remaining classical quadrangles, while the extensions of quadrangles of orthogonal type are precisely the exceptional quadrangles. The algebraic systems of the families of type E_6, E_7 and E_8 involve pairs of vector spaces and the even Clifford algebra of a certain type of quadratic form. We have already commented about the Moufang quadrangles of type F_4, see [26] for a detailed description.

References

[1] P. Abramenko and H. Van Maldeghem, On opposition in spherical buildings and twin buildings, *Annals of Combinatorics* **4** (2000), 125 – 137.

[2] P. Abramenko and H. Van Maldeghem, Characterizations of generalized polygons and opposition in rank 2 twin buildings, submitted.

[3] R. Baer, Homogeneity of projective planes, *Amer. J. Math.* **64** (1942), 137 – 152.

[4] R. Bödi and L. Kramer, On homomorphisms between generalized polygons, *Geom. Dedicata*, **58** (1995), 1 – 14.

[5] L. Brouns, K. Tent and H. Van Maldeghem, Groups of projectivities of generalized quadrangles, *Geom. Dedicata* **73** (1998), 165 – 180.

[6] R. H. Bruck and E. Kleinfeld, The structure of alternative division rings, *Proc. Amer. Math. Soc.* **2** (1951), 878 – 890.

[7] F. Bruhat and J. Tits, Groupes réductifs sur un corps local, I. Données radicielles valuées, *Inst. Hautes Études Sci. Publ. Math.* **41** (1972), 5 – 252.

[8] F. Buekenhout and H. Van Maldeghem, Finite distance transitive generalized polygons, *Geom. Dedicata* **52** (1994), 41 – 51.

[9] A. M. Cohen and J. Tits, On generalized hexagons and a near octagon whose lines have three points, *European J. Combin.* **6** (1985), 13 – 27.

[10] J. R. Faulkner, Groups with Steinberg relations and coordinatization of polygonal geometries, *Mem. Amer. Math. Soc.* (10) **185** (1977).

[11] W. Feit and G. Higman, The nonexistence of certain generalized polygons, *J. Algebra* **1** (1964), 114 – 131.

[12] P. Fong and G. Seitz, Groups with a BN-pair of rank 2, I., *Invent. Math.* **21** (1973), 1 – 57.

[13] P. Fong and G. Seitz, Groups with a BN-pair of rank 2, II., *Invent. Math.* **24** (1974), 191 – 239.

[14] E. Govaert and H. Van Maldeghem, Distance-preserving maps of generalized polygons, I: Maps defined on the the set of flags, to appear in *Contr. Alg. Geom.*

[15] E. Govaert and H. Van Maldeghem, Distance-preserving maps of generalized polygons, II: Maps defined on the the set of points and/or lines, to appear in *Contr. Alg. Geom.*

[16] R. Gramlich and H. Van Maldeghem, Epimorphisms of generalized polygons, Part I: Geometrical characterizations, *Des. Codes Cryptogr.* **21** (2000), 99 – 111.

[17] R. Gramlich and H. Van Maldeghem, Epimorphisms of generalized polygons, Part II: Some existence and nonexistence results, to appear.

[18] T. Grundhöfer and N. Knarr, Topology in generalized quadrangles, *Topology Appl.* **34** (1990) , 139 – 152.

[19] C. Hering, On subgroups with trivial normaliser intersection, *J. Algebra* **20** (1972), 622 – 629.

[20] E. Kleinfeld, Alternative division rings of characteristic 2, *Proc. Nat. Acad. Sci. USA* **37** (1951), 818 – 820.

[21] N. Knarr, The nonexistence of certain topological polygons, *Forum Math.* **2** (1990), 603 – 612.

[22] L. Kamer, *Compact polygons*, Ph. D. thesis, Universität Tübingen, Germany, 1994.

[23] L. Kramer and K. Tent, Algebraic polygons, *J. Algebra* **182** (1996), 435 – 447.

[24] L. Kramer, K. Tent and H. Van Maldeghem, Simple groups of finite Morley rank and Tits-buildings, *Israel J. Math.* **109** (1999), 189 – 224.

[25] R. Moufang, Alternativkörper und der Satz vom vollständigen Vierseit, *Abh. Math. Sem. Univ. Hamburg* **9** (1933), 207 – 222.

[26] B. Mühlherr and H. Van Maldeghem, Exceptional Moufang quadrangles of type F_4, *Canad. J. Math.* **51** (1999), 347 – 371.

[27] H. P. Petersson and M. L. Racine, Jordan algebras of degree 3 and the Tits process, *J. Algebra* **98** (1986), 211 – 243.

[28] H. P. Petersson and M. L. Racine, Classification of algebras arising from the Tits process, *J. Algebra* **98** (1986), 244 – 279.

[29] M. A. Ronan, A geometric characterization of Moufang hexagons, *Invent. Math.* **57** (1980), 227 – 262.

[30] A. E. Schroth, Characterizing symplectic quadrangles by their derivations, *Arch. Math.* **58** (1992), 98 – 104.

[31] A. E. Schroth, *Topologische Laguerreebenen und Topologische Vierecke*, Dissertation, Braunschweig, 1992.

[32] K. Tent, Very homogeneous generalized n-gons of finite Morley rank, *J. London Math. Soc.* (2) **62** (2000), 1 – 15.

[33] K. Tent, Generalized polygons with split Levi factors, to appear in *Arch. Math.*

[34] K. Tent, Generalized polygons and model theory, to appear in "Generalized Polygons and Related Structures", *Acad. Wet. Lett. Sch. K. België.*

[35] K. Tent and H. Van Maldeghem, On irreducible BN-pairs of rank 2, to appear in *Forum Math.*

[36] K. Tent and H. Van Maldeghem, The Cherlin-Zil'ber conjecture for BN-pairs with panels of Morley rank 1, submitted.

[37] K. Tent and H. Van Maldeghem, Moufang polygons and irreducible spherical BN-pairs of rank 2, submitted.

[38] J. A. Thas, Characterization of generalized quadrangles by generalized homologies, *J. Combin. Theory Ser.* A **40** (1985), 331 – 341.

[39] J. A. Thas, The classification of all (x, y)-transitive generalized quadrangles, *J. Combin. Theory Ser.* A **42** (1986), 154 – 157.

[40] J. A. Thas, S. E. Payne and H. Van Maldeghem, Half Moufang implies Moufang for finite generalized quadrangles, *Invent. Math.* **105** (1991), 153 – 156.

[41] J. A. Thas ad H. Van Maldeghem, Generalized Desargues configurations in generalized quadrangles, *Bull. Soc. Math. Belg.* **42** (1990), 713 – 722.

[42] J. A. Thas and H. Van Maldeghem, The classification of all finite generalized quadrangles admitting a group acting transitively on ordered pentagons, *J. London Math. Soc.* (2) **51** (1994), 209 – 218.

[43] J. Tits, Sur la trialité et certains groupes qui s'en déduisent, *Inst. Hautes Études Sci. Publ. Math.* **2** (1959), 13 – 60.

[44] J. Tits, *Buildings of Spherical Type and Finite BN-pairs*, Lect. Notes in Math. **386**, Springer-Verlag, Berlin, Heidelberg, New York, 1974.

[45] J. Tits, Classification of buildings of spherical type and Moufang polygons: a survey, **in** *"Coll. Intern. Teorie Combin. Acc. Naz. Lincei,* Proceedings Roma 1973, *Atti dei convegni Lincei* **17** (1976), 229 – 246.

[46] J. Tits, Non-existence de certains polygones généralisés, I, *Invent. Math.* **36** (1976), 275 – 284.

[47] J. Tits, Endliche Spiegelungsgruppen, die als Weylgruppen auftreten, *Invent. Math.* **43** (1977), 283 – 295.

[48] J. Tits, Non-existence de certains polygones généralisés, II, *Invent. Math.* **51** (1979), 267 – 269.

[49] J. Tits, Moufang octagons and the Ree groups of type $^2F_4$, *Amer. J. Math.* **105** (1983), 539 – 594.

[50] J. Tits, Moufang polygons, I. Root data, *Bull. Belg. Math. Soc. Simon Stevin* **1** (1994), 455 – 468

[51] J. Tits and R. Weiss, *The classification of Moufang polygons*, to appear.

[52] J. van Bon, H. Cuypers and H. Van Maldehem, Hyperbolic lines in generalized polygons, *Forum Math.* **8** (1994), 343 – 362.

[53] H. Van Maldegem, Generalized polygons with valuation, *Arch. Math.* **53** (1989), 513 – 520.

[54] H. Van Maldeghem, A configurational characterization of the Moufang generalized polygons, *European J. Combin.* **11** (1990), 381 – 392.

[55] H. Van Maldeghem, Common characterizations of the finite Moufang polygons, in *Adv. Finite Geom. and Designs*, Proceedings "Third Isle of Thorn Conference on Finite Geometries and Designs", Brighton 1990 (ed. J. W. P. Hirschfeld *et al.*), Oxford University Press, Oxford (1991), 391 – 400.

[56] H. Van Maldeghem, A characterization of the finite Moufang hexagons by generalized homologies, *Pacific J. Math.* **151** (1991), 357 – 367.

[57] H. Van Maldeghem, The nonexistence of certain regular generalized polygons, *Arch. Math.* **64** (1995), 86 – 96

[58] H. Van Maldeghem, A finite generalized hexagon admitting a group acting transitively on ordered heptagons is classical, *J. Combin. Theory Ser.* A **75** (1996), 254 – 269.

[59] H. Van Maldeghem, Some consequences of a result of Brouwer, *Ars Combin.* **48** (1998), 185 – 190.

[60] H. Van Maldeghem, *Generalized Polygons*, Monographs in Math. **93**, Birkhäuser-Verlag, Basel, Boston, Berlin, 1998.

[61] H. Van Maldeghem, Sharply 2-transitive groups of projectivities in generalized polygons, to appear in *Discrete Math.*

[62] H. Van Maldeghem, An elementary construction of the split Cayley hexagon H(2), *Atti Sem. Mat. Fis. Univ. Modena* **XLVIII** (2000), 463 – 471.

[63] H. Van Maldeghem, Generalized polygons in projective spaces, to appear in "Generalized Polygons and Related Structures, *Acad. Wet. Lett. Sch. K. Belgïe.*

[64] H. Van Maldeghem, On a question of Arjeh Cohen: a characterization of Moufang projective planes, to appear in *Bull. Inst. Combin. Appl.*

[65] H. Van Maldeghem and I. Bloemen, Generalized hexagons as amalgamations of generalized quadrangles, *European J. Combin.* **14** (1993), 593 – 604.

[66] H. Van Maldeghem and R. Weiss, On finite Moufang polygons, *Israel J. Math.* **79** (1992), 321 – 330.

[67] M. Walker, On central root automorphisms of finite generalized hexagons, *J. Algebra* **78** (1982), 303 – 340.

[68] M. Walker, On central root automorphisms of finite generalized octagons, *European J. Combin.* **4** (1983), 65 – 86.

[69] R. Weiss, The nonexistence of certain Moufang polygons, *Invent. Math.* **51** (1979), 261 – 266.

[70] A. Yanushka, On order in generalized polygons, *Geom. Dedicata* **10** (1981), 451 – 458.

Buildings and Classical Groups

Linus Kramer*

Mathematisches Institut, Universität Würzburg

Am Hubland, D–97074 Würzburg, Germany

email: kramer@mathematik.uni-wuerzburg.de

In these notes we describe the classical groups, that is, the linear groups and the orthogonal, symplectic, and unitary groups, acting on finite dimensional vector spaces over skew fields, as well as their pseudo-quadratic generalizations. Each such group corresponds in a natural way to a point-line geometry, and to a spherical building. The geometries in question are projective spaces and polar spaces. We emphasize in particular the rôle played by root elations and the groups generated by these elations. The root elations reflect — via their commutator relations — algebraic properties of the underlying vector space.

We also discuss some related algebraic topics: the classical groups as permutation groups and the associated simple groups. I have included some remarks on K-theory, which might be interesting for applications. The first K-group measures the difference between the classical group and its subgroup generated by the root elations. The second K-group is a kind of fundamental group of the group generated by the root elations and is related to central extensions. I also included some material on Moufang sets, since this is an interesting topic. In this context, the projective line over a skew field is treated in some detail, and possibly with some new results. The theory of unitary groups is developed along the lines of Hahn & O'Meara [15]. Other important sources are the books by Taylor [31] and Tits [32], and the classical books by Artin [2] and Dieudonné [9]. The books by Knus [19] and W. Scharlau [29] should also be mentioned here. Finally, I would like to recommend the surveys by Cohen [7] and R. Scharlau [28].

While most of these matters are well-known to experts, there seems to be no book or survey article which contains these aspects simultaneously. Taylor's book [31] is a nice and readable introduction to classical groups, but it is clear that the author secretly thinks of finite fields — the non-commutative theory is almost non-existent in his book. On the other extreme, the book

*Supported by a Heisenberg fellowship by the Deutsche Forschungsgemeinschaft

by Hahn & O'Meara [15] contains many deep algebraic facts about classical groups over skew fields; however, their book contains virtually no geometry (the key words *building* or *parabolic subgroup* are not even in the index!). So, I hope that this survey — which is based on lectures notes from a course I gave in 1998 in Würzburg — is a useful compilation of material from different sources.

There is (at least) one serious omission (more omissions can be found in the last section): I have included nothing about coordinatization. The coordinatization of these geometries is another way to recover some algebraic structure from geometry. For projective planes and projective spaces, this is a classical topic, and I just mention the books by Pickert [23] and Hughes & Piper [17]. For generalized quadrangles (polar spaces of rank 2), Van Maldeghem's book [34] gives a comprehensive introduction. Understanding coordinates in the rank 2 case is in most cases sufficient in order to draw some algebraic conclusions.

I would like to thank Theo Grundhöfer and Bernhard Mühlherr for various remarks on the manuscript, and Katrin Tent. Without her, these notes would have become hyperbolic.

Contents

Preliminaries

We first fix some algebraic terminology. A (left) action of a group G on a set X is a homomorphism $G \longrightarrow \mathsf{Sym}(X)$ of G into the permutation group $\mathsf{Sym}(X)$ of X; the permutation induced by g is (in most cases) also denoted by $g = [x \longmapsto g(x)]$. A *right action* is an anti-homomorphism $G \longrightarrow \mathsf{Sym}(X)$; for right actions, we use exponential notation, $x \longmapsto x^g$. The set of all k-element subsets of a set X is denoted $\binom{X}{k}$.

Given a (not necessarily commutative) ring R (with unit 1), we let $R^\times$ denote the group of multiplicatively invertible elements. If $R^\times = R \backslash \{0\}$, then R is called a *skew field* or *division ring*. Occasionally, we use Wedderburn's Theorem.

Wedderburn's Theorem *A finite skew field is commutative and thus isomorphic to some Galois field $\mathbb{F}_q$, where q is a prime power.*

For a proof see Artin [2] Ch. I Thm. 1.14 or Grundhöfer [13]. □

The *opposite ring* R^{op} of a ring R is obtained by defining a new multiplication $a \cdot b = ba$ on R. An *anti-automorphism* α of R is a ring isomorphism $R \xrightarrow{\alpha} R^{\mathsf{op}}$, i.e. $(xy)^\alpha = y^\alpha x^\alpha$. The group consisting of all automorphisms and anti-automorphisms of D is denoted $\mathsf{AAut}(D)$; it has the automorphism group $\mathsf{Aut}(D)$ of D as a normal subgroup (of index 1 or 2). Let M be an abelian group. A *right R-module structure on M* is a ring homomorphism

$$R^{\mathsf{op}} \xrightarrow{\ \rho\ } \mathsf{End}(M);$$

as customary, we write

$$mrs = \rho(s \cdot r)(m)$$

for $r, s \in R$ and $m \in M$ ('scalars to the right'). The abelian category of all right R-modules is defined in the obvious way and denoted $\mathfrak{Mod}_R$; the subcategory consisting of all finitely generated right R-modules is denoted $\mathfrak{Mod}_R^{\mathsf{fin}}$. If D is a skew field, then $\mathfrak{Mod}_D^{\mathsf{fin}}$ is the category of all finite dimensional *vector spaces* over D. Similarly, we define the category of left R-modules $_R\mathfrak{Mod}$; given a right R-module M, we have the *dual* $M^\vee = \mathsf{Hom}_R(M, R)$ which is in a natural way a left R-module.

A right module over a skew field will be called a *right vector space* or just a vector space. Mostly, we will consider (finite dimensional) right vector spaces (so linear maps act from the left and scalars act from the right), but occasionally we will need both types. Of course, all these distinctions are obsolete over commutative skew fields, but in the non-commutative case one has to be careful.

1   Projective geometry and the general linear group

In this first part we consider the projective geometry over a skew field D and the related groups.

1.1   The general linear group

> *We introduce the projective geometry $\mathsf{PG}(V)$ associated to a finite dimensional vector space V and the general linear group $\mathsf{GL}(V)$, as well as the general semilinear group $\Gamma\mathsf{L}(V)$. We describe the relations between these groups and their projective versions $\mathsf{PGL}(V)$ and $\mathsf{P}\Gamma\mathsf{L}(V)$. Finally, we recall the 'first' Fundamental Theorem of Projective Geometry.*

Let V be a right vector space over a skew field D, of (finite) dimension $\dim(V) = n + 1 \geq 2$. The collection of all k-dimensional subspaces of V is the *Grassmannian*

$$\mathsf{Gr}_k(V) = \{X \subseteq V \mid \dim(X) = k\}.$$

The elements of $\mathsf{Gr}_1(V)$, $\mathsf{Gr}_2(V)$ and $\mathsf{Gr}_n(V)$ are called *points*, *lines*, and *hyperplanes*, respectively. Two subspaces X, Y are called *incident*,

$$X * Y,$$

if either $X \subseteq Y$ or $Y \subseteq X$. The resulting n-sorted structure

$$\mathsf{PG}(V) = (\mathsf{Gr}_1(V), \dots, \mathsf{Gr}_n(V), *)$$

is the *projective geometry* of rank n over D.

It is clear that every linear bijection of V induces an automorphism of $\mathsf{PG}(V)$. More generally, every semilinear bijection induces an automorphism of $\mathsf{PG}(V)$. Recall that a group endomorphism f of $(V, +)$ is called *semilinear* (relative to an automorphism θ of D) if

$$f(va) = f(v)a^\theta$$

holds for all $a \in D$ and $v \in V$. The group of all semilinear bijections of V is denoted $\Gamma\mathsf{L}(V)$; it splits as a semidirect product with the *general linear group* $\mathsf{GL}(V)$, the groups consisting of all linear bijections, as a normal subgroup,

$$1 \longrightarrow \mathsf{GL}(V) \longrightarrow \Gamma\mathsf{L}(V) \overset{\longleftarrow}{\longrightarrow} \mathsf{Aut}(D) \longrightarrow 1.$$

As usual, we write $\mathsf{GL}(D^n) = \mathsf{GL}_n(D)$, and similarly for the groups $\mathsf{PGL}(V)$ and $\mathsf{P\Gamma L}(V)$ induced on $\mathsf{PG}(V)$.

The kernel of the action of $\mathsf{GL}(V)$ on $\mathsf{PG}(V)$ consists of all maps of the form $\rho_c : v \longmapsto vc$, where $c \in \mathsf{Cen}(D^\times)$. Similarly, the kernel of the action of $\mathsf{\Gamma L}(V)$ consists of all maps of the form $\rho_c : v \longmapsto vc$, for $c \in D^\times$. These groups fit together in a commutative diagram with exact rows and columns

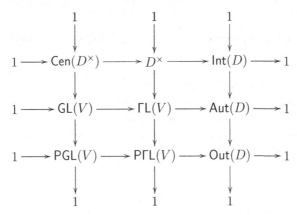

as is easily checked, see Artin [2] II.10, p. 93. Here $\mathsf{Int}(D)$ is the group of inner automorphisms $d \longmapsto d^a = a^{-1}da$ of D, where $a \in D^\times$, and $\mathsf{Out}(D)$ is the quotient group $\mathsf{Aut}(D)/\mathsf{Int}(D)$. The bottom line shows the groups induced on $\mathsf{PG}(V)$, and the top line the kernels of the respective actions. For example, $\mathsf{Out}(\mathbb{H}) = 1$ holds for the quaternion division algebra $\mathbb{H}$ over any real closed field, whence $\mathsf{PGL}_{n+1}(\mathbb{H}) \cong \mathsf{P\Gamma L}_{n+1}(\mathbb{H})$ for $n \geq 1$. On the other hand, $\mathsf{Int}(D) = 1$ if D is commutative. Recall the Fundamental Theorem of Projective Geometry, which basically says that $\mathsf{Aut}(\mathsf{PG}(V)) = \mathsf{P\Gamma L}(V)$. The actual statement is in fact somewhat stronger.

1.1.1 The Fundamental Theorem of Projective Geometry, I
For $i = 1, 2$, let V_i be vector spaces over D_i, of (finite) dimensions $n_i + 1 \geq 3$, and let

$$\mathsf{PG}(V_1) \xrightarrow[\cong]{\phi} \mathsf{PG}(V_2)$$

be an isomorphism. Then $n_1 = n_2$, and there exists a skew field isomorphism $\theta : D_1 \xrightarrow{\cong} D_2$ and a θ-semilinear bijection

$$f : V_1 \longrightarrow V_2$$

(i.e. $f(va) = f(v)a^\theta$)) such that $\phi(X) = f(X)$ for all $X \in \mathsf{Gr}_k(V_1)$, for $k = 1, 2, \ldots, n_1$. In particular,

$$\mathsf{Aut}(\mathsf{PG}(V)) = \mathsf{P\Gamma L}(V).$$

For a proof see Hahn & O'Meara [15] 3.1.C, Artin [2] Ch. II Thm. 2.26, Lüneburg [20] [21], or Faure & Frölicher [10]. □

The theorem is also valid for infinite dimensional projective spaces; this will be important in the second part, when we consider hermitian forms. Note that for $n = 1$, the structure $\mathsf{PG}(V) = (\mathsf{Gr}_1(V), =)$ is rather trivial. We will come back to a refined version of the projective line in Section 1.8.

1.2   Elations, transvections, and the elementary linear group

We introduce elations, transvections, and the elementary linear group $\mathsf{EL}(V)$ generated by all elations. The projective elementary group is a simple group, except for some low dimensional cases over fields of small cardinality; in the commutative case, the elementary linear group coincides with the special linear group $\mathsf{SL}(V)$.

As in the previous section, V is a right vector space over D of finite dimension $n + 1 \geq 2$. Let $A \in \mathsf{Gr}_n(V)$ be a hyperplane, and let $z \in \mathsf{Gr}_1(V)$ be a point incident with A. An automorphism τ of $\mathsf{PG}(V)$ which fixes A pointwise and z linewise is called an *elation* or *translation*, with *axis* A and *center* z. We can choose a base $b_0, \ldots, b_n$ of V such that A is spanned by $b_0, \ldots, b_{n-1}$ and z by b_0, and such that τ is represented by a matrix of the form

$$\begin{pmatrix} 1 & & & & & a \\ & 1 & & & & \\ & & 1 & & & \\ & & & 1 & & \\ & & & & \ddots & \\ & & & & & 1 \end{pmatrix}$$

for some $a \in D$. Such an elation is called a (z, A)-elation; the group $U_{(z,A)}$ consisting of all (z, A)-elations is isomorphic to the additive group $(D, +)$. There is a coordinate-free way to describe elations. Let

$$V^\vee = \mathsf{Hom}_D(V, D)$$

denote the dual of V. This is a left vector space over D, so we can form the tensor product $V \otimes_D V^\vee$; this is an abelian group (even a ring) which is naturally isomorphic to the endomorphism ring of V,

$$\mathsf{End}_D(V) \cong V \otimes_D V^\vee.$$

We write $x \otimes \phi = x\phi = [v \longmapsto x\phi(v)]$. If the elements of V are represented as column vectors and the elements of $V^{\vee}$ as row vectors, then $x\phi$ is just the standard matrix product. Now let $(u, \rho) \in V \times V^{\vee}$ be a pair such that the endomorphism $u\rho \in \mathsf{End}_D(V)$ is nilpotent, i.e. $u\rho u\rho = u\rho(u)\rho = 0$ (which is equivalent to $\rho(u) = 0$). The linear map

$$\tau_{u\rho} = \mathsf{id}_V + u\rho = [\, v \longmapsto v + u\rho(v) \,]$$

is called a *transvection*. Note that

$$\tau_{ua\rho}\tau_{ub\rho} = \tau_{u(a+b)\rho} \qquad \text{whence} \qquad \tau_{u\rho}\tau_{-u\rho} = \mathsf{id}_V.$$

Such a transvection induces an elation on $\mathsf{PG}(V)$, and conversely, every elation in $\mathsf{PG}(V)$ is induced by a transvection. In fact, suppose that $u\rho \neq 0$. The center of $\tau_{u\rho}$ is $z = uD \in \mathsf{Gr}_1(V)$, and the axis is $A = \mathsf{ker}(\rho)$. The group $(D, +) \cong \{\tau_{ua\rho} \mid a \in D\}$ maps isomorphically onto $U_{(z,A)}$. Suppose that $\tau_{u\phi} \neq 1 \neq \tau_{v\psi}$ are commuting transvections,

$$\tau_{u\phi}\tau_{v\psi} = \tau_{v\psi}\tau_{u\phi}.$$

This implies that $u\phi(v)\psi = v\psi(u)\phi$, and thus $\psi(u) = \phi(v) = 0$.

1.2.1 Lemma *Two non-trivial transvections commute if and only if they have either the same axis or the same center.* $\qquad\square$

The group generated by all transvections is the *elementary linear group* $\mathsf{EL}(V)$. It is a normal subgroup of the group $\mathsf{\Gamma L}(V)$ and of $\mathsf{GL}(V)$ (because the conjugate of a transvection is again a transvection). The subgroup in $\mathsf{Aut}(\mathsf{PG}(V))$ generated by all elations is the *little projective group* of $\mathsf{PG}(V)$; it is normal and an epimorphic image of $\mathsf{EL}(V)$. We denote this group by $\mathsf{PEL}(V)$. We collect a few facts about the group $\mathsf{EL}(V)$.

1.2.2 Lemma *The action of $\mathsf{PEL}(V)$ on $\mathsf{Gr}_1(V)$ is 2-transitive and in particular primitive.*

Proof. It suffices to show that given a point $p \in \mathsf{Gr}_1(V)$, the stabilizer $\mathsf{PEL}(V)_p$ acts transitively on $\mathsf{Gr}_1(V) \setminus \{p\}$. Also, it is easy to see that given a line ℓ passing through p, the stabilizer $\mathsf{PEL}(V)_{p,\ell}$ acts transitively on the points lying on ℓ and different from p — choose an axis A passing through the center p, not containing ℓ. It is also not difficult to prove that $\mathsf{PEL}(V)_p$ acts transitively on the lines passing through p (here, one chooses an axis containing p, but with a different center). The result follows from these observations. $\qquad\square$

Mutatis mutandis, one proves that $\mathsf{PEL}(V)$ acts 2-transitively on the hyperplanes.

1.2.3 Lemma *If $n \geq 2$, then $\mathsf{EL}(V)$ is perfect, i.e. $[\mathsf{EL}(V), \mathsf{EL}(V)] = \mathsf{EL}(V)$. The same is true for $n = 1$, provided that $|D| \geq 4$.*

Proof. Assume first that $n \geq 2$. Given $\tau_{u\phi}$, choose ψ linearly independent from ϕ, such that $\psi(u) = 0$ and $v \in \ker(\phi)$ with $\psi(v) = 1$, Then $[\tau_{\psi u}, \tau_{\phi v}] = \tau_{\phi u}$, so every transvection is a commutator of transvections. In the case $n = 1$ one uses some clever matrix identities, and the fact that the equation $x^2 - 1 \neq 0$ has a solution in D if $|D| \geq 4$, see Hahn & O'Meara [15] 2.2.3. □

1.2.4 Proposition *If $n \geq 2$ or if $n = 1$ and $|D| \geq 4$, then $\mathsf{PEL}(V)$ is a simple group.*

Proof. The proof uses Iwasawa's simplicity criterion, see Hahn & O'Meara [15] 2.2.B. The group $\mathsf{PEL}(V)$ is perfect, and the stabilizer of a point z has an abelian normal subgroup, the group consisting of all elations with center z. These are the main ingredients of the proof, see *loc.cit.* 2.2.13. See also Artin [2] Ch. IV Thm. 4.10. □

We collect some further results about the group $\mathsf{EL}(V)$. Let $H \cong \mathsf{GL}_1(D)$ denote the subgroup of $\mathsf{GL}(V)$ consisting of all matrices of the form

$$\begin{pmatrix} a & & & & & \\ & 1 & & & & \\ & & 1 & & & \\ & & & 1 & & \\ & & & & \ddots & \\ & & & & & 1 \end{pmatrix}$$

for $a \in D^\times$. One can show that $\mathsf{GL}(V) = H\mathsf{EL}(V)$, see Hahn & O'Meara [15] 1.2.10 (the theorem applies, since a skew field is a euclidean ring in the terminology of *loc.cit.*). An immediate consequence is the following.

1.2.5 *If D is commutative, then $\mathsf{EL}(V) = \mathsf{SL}(V)$.* □

(Artin [2] denotes the group $\mathsf{EL}(V)$ by $\mathsf{SL}(V)$, even if D is not commutative. Hahn & O'Meara [15] — and other modern books — have a different terminology; in their book, $\mathsf{SL}(V)$ is the kernel of the reduced norm.)

The next result follows from the fact that $\mathsf{EL}(V)$ acts strongly transitively on the building $\Delta(V)$; such an action is always primitive on the vertices of a fixed type, because the maximal parabolics are maximal subgroups. In our case, the vertices of the building $\Delta(V)$ are precisely the subspaces of V, see Section 1.5.

1.2.6 *The action of $\mathsf{EL}(V)$ on $\mathsf{Gr}_k(V)$ is primitive, for $1 \leq k \leq n$. It is two-transitive if and only if $k = 1, n$.* □

Finally, we mention some exceptional phenomena.

1.2.7 Suppose that $D \cong \mathbb{F}_q$ is finite, and let $\mathsf{PSL}_m(q) = \mathsf{PEL}_m(\mathbb{F}_q)$. There are the following isomorphisms (and no others, see Hahn & O'Meara [15] p. 81).

$$\mathsf{PSL}_2(2) \cong \mathsf{Sym}(3)$$
$$\mathsf{PSL}_2(3) \cong \mathsf{Alt}(4)$$
$$\mathsf{PSL}_2(4) \cong \mathsf{PSL}_2(5) \cong \mathsf{Alt}(5)$$
$$\mathsf{PSL}_2(7) \cong \mathsf{PSL}_3(2)$$
$$\mathsf{PSL}_2(9) \cong \mathsf{Alt}(6)$$
$$\mathsf{PSL}_4(2) \cong \mathsf{Alt}(8)$$

In particular, the groups $\mathsf{PSL}_2(q)$ are not perfect for $q = 2, 3$. Note that the groups $\mathsf{PSL}_3(4)$ and $\mathsf{PSL}_4(2)$ have the same order $20\,160$ without being isomorphic. □

1.3 K_1 and the Dieudonné determinant

We explain the connection between the elementary linear group,
the first K-group $\mathsf{K}_1(D)$ and the Dieudonné determinant.

Recall the category $\mathfrak{Mod}_D^{\mathsf{fin}}$ of all finite dimensional vector spaces over D. Let $\mathcal{M} = \{D^n \mid n \geq 0\}$; these vector spaces together with the linear maps between them form a small and full subcategory of $\mathfrak{Mod}_D^{\mathsf{fin}}$. Every vector space V in $\mathfrak{Mod}_D^{\mathsf{fin}}$ is isomorphic to a unique element $[V] \in \mathcal{M}$. We make $\mathcal{M}$ into an additive semigroup with addition $[V] + [W] = [V \oplus W]$, and neutral element $[0]$. The dimension functor yields an isomorphism

$$(\mathcal{M}, +) \xrightarrow{\ \mathsf{dim}\ } (\mathbb{N}, +).$$

Now $\mathsf{K}_0(D) = \mathsf{K}_0(\mathfrak{Mod}_D^{\mathsf{fin}})$ is defined to be the Grothendieck group generated by the additive semigroup of isomorphism classes of finite dimensional vector spaces over D, i.e. $\mathsf{K}_0(D) \cong \mathbb{Z}$. (The Grothendieck group of a commutative semigroup S is the universal solution $G(S)$ of the problem

where H is any group and f is any semigroup homomorphism.) While all this is rather trivial general nonsense (but only since D is a skew field!), there are higher-rank K-groups which bear more information even for skew fields.

For $n \leq m$, there is a natural inclusion $\mathsf{GL}_n(D) \hookrightarrow \mathsf{GL}_m(D)$ (as block matrices in the upper left, with 1s on the diagonal in the lower right). Let $\mathsf{GL}_{\mathsf{stb}}(D)$ denote the direct limit over these inclusions (this is called the *stable linear group* — not to be confused with stability theory in the model theoretic sense), and let $\mathsf{EL}_{\mathsf{stb}}(D)$ denote the corresponding direct limit over the groups $\mathsf{EL}_n(D)$. For $n \geq 1$, there are exact sequences

$$
\begin{array}{ccccccccc}
1 & \longrightarrow & \mathsf{EL}_n(D) & \longrightarrow & \mathsf{GL}_n(D) & \longrightarrow & \mathsf{GL}_n(D)/\mathsf{EL}_n(D) & \longrightarrow & 1 \\
& & \downarrow & & \downarrow & & \downarrow {\scriptstyle \cong} & & \\
1 & \longrightarrow & \mathsf{EL}_{\mathsf{stb}}(D) & \longrightarrow & \mathsf{GL}_{\mathsf{stb}}(D) & \longrightarrow & \mathsf{GL}_{\mathsf{stb}}(D)/\mathsf{EL}_{\mathsf{stb}}(D) & \longrightarrow & 1
\end{array}
$$

We denote these quotients

$$
\mathsf{K}_{1,n}(D) = \mathsf{GL}_n(D)/\mathsf{EL}_n(D)
$$

and put $\mathsf{K}_1(D) = \mathsf{GL}_{\mathsf{stb}}(D)/\mathsf{EL}_{\mathsf{stb}}(D)$. The groups $\mathsf{K}_{1,n}(D)$ are *stable*, i.e. independent of n; there are isomorphisms $\mathsf{K}_{1,2}(D) \cong \mathsf{K}_{1,3}(D) \cong \cdots \cong \mathsf{K}_1(D)$, see Hahn & O'Meara [15] 2.2.4. In this way we have obtained the *first K-group* $\mathsf{K}_1(D)$.

1.3.1 Proposition *There is an isomorphism* $\mathsf{K}_1(D) \cong D^\times/[D^\times, D^\times]$. *The composite*

$$
\begin{array}{ccc}
\mathsf{GL}_n(D) & \longrightarrow & \mathsf{GL}_{\mathsf{stb}}(D) \\
& \searrow & \downarrow {\scriptstyle \mathsf{det}} \\
& & \\
\mathsf{K}_1(D) & \xrightarrow{\ \cong\ } & D^\times/[D^\times, D^\times]
\end{array}
$$

is precisely the Dieudonné determinant det, *see Hahn & O'Meara [15] 2.2.2 and Artin [2] Ch. IV Thm. 4.6.*   □

Since the determinant takes values in an abelian group, it is invariant under base change (matrix conjugation); in particular, there is a well-defined (base independent) determinant map

$$
\mathsf{GL}(V) \xrightarrow{\ \mathsf{det}\ } D^\times/[D^\times, D^\times].
$$

We end this section with a commutative diagram which compares the linear and the projective actions. Given $c \in D^\times$, we denote its image in

$D^\times/[D^\times, D^\times]$ by $\bar{c}$. Let $C = \mathsf{Cen}(D^\times)$.

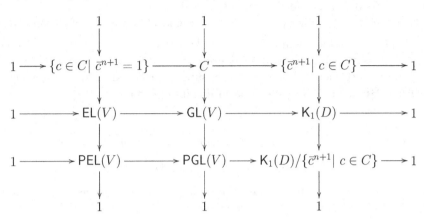

Here are some examples. If D is commutative, then $\mathsf{K}_1(D) = D^\times$. Now let $\mathbb{H}$ denote the quaternion division algebra over a real closed field $\mathbb{R}$. The *norm* N is defined as $N(x_0 + \mathbf{i}x_1 + \mathbf{j}x_2 + \mathbf{ij}x_3) = x_0^2 + x_1^2 + x_2^2 + x_3^2$. Then $\mathbb{H}^\times \xrightarrow{N} \mathbb{R}_{>0}$ has kernel $[\mathbb{H}^\times, \mathbb{H}^\times] = \mathbb{S}^3$, whence $\mathsf{K}_1(\mathbb{H}) \cong \mathbb{R}_{>0}$.

1.4 Steinberg relations and $\mathbf{K}_2$

We introduce the Steinberg relations, which are the basic commutator relations for projective spaces, and indicate briefly the connection with higher K-theory.

In this section we assume that $\dim(V) = n + 1$ is finite, and that $n \geq 2$. Let $b_0, \dots, b_n$ be a base for V, and let $\beta_0, \dots, \beta_n \in V^\vee$ be the dual base (i.e. $\beta_i(b_j) = \delta_{ij}$). Let

$$\tau_{ij}(a) = \tau_{b_i a \beta_j}, \qquad \text{for } i \neq j.$$

Thus $\tau_{ij}(a)$ can be pictured as a matrix with 1s on the diagonal, an entry a at position (i, j) (ith row, jth column), and 0s else. We claim that the elations $\{\tau_{ij}(a) | \ a \in D, \ i \neq j\}$ generate $\mathsf{EL}(V)$. Indeed, it is easy to see that the group generated by these elations acts transitively on incident point-hyperplane pairs; therefore, it contains all elations. The maps $\tau_{ij}(a)$ satisfy the following relations, as is easily checked.

SR1 $\tau_{ij}(a)\tau_{ij}(b) = \tau_{ij}(a + b)$ for $i \neq j$.

SR2 $[\tau_{ij}(a), \tau_{kl}(b)] = 1$ for $i \neq k$ and $j \neq l$.

SR3 $[\tau_{ij}(a), \tau_{jk}(b)] = \tau_{ik}(ab)$ for i, j, k pairwise distinct.

These are the *Steinberg relations*. They show that the algebraic structure of the skew field D is encoded in the little projective group $\mathsf{PEL}_n(D)$.

For each pair (i,j) with $i \neq j$ we fix an isomorphic copy U_{ij} of the additive group $(D,+)$, and an isomorphism $\tau_{ij} : (D,+) \overset{\cong}{\longrightarrow} U_{ij}$. For $n \geq 2$, we define $\mathsf{St}_{n+1}(D)$ as the free amalgamated product the of $n(n+1)$ groups $U_{ij} = \{\tau_{ij}(a)| \; a \in D\}$, factored by the normal subgroup generated by the Steinberg relations **SR2**, **SR3**. There is a natural epimorphism

$$\mathsf{St}_{n+1}(D) \longrightarrow \mathsf{EL}_{n+1}(D)$$

whose kernel is denoted $\mathsf{K}_{2,n+1}(D)$. Again, there are natural maps

$$\mathsf{K}_{2,n+1}(D) \longrightarrow \mathsf{K}_{2,m+1}(D)$$

for $m \geq n$, and one can consider the limits, the *stable groups* $\mathsf{St}_{\mathsf{stb}}(D)$ and $\mathsf{K}_2(D)$. Clearly, there are exact sequences

$$1 \longrightarrow \mathsf{K}_{2,n+1}(D) \longrightarrow \mathsf{St}_{n+1}(D) \longrightarrow \mathsf{GL}_{n+1}(D) \longrightarrow \mathsf{K}_{1,n+1}(D) \longrightarrow 1$$

(and similarly in the limit). The groups $\mathsf{K}_2(D)$ bear some information about the skew field D. See Milnor [22] for $\mathsf{K}_2(D)$ of certain fields D; for quaternion algebras, $\mathsf{K}_2(D)$ is determined in Alperin-Dennis [1]. One can prove that $\mathsf{K}_{2,n+1}(D) \cong \mathsf{K}_2(D)$ for $n > 1$, see Hahn & O'Meara [15] 4.2.18 and that $\mathsf{St}_{n+1}(D)$ is a universal central extension of $\mathsf{EL}_{n+1}(D)$, provided that $n \geq 4$, see *loc.cit.* 4.2.20, or that $n \geq 3$ and that $\mathsf{Cen}(D)$ has at least 5 elements, see Strooker [30] Thm. 1. We just mention the following facts.

(1) If D is finite, then $\mathsf{K}_2(D) = 0$, so the groups $\mathsf{SL}_{n+1}(q)$ are centrally closed for $n \geq 4$ (they don't admit non-trivial central extensions), see Hahn & O'Meara [15] 2.3.10 (in low dimensions over small fields, there are exceptions, see *loc.cit.*). Also, the Steinberg relations yield a presentation of the groups $\mathsf{SL}_{n+1}(q)$ for $n \geq 4$.

(2) Suppose that D is a field with a primitive mth root of unity. Let $\mathsf{Br}(D)$ denote its Brauer group. Then there is an exact sequence of abelian groups

$$1 \longrightarrow [\mathsf{K}_2(D)]^m \longrightarrow \mathsf{K}_2(D) \longrightarrow \mathsf{Br}(D) \longrightarrow [\mathsf{Br}(D)]^m \longrightarrow 1 \, .$$

(where we write $[A]^m = \{a^m| \; a \in A\}$ for an abelian group $(A, \cdot)$), see Hahn & O'Meara [15] 2.3.12.

Applications of K_2, e.g. in number theory, are mentioned in Milnor [22] and in Rosenberg [27].

1.5 Different notions of projective space, characterizations

We introduce the point-line geometry and the building obtained from a projective geometry and compare the resulting structures.

Then we mention the 'second' Fundamental Theorem of Projective Geometry which characterizes projective geometries of rank at least 3. We describe the Tits system (BN-pair) for the projective geometry and the root system, and we explain how the root system reflects properties of commutators of root elations.

A *point-line geometry* is a structure

$$(\mathcal{P}, \mathcal{L}, *),$$

where $\mathcal{P}$ and $\mathcal{L}$ are non-empty disjoint sets, and $* \subseteq (\mathcal{P} \cup \mathcal{L}) \times (\mathcal{P} \cup \mathcal{L})$ is a symmetric and reflexive binary relation, such that $*|_{\mathcal{P} \times \mathcal{P}} = \mathrm{id}_{\mathcal{P}}$ and $*|_{\mathcal{L} \times \mathcal{L}} = \mathrm{id}_{\mathcal{L}}$. Given a projective geometry $\mathsf{PG}(V) = (\mathsf{Gr}_1(V), \dots, \mathsf{Gr}_n(V), *)$ of rank $n \geq 2$, we can consider the point-line geometry

$$\mathsf{PG}(V)_{1,2} = (\mathsf{Gr}_1(V), \mathsf{Gr}_2(V), *).$$

It is easy to recover the whole structure $\mathsf{PG}(V)$ from this; call a set X of points a *subspace* if it has the following property: for every triple of pairwise distinct collinear points p, q, r (i.e. there exists a line $\ell \in \mathsf{Gr}_2(V)$ with $p, q, r * \ell$), we have the implication

$$(p, q \in X) \quad \implies \quad (r \in X).$$

We define the *rank* of a subspace inductively as $\mathrm{rk}(\varnothing) = -1$, and $\mathrm{rk}(X) \geq k+1$ if X contains a proper subspace $Y \subsetneq X$ with $\mathrm{rk}(Y) \geq k$. Clearly, $\mathsf{Gr}_{k+1}(V)$ can be identified with the set of all subspaces of rank k.

The point-line geometry $(\mathcal{P}, \mathcal{L}, *) = (\mathsf{Gr}_1(V), \mathsf{Gr}_2(V), *)$ has the following properties.

PG1 Every line is incident with at least 3 distinct points.

PG2 Any two distinct points p, q can be joined by a unique line which we denote by $p \vee q$.

PG3 There exist at least 2 distinct lines.

PG4 If p, q, r are three distinct points, and if ℓ is a line which meets $p \vee q$ and $p \vee r$ in two distinct points, then ℓ meets $q \vee r$.

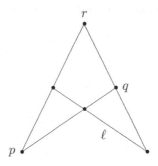

Axiom **PG4** is also called *Veblen's axiom* or the *Veblen-Young property*. A point-line geometry which satisfies these axioms is called a *projective (point-line) geometry*. If there exist two lines which don't intersect, then it is called a *projective space*, otherwise a *projective plane*.

Veblen's axiom PG4 is the important 'geometric' axiom in this list; the axioms PG1–PG3 exclude only some obvious pathologies. It is one of the marvels of incidence geometry that this simple axiom encodes — by the Fundamental Theorem of Projective Geometry stated below — the whole theory of skew fields, vector spaces, and linear algebra.

1.5.1 The Fundamental Theorem of Projective Geometry, II
Let $(\mathcal{P}, \mathcal{L}, *)$ be a projective (point-line) space which is not a projective plane. Then there exists a skew field D, unique up to isomorphism, and a right vector space V over D, unique up to isomorphism, such that

$$(\mathcal{P}, \mathcal{L}, *) = (\mathsf{Gr}_1(V), \mathsf{Gr}_2(V), *).$$

Here, the vector space dimension can be infinite. In fact, the dimension is finite if and only if one the following holds:
(1) Every subspace has finite rank.
(2) There exists no subspace U and no automorphism ϕ such that $\phi(U)$ is a proper subset of U.

The theorem is folklore; we just refer to Lüneburg [20] [21], or to Faure & Frölicher [10] for a category-theoretic proof. □

Let $V^{\vee}$ denote the dual of V. This is a right vector space over D^{op}; it is easy to see that there is an isomorphism

$$\mathsf{PG}(V^{\vee})_{1,2} \cong (\mathsf{Gr}_n(V), \mathsf{Gr}_{n-1}(V), *).$$

Now consider the following structure (for finite dimension $\dim(V) = n + 1$). Let $\mathcal{V} = \mathsf{Gr}_1(V) \cup \mathsf{Gr}_2(V) \cup \cdots \cup \mathsf{Gr}_n(V)$ and let Δ denote the collection of all subsets of $\mathcal{V}$ which consist of pairwise incident elements. Such a set is finite and has at most n elements. Then $\Delta(V) = (\Delta, \subseteq)$ is a poset, and in fact an abstract simplicial complex of dimension $n - 1$. The set $\mathcal{V}$ can be identified with the minimal elements (the vertices) of Δ. There is an exact sequence

$$1 \longrightarrow \mathsf{Aut}(\mathsf{PG}(V)) \longrightarrow \mathsf{Aut}(\Delta(V)) \longrightarrow \mathsf{AAut}(D)/\mathsf{Aut}(D) \longrightarrow 1$$

The poset $\Delta(V)$ is the *building* associated to the projective space $\mathsf{PG}(V)$. See Brown [4], Garrett [12], Grundhöfer [14] (these proceedings), Ronan [26], Taylor [31], Tits [32]. Here, we view a building as a simplicial complex, without a type function (the type function **type** would associate to a vertex $v \in \mathsf{Gr}_k(V)$ the number k). Such a type function can always be defined and is unique up to automorphisms; if we consider only type-preserving automorphisms, then we obtain $\mathsf{Aut}(\mathsf{PG}(V))$ as the automorphism group. (In Grundhöfer [14], the buildings are always endowed with a type funtion.) There is no natural way to recover $\mathsf{PG}(V)$ from $\Delta(V)$, but we can recover both $\mathsf{PG}(V)$ and $\mathsf{PG}(V^\vee)$ simultaneously. The following diagram shows the various 'expansions' and 'reductions'. Only the solid arrows describe natural constructions; the dotted arrows require the choice of a type function, i.e. one has to choose which elements are called points, and which ones are called hyperplanes.

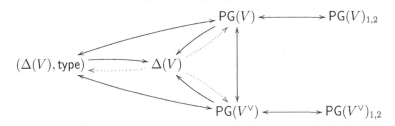

This is probably the right place to introduce the *Tits system* (or *BN-pair*) of $\mathsf{PEL}(V)$. Actually, it is easier if we lift everything into the group $\mathsf{EL}(V)$ (we could equally well work with the Steinberg group $\mathsf{St}_n(D)$, or the general linear group $\mathsf{GL}(V)$).

1.5.2 The Tits system for $\mathsf{EL}(V)$ Let $b_0, \ldots, b_n$ be a base for V, and let $p_i = b_i D$. Every proper subset of $\{p_1, \ldots, p_n\}$ spans a subspace; in this way, we obtain a collection

$$\Sigma^{(0)} = \{V_J = \mathsf{span}\{b_j \mid j \in J\} \mid \varnothing \neq J \subsetneq \{0, \ldots, n\}\}$$

of $2(2^n - 1)$ subspaces. With the natural inclusion '$\subseteq$', this becomes a poset and an abstract simplicial complex; combinatorially, this is the complex $\partial\Delta^{n+1}$ of all proper faces of a $n+1$-simplex; for $n = 2$, we have the points and sides of a triangle, and for $n = 3$ the points, edges and sides of a tetrahedron.

Now we consider a different simplicial complex, the *apartment* Σ. The vertices of Σ are the elements of $\Sigma^{(0)}$, and the higher rank elements are sets of pairwise incident elements. This complex can be pictured as follows: consider the first barycentric subdivision $\mathsf{Sd}\Sigma^{(0)}$ of $\Sigma^{(0)}$; the barycentric subdivision adds a vertex in every face of $\Sigma^{(0)}$. This flag complex

$$\Sigma = \mathsf{Sd}\partial\Delta^{n+1}$$

is the *apartment* spanned by $p_0, \ldots, p_n$.

Let $T \subseteq \mathsf{EL}(V)$ denote the pointwise stabilizer of $p_0, \ldots, p_n$ (equivalently, the elementwise stabilizer of $\Sigma^{(0)}$ or Σ), and N the setwise stabilizer of $\{p_0, \dots, p_n\}$ (or $\Sigma^{(0)}$, or Σ). Thus $N/T \cong \mathsf{Sym}(n+1)$; this quotient is the *Weyl group* for $\mathsf{PGL}(V)$. Let $B \subseteq \mathsf{EL}(V)$ denote the stabilizer of the flag

$$C = (p_0, p_0 \oplus p_2, p_0 \oplus p_1 \oplus p_2, \dots, p_0 \oplus \cdots \oplus p_{n-1})$$

With respect to the base $b_0, \dots, b_n$, the group B consists of the upper tri-angular matrices in $\mathsf{EL}(V)$, and N consists of permutation (or monomial) matrices (a permutation matrix has precisely one non-zero entry in every row and column), and $T = B \cap N$ consists of diagonal matrices with the property that the product of the entries lies in the commutator group $[D^\times, D^\times]$. For $1 \leq i \leq n$, let s_i be the T-coset of the matrix

$$\begin{pmatrix} 1 & & & & & & \\ & \ddots & & & & & \\ & & 1 & & & & \\ & & & 0 & 1 & & \\ & & & -1 & 0 & & \\ & & & & & 1 & \\ & & & & & & \ddots \\ & & & & & & & 1 \end{pmatrix}$$

which interchanges p_i and p_{i-1}. Then s_i is an involution in W, and

$$(W, \{s_1, \dots, s_n\})$$

presents W as a *Coxeter group*. It is a routine matter to check that these data $(G, B, N, \{s_1, \dots, s_n\})$ satisfy the axioms of a *Tits system*:

TS1 B and N generate G.

TS2 $T = B \cap N$ is normalized by N.

TS3 The set $S = \{s_1, \dots, s_n\}$ generates N/T and has the following properties.

TS4 $sBs \neq s$ for all $s \in S$.

TS5 $BsBwB \subseteq BwB \cup BswB$ for all $s \in S$ and $w \in W = N/T$.

The group $T \subseteq B$ has a normal complement, the group U generated by all elations $\tau_{ij}(a)$, for $i < j$. Thus B is a semidirect product $B = TU$, and $(\mathsf{EL}(V), B, N, S)$ is what is called a (strongly split) *Tits system* (or *BN-pair*). (A Tits system is called *split* if B can be written as a not necessarily semidirect product $B = TU$, such that $U \triangleleft B$ is normal in B. Usually, one also requires U to be nilpotent — this is the case in our example. The group U acts transitively on the apartments containing the chamber corresponding to B. If $U \cap T = 1$, the Tits system is said to be *strongly split*.) Note also that B is *not* solvable if D is not commutative.

Now we describe the root system for $\Delta(V)$.

1.5.3 Thin projective spaces and the root system Let $\mathcal{P} = \{0, \dots, n\}$ and let $\mathcal{L} = \binom{\mathcal{P}}{2}$ denote the collection of all 2-element subsets of $\mathcal{P}$. The incidence $*$ is the symmetrized inclusion relation. Then $(\mathcal{P}, \mathcal{L}, *)$ is a thin projective geometry of rank n, i.e. a projective space where every line is incident with precisely two points. The corresponding thin projective space (the projective geometry over the 'field 0 with one element') is

$$\mathsf{PG}_n(0) = \left(\binom{\mathcal{P}}{1}, \binom{\mathcal{P}}{2}, \dots, \binom{\mathcal{P}}{n}, \subseteq \right).$$

Now we construct a 'linear model' for this geometry. Consider the real euclidean vector space $\mathbb{R}^{n+1}$ with its standard inner product $\langle -, - \rangle$ and standard base $e_0, e_1, \dots, e_n$. Let E denote the orthogonal complement of the vector $v = e_0 + \cdots + e_n = (1, 1, \dots, 1) \in \mathbb{R}^{n+1}$. Let

$$p_i = e_i - v\frac{1}{n} \in E, \text{ for } i = 0, \dots, n.$$

We identify $p_0, \dots, p_n$ with the points of $\mathsf{PG}_n(0)$. The subspaces of rank k correspond precisely to (linearly independent) subsets $p_{i_1}, \dots, p_{i_k}$ of $\mathcal{P}$. We identify such a subspace with the vector $\frac{1}{k}(p_{i_1} + \cdots + p_{i_k})$; this is the barycenter of the convex hull of $\{p_{i_1}, \dots, p_{i_k}\}$. This is our model of $\mathsf{PG}_n(0)$; the picture shows the case $n = 2$.

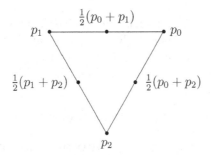

Now we construct the Weyl group. For $i \neq j$ put $\varepsilon_{ij} = e_i - e_j$. The reflection r_{ij} at the hyperplane $\varepsilon_{ij}^{\perp}$ in E,

$$x \overset{r_{ij}}{\longmapsto} x - \langle x, \varepsilon_{ij} \rangle \varepsilon_{ij},$$

permutes $p_0, \dots, p_n$; the isometry group W generated by these reflections is the *Weyl group* of type $\mathbf{A}_n$ (which is isomorphic to the Coxeter group N/T of the Tits system).

These vectors ε_{ij} form a *root system of type* $\mathbf{A}_n$ in E, with $\Phi = \{\varepsilon_{ij} | i \neq j\}$ as set of *roots*. We call the set $\Phi^+ = \{\varepsilon_{ij} | i < j\}$ the set of *positive roots*; this determines the n *fundamental roots* $\{\varepsilon_{01}, \varepsilon_{12}, \ldots, \varepsilon_{n-1,n}\}$. The fundamental roots form a base of E, and every root is an integral linear combination of fundamental roots, such that either all coefficients are non-negative (this yields the positive roots) or non-positive.

To each root ε_{ij}, we attach the group U_{ij} as defined in Section 1.4. From the Steinberg relations, we see the following: if $i < j$ and $k < l$, then $[U_{ij}, U_{kl}] = 0$ if there exists no positive root which is a linear combination $\varepsilon_{ij}a + \varepsilon_{kl}b$, with $a, b \in \mathbb{Z}_{>0}$. For our root system, the only instance where such a linear combination is a positive root is when $\mathsf{card}\{i, j, k, l\} = 3$, and in this case the Steinberg relations show that the commutator is indeed not trivial. The picture below shows the root system $\mathbf{A}_2$ (the case $n = 2$); the fundamental roots are ε_{01} and ε_{12}, and the positive roots are ε_{01}, ε_{12}, and $\varepsilon_{02} = \varepsilon_{01} + \varepsilon_{12}$.

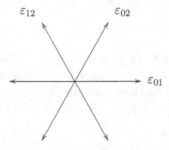

The hyperplanes $\varepsilon_{ij}^{\perp} \subseteq E$ yield a triangulation of the unit sphere $\mathbb{S}^{n-1} \subseteq E$; as a simplicial complex, this is precisely the apartment Σ. The half-apartments correspond to the half-spaces $\{v \in E | \ \langle v, \varepsilon_{ij} \rangle \geq 0\}$. From this, it is not hard to see that the groups U_{ij} are root groups in the building-theoretic sense (as defined in Grundhöfer's article [14] in these proceedings), see also Ronan [26] Ch. 6.

1.6 The little projective group as a 2-transitive group

We show that the projective space is determined by (and can be recovered from) the action of the elementary linear group on the point set.

Let $\mathsf{PEL}(V) \subseteq H \subseteq \mathsf{PGL}(V)$ be a subgroup, and assume that $\dim(V) \geq 3$. Then $(H, \mathsf{Gr}_1(V))$ is a 2-transitive permutation group. Let $L \subseteq \mathsf{Gr}_1(V)$ be a point row, i.e. the set of all points lying on a line $\ell \in \mathsf{Gr}_2(V)$. Let $p, q \in L$ be distinct points. Then $H_{p,q}$ has precisely four orbits in $\mathsf{Gr}_1(V)$: the two singletons $\{p\}, \{q\}$, the set $L \setminus \{p, q\}$, and $X = \mathsf{Gr}_1(V) \setminus L$. The set X has the property that every h in H which fixes X pointwise fixes $\mathsf{Gr}_1(V)$ pointwise.

None of the other three orbits has this property. Thus one can see the line $L \subset \mathsf{Gr}_1(V)$ from the H-action, we have a canonical (re)construction

$$(H, \mathsf{Gr}_1(V)) \longmapsto (H, \mathsf{Gr}_1(V), \mathsf{Gr}_2(V), *).$$

Combining this with the Fundamental Theorem of Projective Geometry 1.1.1, we have the next result.

1.6.1 Proposition *For $i = 1, 2$, let $\mathsf{PG}(V_i)$ be projective geometries (of possibly different ranks $n_i \geq 2$) over skew fields D_1, D_2. Let $\mathsf{PEL}(V_i) \subseteq H_i \subseteq \mathsf{PGL}(V_i)$ be subgroups. If there exists an isomorphism of permutation groups*

$$(H_1, \mathsf{Gr}_1(V_1)) \xrightarrow[\cong]{\phi} (H_2, \mathsf{Gr}_1(V_2))$$

then there exists a semilinear bijection $F : V_1 \longrightarrow V_2$ which induces ϕ (and $n_1 = n_2$). □

The result is also true in dimension 2, but the proof is more complicated, as we will see in Section 1.8. The problem whether an abstract group isomorphism $H_1 \xrightarrow[\cong]{\phi} H_2$ is always induced by a semilinear map is much more subtle. The result is indeed that such an isomorphism is induced by a linear map, composed with an isomorphism or anti-isomorphism of the skew fields in question, provided that the vector space dimensions are large enough (at least 3), see Hahn & O'Meara [15] 2.2D. The crucial (and difficult) step is to show that ϕ maps transvections to transvections.

1.7 Projective planes

We mention the classification of Moufang planes.

The Fundamental Theorem of Projective Geometry does not apply to projective planes. However, there is the following result. Suppose that $(\mathcal{P}, \mathcal{L}, *)$ is a projective plane. Given a flag (p, ℓ), (i.e. $p * \ell$)), the group $G_{[p,\ell]}$ is defined to be the set of all automorphisms which fix p linewise and ℓ pointwise (so for $\mathsf{PG}(D^3)$, we have $G_{[p,\ell]} = U_{(p,\ell)}$ in our previous notation). Let h be a line passing through p and different from ℓ. It is easy to see that $G_{[p,\ell]}$ acts freely on the set $H' = \{q \in \mathcal{P} | \ q \neq p, \ q * \ell\}$. If this action is transitive, then $(\mathcal{P}, \mathcal{L}, *)$ is called (p, ℓ)-*homogeneous*. The projective plane is called a *Moufang plane* if it is (p, ℓ)-homogeneous for any flag (p, ℓ). If $(\mathcal{P}, \mathcal{L}, *) = \mathsf{PGL}(V)$ for some 3-dimensional vector space V, then we have a Moufang plane.

Recall that an *alternative field* is a (not necessarily associative) algebra with unit, satisfying the following relations.

AF1 If $a \neq 0$, then the equations $ax = b$ and $ya = b$ have unique solutions x, y.

AF2 The equalities $x^2 y = x(xy)$ and $yx^2 = (yx)x$ hold for all x, y.

Clearly, every field or skew field is an alternative field. The structure theorem of alternative fields says that every non-associative alternative field is a central 8-dimensional algebra over a field K, a so-called *Cayley division algebra*. Not every field K admits a Cayley division algebra; it is necessary that K admits an anisotropic quadratic form of dimension 8, so finite fields or algebraically closed fields do not admit Cayley division algebras. Every real closed (or ordered) field admits a Cayley division algebra.

Given an alternative field A, we construct a projective plane $\mathsf{PG}_2(A)$ as follows. Let ∞ be a symbol which is not an element of A. Let

$$\mathcal{P} = \{(\infty)\} \cup \{(a) \mid a \in D\} \cup \{(x, y) \mid x, y \in A\}$$
$$\mathcal{L} = \{[\infty]\} \cup \{[a] \mid a \in D\} \cup \{[x, y] \mid x, y \in A\}$$

The incidence $*$ is defined as

$$(\infty) * [a] * (a, sa + t) * [s, t] * (s) * [\infty] * (\infty)$$

Note that for a field or skew field D, this is precisely $\mathsf{PG}(D^3)$.

1.7.1 Theorem (Moufang Planes)
Let $(\mathcal{P}, \mathcal{L}, *)$ be a Moufang plane. The there exists an alternative field A, unique up to isomorphism, such that $(\mathcal{P}, \mathcal{L}, *)$ is isomorphic to $\mathsf{PG}_2(A)$.

For a proof see Hughes & Piper [17] or Pickert [23]; the structure theorem for non-associative alternative fields is proved in Van Maldeghem [34].   □

The Moufang planes can also be described by means of Steinberg relations: for each pair (i, j) with $i \neq j$ and $i, j \in \{1, 2, 3\}$, fix a copy $U_{ij} \cong (A, +)$. Let G denote the free product of these six groups, factored by the Steinberg relations as given in 1.4 (note that the Steinberg relations make sense even in the non-associative case). The group G has a natural Tits system which yields the Cayley plane $\mathsf{PG}_2(A)$. The group induced by G on $\mathsf{PG}_2(A)$ is a K-form of a simple adjoint algebraic group of type $\mathbf{E}_6$.

1.8   The projective line and Moufang sets

We investigate the action of the linear group as a permutation group on the projective line. This is a special case of a Moufang set.

The projective line $\mathcal{P} = \mathsf{Gr}_1(V)$, for $V \cong D^2$, is a set without further structure. We add structure by specifying properties of the group $G = \mathsf{PGL}(V)$ acting on it. This group has two remarkable properties: (1) G acts 2-transitively on $\mathcal{P}$. (2) The stabilizer $B = G_p$ of a point $p = vD \in \mathcal{P}$ has a regular normal subgroup U_p, the group induced by maps of the form $\mathrm{id}_V + v\rho$, where ρ runs through the collection of all non-zero elements of $V^\vee$ which annihilate v.

Moufang sets were first defined by Tits in [33]; our definition given below is stated in a slightly different way. We define a *Moufang set* as a triple (G, U, X), where G is a group acting on a set X (with at least 3 elements), and U is a subgroup of G. We require the following properties.

MS1 The action of G on X is 2-transitive.

MS2 The group U fixes a point x and acts regularly on $X \setminus \{x\}$.

MS3 The group U is normal in the stabilizer G_x.

The properties **MS2** and **MS3** will be summarized in the sequel as 'G_x *has a regular normal subgroup*'; we will also say that 'U *makes* (G, X) *into a Moufang set*'. Let $y \in X \setminus \{x\}$, and put $T = G_{x,y}$. Then clearly $G_x = TU$ is a semidirect product. (The pair (G_x, T) is a what is called a (strongly) split Tits system (BN-pair) of rank 1 for the group G.)

If $U = G_x$, then $T = 1$, so G is sharply 2-transitive. This case has its own, special flavor. Note that in general, U is *not* determined by G and x. As a counterexample, let $\mathbb{H}$ denote the quaternion division algebra over a real closed field, let $X = \mathbb{H}$, and consider the group consisting of maps of the form $[x \longmapsto axb + t]$, for $a, b \in \mathbb{H}^\times$ and $t \in \mathbb{H}$. The stabilizer of 0 consists of the maps $[x \longmapsto axb]$, and it has two regular normal subgroups isomorphic to $\mathbb{H}^\times$, consisting of the maps $[x \longmapsto ax]$ or $[x \longmapsto xb]$.

Let $H \subseteq \mathsf{PGL}_2(D)$ be a subgroup containing $\mathsf{PEL}_2(D)$. We identify the projective line $\mathsf{Gr}_1(D^2)$ with $D \cup \{\infty\}$, identifying $\binom{x}{1}D$ with x and $\binom{1}{0}D$ with ∞. Let U_∞ denote the group consisting of the maps $[x \longmapsto x + t]$, for $t \in D$. So

$$(H, U_\infty, D \cup \{\infty\})$$

is a Moufang set. The stabilizer $T = H_{0,\infty}$ contains all maps $[x \longmapsto axa]$, for $a \in D^\times$. In particularly, T is commutative if and only if D is commutative. Now we consider the following problem:

Is it possible to recover U_∞ from the action of H?

In the commutative case, the answer is easy: U_∞ is the commutator group of H_∞,

$$T \text{ commutative} \quad \Longrightarrow \quad U_\infty = [H_\infty, H_\infty].$$

Also, an element $1 \neq g \in H_\infty$ is contained in U_∞ if and only if g has no fixed point in D. Thus, U_∞ is the *only* regular normal subgroup of H_∞.

Now suppose that D is not commutative. In this case, we will prove first that the action of H_∞ on D is primitive. We have to show that T is a maximal subgroup of H_∞. Let $g \in H_\infty \setminus T$, and consider the group K generated by T and g. Since H_∞ splits as a semidirect product $H_\infty = TU_\infty$, we may assume that $g = [x \longmapsto x + t] \in U$, with $t \neq 1$. Since K contains T, it contains all maps of the form $[x \longmapsto x + taba^{-1}b^{-1}]$. Using some algebraic identities for multiplicative commutators as in Cohn [8] Sec. 3.9, one shows that the set of all multiplicative commutators generates D additively. Thus $U_\infty \subseteq K$, whence $K = H_\infty$,

$$T \text{ not commutative} \quad \Longrightarrow \quad H_\infty \text{ primitive}$$
$$\Longrightarrow \quad U_\infty \text{ unique abelian normal subgroup of } H_\infty$$

(for the last implication see Robinson [25] 7.2.6). As in the commutative case, there is no other way of making the projective line into a Moufang set. Indeed, suppose that $U_\infty \neq U' \trianglelefteq H_\infty$ is another regular normal subgroup. Then $U_\infty \cap U' \trianglelefteq H_\infty$ is also normal, so either $U_\infty \cap U' = 1$, or $U' \supseteq U_\infty$. In the latter case, $U_\infty = U'$ since we assumed the action to be regular. So suppose $U_\infty \cap U' = 1$. Then $U'U_\infty$ is a direct product. Define a map $\phi : U_\infty \longrightarrow U'$ by putting $\phi(u) = u'$ if and only if $u(0) = u'(0)$. Then $\phi(u_1 u_2)(0) = (u_1 u_2)'(0) = u_1 u_2'(0) = u_2' u_1(0) = u_2' u_1'(0)$, so ϕ is an anti-isomorphism. In particular, U' is abelian, whence $U_\infty = U'$, a contradiction.

1.8.1 Proposition *Let V be a two-dimensional vector space over a skew field D, and assume that $\mathsf{PEL}(V) \subseteq H \subseteq \mathsf{PGL}(V)$. Then G_x contains a unique regular normal subgroup U_∞, i.e. there is a unique way of making $(H, \mathsf{Gr}_1(V))$ into a Moufang set.* □

We combine this with Hua's Theorem.

1.8.2 Theorem (Hua) *Let*

$$(\mathsf{PEL}(V), U_\infty, D \cup \{\infty\}) \xrightarrow[\cong]{\alpha} (\mathsf{PEL}(V'), U'_\infty, D' \cup \{\infty'\})$$

be an isomorphism of Moufang sets. Then α is induced by an isomorphism or anti-isomorphism of skew fields.

For a proof see Tits [32] 8.12.3 or Van Maldeghem [34] p. 383–385. □

1.8.3 Corollary *Let V, V' be 2-dimensional vector spaces over skew fields D, D', let $\mathsf{PEL}(V) \subseteq H \subseteq \mathsf{PGL}(V)$ and $\mathsf{PEL}(V') \subseteq H' \subseteq \mathsf{PGL}(V')$ and assume that*

$$(H, \mathsf{Gr}_1(V)) \xrightarrow[\cong]{\alpha} (H', \mathsf{Gr}_1(V'))$$

is an isomorphism of permutation groups. Then there exists an isomorphism or anti-isomorphism $D \xrightarrow[\cong]{\theta} D'$ *which induces* α.

Proof. Clearly, α induces an isomorphism of the commutator groups $[H, H] = \mathsf{PEL}(V)$ and $[H', H'] = \mathsf{PEL}(V')$ (we can safely disregard the small fields $\mathbb{F}_2$ and $\mathbb{F}_3$, since here, counting suffices). There is a unique way of making these permutation groups into Moufang sets, and to these Moufang sets, we apply Hua's Theorem. $\square$

The following observation is due to Hendrik Van Maldeghem. Let V be a vector space of dimension at least 3, let H be a group of automorphisms of $\mathsf{PG}(V)$ containing $\mathsf{PEL}(V)$. Then $(H, \mathsf{Gr}_1(V))$ cannot be made into a Moufang set. To see this, let $p \in \mathsf{Gr}_1(V)$ and assume that $U \trianglelefteq H_p$ is a normal subgroup acting regularly on $\mathsf{Gr}_1(V) \setminus \{p\}$. Let $u \in U$, and let τ be an elation with center p. Then $u\tau u^{-1}$ is also an elation with center p, and so is $[u, \tau]$. If we choose u, τ in such a way that u doesn't fix the axis of τ (which is possible, since $\dim(V) \geq 3$), then $[u, \tau] \in U$ is a non-trivial elation with center p. Since U is normal, U contains all elations with center p. These elations form an abelian normal subgroup of H_p which is, however, not regular on $\mathsf{Gr}_1(V)$.

1.8.4 Lemma *Let* $\mathsf{PEL}(V) \subseteq H \subseteq \mathsf{PGL}(V)$ *and assume that* $\dim(V) \geq 3$. *Then* H_p *contains no regular normal subgroup; in particular,* $(H, \mathsf{Gr}_1(V))$ *cannot be made into a Moufang set.* $\square$

Combining the results of this section with Proposition 1.6.1, we have the following final result about actions on the point set.

1.8.5 Corollary *For* $i = 1, 2$, *let* V_i *be vector spaces over skew fields* D_i, *of (finite) dimensions* $n_i \geq 2$. *Let* $\mathsf{PEL}(V_i) \subseteq H_i \subseteq \mathsf{PGL}(V_i)$ *and assume that*

$$(H_1, \mathsf{Gr}_1(V_1)) \xrightarrow[\cong]{\phi} (H_2, \mathsf{Gr}_1(V_2))$$

is an isomorphism of permutation groups. Then $n_1 = n_2$, *and* ϕ *is induced by a semilinear isomorphism, except if* $n_1 = 2$, *in which case* ϕ *may also be induced by an anti-isomorphism of skew fields.* $\square$

These results are also true for infinite dimensional vector spaces.

2 Polar spaces and quadratic forms

In this second part we consider (σ, ε)-hermitian forms and their generalizations, pseudo-quadratic forms. From now on, we consider also infinite dimensional vector spaces. As a finite dimensional motivation, we start with dualities. Suppose that $\dim(V) = n + 1$ is finite, and that there is an isomorphism ϕ between the projective space $\mathsf{PG}(V)$ and its dual,

$$\mathsf{PG}(V) \xrightarrow[\cong]{\phi} \mathsf{PG}(V^\vee).$$

Now $V^\vee$ is in a natural way a right vector space over the opposite skew field D^{op}. By the Fundamental Theorem of Projective Geometry 1.1.1, ϕ is induced by a σ-semilinear bijection $f : V \xrightarrow{\cong} V^\vee$, relative to an isomorphism $\sigma : D \longrightarrow D^{\mathrm{op}}$, i.e. σ is an anti-automorphism of D. Note also that there is a natural isomorphism

$$\mathsf{PG}(V^\vee) \cong (\mathsf{Gr}_n(V), \mathsf{Gr}_{n-1}(V), \dots, \mathsf{Gr}_1(V), *).$$

Thus, we may view ϕ as an non type-preserving automorphism of the building $\Delta(V)$. Such an isomorphism is called a *duality*. If $\phi^2 = \mathrm{id}_{\Delta(V)}$ (this makes sense in view of the identification above), then ϕ is called a *polarity*. In this section we study polarities and the related geometries, *polar spaces*.

2.1   Forms and polarities

We study some basic properties of forms (sesquilinear maps) and their relation with dualities and polarities.

In this section, V is a (possibly infinite dimensional) right vector space over D. We fix an anti-automorphism σ of D. Using σ, we make the dual space $V^\vee$ into a *right* vector space over D, denoted V^σ, by defining $\lambda a = [v \longmapsto a^\sigma \lambda(v)]$, for $v \in V$, $\lambda \in V^\vee$ and $a \in D$. We put

$$\mathsf{Form}_\sigma(V) = \mathsf{Hom}_D(V, V^\sigma).$$

Given an element $f \in \mathsf{Form}_\sigma(V)$, we write $f(u, v) = f(u)(v)$; the map $(u, v) \longmapsto f(u, v)$ is *biadditive* ($\mathbb{Z}$-linear in each argument) and σ-*sesquilinear*, $f(ua, vb) = a^\sigma f(u, v)b$. Note that σ is uniquely determined by the map f, provided that $f \neq 0$. Suppose that that

$$F : V \longrightarrow V'$$

is a linear map of vector spaces over D. We define

$$\mathsf{Form}_\sigma(V) \xleftarrow{F^*} \mathsf{Form}_\sigma(V')$$

by $F^*(f')(u, v) = f'(F(u), F(v))$. A similar construction works if $V \xrightarrow{F} V'$ is θ-semilinear relative to a skew field isomorphism $D \xrightarrow[\cong]{\theta} D'$; here, we define $F^*(f')(u, v) = f'(F(u), F(v))^{\theta^{-1}}$ to obtain

$$\mathsf{Form}_\sigma(V) \xleftarrow{F^*} \mathsf{Form}_{\sigma'}(V'),$$

where $\sigma = \theta\sigma'\theta^{-1}$. The group $\mathsf{GL}(V)$ acts thus in a natural way from the right on $\mathsf{Form}_\sigma(V)$, by putting

$$fg = g^*f.$$

Forms in the same $\mathsf{GL}(V)$-orbit are called *equivalent*; the stabilizer of a form f is denoted

$$\mathsf{GL}(V)_f = \{g \in \mathsf{GL}(V)|\ f(g(u), g(v)) = f(u, v) \text{ for all } u, v \in V\}.$$

The assignment $V \longmapsto V^\sigma$ is a natural cofunctor on $\mathfrak{Mod}_D^{\text{fin}}$ (or $\mathfrak{Mod}_D$, if we allow infinite dimensional vector spaces) which we also denote by σ,

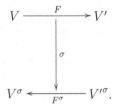

Let $V^\vee \xleftarrow{F^\vee} V'^\vee$ be the dual or *adjoint* of $V \xrightarrow{F} V'$ (i.e. $F^\vee(\lambda) = \lambda F$). Then set-theoretically, $F^\sigma = F^\vee$. There is a canonical linear injection

$$V \xrightarrow{\text{can}} V^{\sigma\sigma}$$

which sends $v \in V$ to the map $\mathsf{can}(v) = [\lambda \longmapsto \lambda(v)^{\sigma^{-1}}]$. If the dimension of V is finite, then can is an isomorphism (and the data σ and can make the abelian category $\mathfrak{Mod}_D^{\text{fin}}$ into a *hermitian category* $\mathfrak{Herm}_{D,\sigma}^{\text{fin}}$). Given a form f, we have a diagram

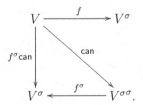

Note that $f^\sigma\mathsf{can}(v) = [u \longmapsto f(u, v)^{\sigma^{-1}}]$. If both f and $f^\sigma\mathsf{can}$ are injective, then we call the form f *non-degenerate* (if $\dim(V)$ is finite, then it suffices to require that f is injective). For a subspace $U \subseteq V$, put

$$U^{\perp_f} = \bigcap\{\mathsf{ker}(f(u))|\ u \in U\} = \{v \in V|\ f(u, v) = 0 \text{ for all } u \in U\}$$

$$^{\perp_f}U = \bigcap\{\mathsf{ker}(f^\sigma\mathsf{can}(u))|\ u \in U\} = \{v \in V|\ f(v, u) = 0 \text{ for all } u \in U\}.$$

Thus f is non-degenerate if and only if $V^{\perp_f} = 0 = {}^{\perp_f}V$.

In the finite dimensional case, a non-degenerate form defines a duality of $\mathsf{PG}(V)$ (by $U \longmapsto U^{\perp_f}$, and by the Fundamental Theorem of Projective Geometry 1.1.1, every duality of $\mathsf{PG}(V)$ is obtained in this way). Also, $\perp_f$ determines the anti-automorphism σ up to conjugation with elements of $\mathsf{Int}(D)$. The form f itself is, however, not determined by $\perp_f$. Therefore, we introduce another equivalence relation on forms. If f is σ-sesquilinear, and if $s \in D^\times$, then $sf : (u, v) \longmapsto sf(u, v)$ is σs^{-1}-sesquilinear,

$$sf(ua, vb) = sa^\sigma f(u, v)b = (sa^\sigma s^{-1})sf(u, v)b = a^{\sigma s^{-1}}(sf)(u, v)b.$$

The forms f and sf are called *proportional*, and we say that sf is obtained from f by *scaling* with s; proportional forms induce the same dualities. Scaling with s yields an isomorphism $\mathsf{Form}_\sigma(V) \overset{\cong}{\longrightarrow} \mathsf{Form}_{\sigma s^{-1}}(V)$.

2.1.1 Proposition *Suppose that* $\dim(V)$ *is finite. Let* $f \in \mathsf{Form}_\sigma(V)$ *and* $f' \in \mathsf{Form}_{\sigma'}(V)$ *be non-degenerate forms. If* f *and* f' *induce the same duality, then* $\sigma'\sigma^{-1} \in \mathsf{Int}(D)$, *and* f *and* f' *are proportional. There is a 1-1 correspondence*

$$\left\{ \begin{matrix} \text{Dualities in} \\ \mathsf{PG}(V) \end{matrix} \right\} \longleftrightarrow \left\{ \begin{matrix} \text{Proportionality classes of} \\ \text{non-degenerate forms} \end{matrix} \right\}.$$

$\square$

A non-degenerate sesquilinear form which induces a duality has the property that $(U^{\perp_f})^{\perp_f} = U$ holds for all subspaces U. This condition makes also sense in the infinite dimensional case if we restrict it to finite dimensional subspaces (although there, no dualities exist), and boils down to f being *reflexive*;

$$f(u, v) = 0 \quad \Longleftrightarrow \quad f(v, u) = 0.$$

In other words, $U^{\perp_f} = {}^{\perp_f}U$ holds for all subspaces $U \subseteq V$. If $f \neq 0$ is reflexive, then there exists a unique element $\varepsilon \in D^\times$ such that $f(u, v) = f(v, u)^\sigma \varepsilon$ for all $u, v \in V$, see Dieudonné [9] Ch. I §6. Furthermore, this implies that $\varepsilon^\sigma = \varepsilon^{-1}$ (choose u, v with $f(u, v) = 1$, then $f(v, u) = \varepsilon$), and $a^{\sigma^2} = a^{\varepsilon^{-1}}$ for all $a \in D$ (consider $f(u, va)$). A form h which satisfies the identity

$$h(u, v) = h(v, u)^\sigma \varepsilon \quad \text{for all } u, v \in V$$

is called (σ, ε)-*hermitian*. The collection of all (σ, ε)-hermitian forms is a subgroup of $\mathsf{Form}_\sigma(V)$ which we denote $\mathsf{Herm}_{\sigma,\varepsilon}(V)$. A *non-degenerate* (σ, ε)-*hermitian form* h induces thus an involution $\perp_h$ on the building $\Delta(V)$, for *finite dimensional* V. In the finite dimensional setting, we have thus a 1-1 correspondence

$$\left\{\begin{array}{c}\text{Polarities in}\\ \text{PG}(V)\end{array}\right\} \longleftrightarrow \left\{\begin{array}{c}\text{Proportionality classes}\\ \text{of non-degenerate}\\ (\sigma, \varepsilon)\text{-hermitian forms}\end{array}\right\}$$

If a (σ, ε)-hermitian form h is scaled with $s \in D^{\times}$, then the resulting form sh is $(\sigma s^{-1}, ss^{\sigma}\varepsilon)$-hermitian.

2.2 Polar spaces

We introduce polar spaces as certain point-line geometries.

Assume that h is non-degenerate (σ, ε)-hermitian. An element $U \in \text{Gr}_k(V)$ is called *absolute* if it is incident with its image $U^{\perp h}$. If $2k \le \dim(V)$, then this means that $U \subseteq U^{\perp h}$, or, in other words, that $h|_{U \times U} = 0$. A subspace with this property is called *totally isotropic* (with respect to h). Let $\text{Gr}_k^h(V)$ denote the collection of all totally isotropic k-dimensional subspaces. The maximum number k for which this set is non-empty is called the *Witt index* $\text{ind}(h)$ of h (If the vector space dimension is infinite, then $\text{ind}(h)$ can of course be infinite). This makes sense also for possibly degenerate (σ, ε)-hermitian forms:

$$\text{ind}(h) = \max\{\dim(U)| \ U \subseteq V, \ h|_{U \times U} = 0\}$$

If h is non-degenerate, then $2\,\text{ind}(V) \le \dim(V)$. Suppose that h is non-degenerate and of finite index $m \ge 2$. Let $\text{PG}^h(V)$ denote the structure

$$\text{PG}^h(V) = (\text{Gr}_1^h(V), \dots, \text{Gr}_m^h(V), *),$$

and let $\text{PG}^h(V)_{1,2} = (\text{Gr}_1^h(V), \text{Gr}_2^h(V), *)$ denote the corresponding point-line geometry. This is an example of a *polar space*. This geometry has one crucial property: given a point $p \in \text{Gr}_1^h(V)$ and a line $L \in \text{Gr}_2^h(V)$ which are not incident, $p \not\subseteq L$, there are two possibilities: either there exists an element $H \in \text{Gr}_3^h(V)$ containing both p and L, or there exists precisely one point q incident with L, such that $p \oplus q \in \text{Gr}_2^h(V)$. Algebraically, this means that either $L \subseteq p^{\perp h}$ or $q = L \cap p^{\perp h}$ (note that $p^{\perp h}$ is a hyperplane, so the intersection has at least dimension 1).

A point-line geometry $(\mathcal{P}, \mathcal{L}, *)$ is called a (non-degenerate) *polar space* if it satisfies the following properties.

PS1 There exist two distinct lines. Every line is incident with at least 3 points. Two lines which have more than one point in common are equal.

PS2 Given $p \in \mathcal{P}$ there exists $q \in \mathcal{P}$ such that p and q are not joined by a line.

PS3 Given a point $p \in \mathcal{P}$ and a line $\ell \in \mathcal{L}$, either p is collinear with every point which is incident with ℓ, or with precisely one point which is incident with ℓ.

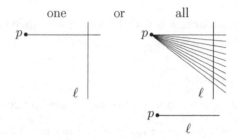

A *subspace* of a geometry satisfying PS1–PS3 is a set X of points with the following two properties: given two distinct points $p, q \in X$, there exists a line ℓ incident with p, q, and if r is also incident with ℓ, then $r \in X$. It is a (non-trivial) fact that every subspace which contains 3 non-collinear points is a projective space. We define the *rank* of $(\mathcal{P}, \mathcal{L}, *)$ as the maximum of the ranks of the subspaces minus 1.

PS4 The rank m is finite.

PS5 Every subspace of rank $m - 2$ is contained in at least 3 subspaces of rank $m - 1$.

A structure satisfying the axioms PS1–PS5 is called a *thick polar space*. It is easy to see that the structure

$$\mathsf{PG}^h(V) = (\mathsf{Gr}_1^h(V), \ldots, \mathsf{Gr}_m^h(V), *),$$

satisfies PS1–PS4, provided that h is non-degenerate and $\mathrm{ind}(h) = m \geq 2$. Axiom PS5 is more subtle, but is is easy to see that every subspace of rank $m - 1$ is contained in at least 2 subspaces of rank m. We call such a structure a *weak polar space* (with thick lines). Our set of axioms is a variation of the Buekenhout-Shult axiomatization of polar spaces given in Buekenhout & Shult [6]. Similarly as Veblen's axiom in the definition of a projective geometry in Section 1.5, the one axiom which is geometrically important is the Buekenhout-Shult 'one or all' axiom PS3. By the Fundamental Theorem of Polar Spaces 2.6.3, this simple axiom encodes the whole body of geometric algebra!

We mention some examples of (weak) polar spaces which do not involve hermitian forms.

2.2.1 Examples

(0) Let X, Y be disjoint sets of cardinality at least 3, put $\mathcal{P} = X \times Y$, and $\mathcal{L} = X \cup Y$. By definition, a point (x, y) is incident with the lines x and y. The resulting geometry is a weak polar space of rank 2; every point is incident with precisely 2 lines. This in fact an example of a weak generalized quadrangle, see Van Maldeghem's article [35] in these proceedings.

(1) Every (thick) generalized quadrangle (see Van Maldeghem's article [35]) is a thick polar space of rank 2.

(2) Let V be a 4-dimensional vector space over a skew field D, put $\mathcal{P} = \mathsf{Gr}_2(V)$ and $\mathcal{L} = \{(p, A) \in \mathsf{Gr}_1(V) \times \mathsf{Gr}_3(V) |\ p \subseteq A\}$. The incidence is the natural one (inclusion of subspaces). The resulting polar space which we denote by $\mathsf{A}_{3,2}(D)$ has rank 3; the planes of the polar space are the points and planes of $\mathsf{PG}(V)$, and every line of this polar space is incident with precisely two planes, so $\mathsf{A}_{3,2}(D)$ is a weak polar space.

The next theorem is an important step in the classification of polar spaces of higher rank (the full classification will be stated in Section 2.6).

2.2.2 Theorem (Tits) *Let X be a subspace of a polar space $(\mathcal{P}, \mathcal{L}, *)$. If X contains 3 non-collinear points, then X, together with the set of all lines which intersect X in more than one point, is a self-dual projective space. This projective space is either a Moufang plane over some alternative field A, or a desarguesian projective space over some skew field D.*

For a proof see Tits [32] 7.9, 7.10, and 7.11. □

2.3 Hermitian forms

We continue to study properties of hermitian forms.

We fix the following data: D is a skew field, σ is an anti-automorphism of D, and $\varepsilon \in D^\times$ is an element with $\varepsilon^\sigma \varepsilon = 1$ and $x^{\sigma^2} = x^{\varepsilon^{-1}}$ as in Section 2.1, and $\mathsf{Herm}_{\sigma,\varepsilon}(V)$ is the group of all (σ, ε)-hermitian forms on V. We define the set of (σ, ε)-*traces* as

$$D_{\sigma,\varepsilon} = \{c + c^\sigma \varepsilon |\ c \in D\}$$

This is an additive subgroup of $(D, +)$, with the property that $c^\sigma D_{\sigma,\varepsilon} c \subseteq D_{\sigma,\varepsilon}$, for all $c \in D$. A form $h \in \mathsf{Herm}_{\sigma,\varepsilon}(V)$ is called *trace valued* if $h(v, v) \in D_{\sigma,\varepsilon}$ holds for all $v \in V$. We call such a form h *trace* (σ, ε)-*hermitian*. (If $\mathsf{char}(D) \neq 2$, then it's easy to show that every hermitian form is trace hermitian.) A hermitian form h is trace hermitian if and only if it can be written as

$$h(u, v) = f(u, v) + f(v, u)^\sigma \varepsilon,$$

for some $f \in \mathsf{Form}_\sigma(V)$. We denote the group of all trace hermitian forms by $\mathsf{TrHerm}_{\sigma,\varepsilon}(V)$.

2.3.1 Lemma *Let* $V_0 = \{v \in V \mid h(v,v) \in D_{\sigma,\varepsilon}\}$. *Then* V_0 *is a subspace of* V *containing all totally isotropic subspaces of* V. □

The form induced by h on $V_0/(V_0 \cap V_0^{\perp_h})$ is thus trace hermitian and non-degenerate. Since we are only interested in the polar space arising from h, we can thus safely assume that h is trace hermitian.

Suppose now that we scale the form with an element $s \in D^\times$. Let $h' = sh$ and put $\sigma' = \sigma s^{-1}$, and $\varepsilon' = ss^{-\sigma}\varepsilon$. Then $D_{\sigma',\varepsilon'} = sD_{\sigma,\varepsilon}$. Thus we can achieve that either $D_{\sigma,\varepsilon} = 0$, or that $1 \in D_{\sigma,\varepsilon}$. In the first case we have $\varepsilon' = -1$ and $\sigma = \sigma' = \mathsf{id}_D$ (and then D is commutative), and in the second case $\varepsilon = 1$ and $\sigma'^2 = \mathsf{id}_D$. The study of (non-degenerate) trace hermitian forms is thus — by means of scaling — reduced to the following subcases:

Trace σ-hermitian forms

$\sigma^2 = \mathsf{id}_D$ (here $\sigma = \mathsf{id}_D$ is allowed if D is commutative) and $h(u,v) = h(v,u)^\sigma$ for all $u,v \in V$.

If $\sigma = \mathsf{id}_D$, then we call h *symmetric*, and (in the non-degenerate case) $\mathsf{GL}(V)_h = \mathsf{O}(V,h)$ is the *orthogonal group* of h; if $\sigma \neq \mathsf{id}_D$ (and if h is non-degenerate), then $\mathsf{U}(V,h) = \mathsf{GL}(V)_h$ is called the *unitary group* of h.

Symplectic forms

$\sigma = \mathsf{id}_D$ and $h(v,v) = 0$ for all $v \in V$ (so h is symplectic). The group $\mathsf{GL}(V)_h = \mathsf{Sp}(V,h)$ is called the *symplectic group* (in the non-degenerate case).

Before we consider these groups and forms in more detail, we extend the whole theory to include (pseudo-)quadratic forms.

2.4　Pseudo-quadratic forms

We introduce pseudo-quadratic forms, which are certain cosets of sesquilinear forms. In characteristic 2, this generalizes trace hermitian forms.

In the course of the classification of polar spaces, it turns out that in characteristic 2, (σ,ε)-trace hermitian forms are not sufficient; one needs the notion of a pseudo-quadratic form which is due to Tits. We follow the treatment which is now standard and which is based on Bak's concept [3] of form parameters. This is a modification of Tits' original approach; however, the characteristic 2 theory of unitary groups over skew fields is entirely due to Tits, a fact which is not always properly reflected in books on classical groups (see e.g. the footnote on p. 190 in Hahn & O'Meara [15]). Let Λ be an additive subgroup of D, with

$$D_{\sigma,-\varepsilon} = \{c - c^\sigma \varepsilon \mid c \in D\} \subseteq \Lambda \subseteq D^{\sigma,-\varepsilon} = \{c \in D \mid c^\sigma \varepsilon = -c\},$$

and with the property that

$$s^\sigma \Lambda s \subseteq \Lambda$$

for all $s \in D$. Such a subset Λ is called a *form parameter*. Given an element $f \in \mathsf{Form}_\sigma(V)$, we define the *pseudo-quadratic form* $[f] = (q_f, h_h)$ to be the pair of maps

$$q_f : V \longrightarrow D/\Lambda \qquad\qquad h_f : V \times V \longrightarrow D$$

$$v \longmapsto f(v,v) + \Lambda \qquad\qquad (u,v) \longmapsto f(u,v) + f(v,u)^\sigma \varepsilon.$$

It is not difficult to see that the map $f \longmapsto [f]$ is additive; the kernel is

$$\Lambda\text{-}\mathsf{Herm}_{\sigma,-\varepsilon}(V) = \{f \in \mathsf{TrHerm}_{\sigma,-\varepsilon}(V) \mid f(v,v) \in \Lambda \text{ for all } v \in V\}.$$

The resulting group of *pseudo-quadratic forms* is denoted $\Lambda\text{-}\mathsf{Quad}_{\sigma,\varepsilon}(V)$, and we have an exact sequence

$$0 \longrightarrow \Lambda\text{-}\mathsf{Herm}_{\sigma,-\varepsilon}(V) \longrightarrow \mathsf{Form}_\sigma(V) \longrightarrow \Lambda\text{-}\mathsf{Quad}_{\sigma,\varepsilon}(V) \longrightarrow 0$$

which is compatible with the $\mathsf{GL}(V)$-action on $\mathsf{Form}_\sigma(V)$; furthermore, scaling is a well-defined process on pseudo-quadratic forms. Let $\psi(c) = c + c^\sigma \varepsilon$. Then $\ker(\psi) = D^{\sigma,-\varepsilon} \supseteq \Lambda$; consequently, there is a well-defined map $D/\Lambda \overset{\bar\psi}{\longrightarrow} D$ sending $c + \Lambda$ to $\psi(c)$,

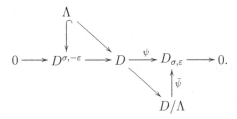

Thus we have $\bar\psi(q_f(v)) = h(v,v)$; in particular,

$$q_f(v) = 0 \quad\Longrightarrow\quad h_f(v,v) = 0.$$

We call a pseudo-quadratic form *non-degenerate* if h_f is non-degenerate (this differs from Tits' notion of non-degeneracy [32], [5]; we'll come back to that point later). Similarly as before, a subspace U is called *totally isotropic* if q_f and h_f vanish on U; the collection of all k-dimensional totally isotropic subspaces is denoted $\mathsf{Gr}_k^{[f]}(V)$, and the Witt index $\mathsf{ind}[f]$ is defined in the obvious way. Note that

$$\mathsf{Gr}_k^{[f]}(V) \subseteq \mathsf{Gr}_k^{h_f}(V).$$

Suppose that V, V' are vector spaces over D, and that

$$F : V \longrightarrow V'$$

is linear. We define a map

$$\Lambda\text{-}\mathsf{Quad}_{\sigma,\varepsilon}(V) \xleftarrow{\ F^*\ } \Lambda\text{-}\mathsf{Quad}_{\sigma,\varepsilon}(V'),$$

by $[f'] \longmapsto [F^*(f')]$ which we denote also by F^*. A similar construction works if $V \xrightarrow{\ F\ } V'$ is θ-semilinear relative to $D \xrightarrow{\ \theta\ }_{\cong} D'$. The group

$$\mathsf{U}([f]) = \mathsf{GL}(V)_{[f]}$$

is the group of all isometries of $(V, [f])$. Let $g \in \mathsf{\Gamma L}(V)$. If there exists an element $s \in D$ such that $g^*[f] = [sf]$, then g is called a *semi-similitude*; if g is linear, then it is called a *similitude*. The corresponding groups are denoted

$$\mathsf{U}([f]) \overset{\trianglelefteq}{\hookrightarrow} \mathsf{GU}([f]) \overset{\trianglelefteq}{\hookrightarrow} \mathsf{\Gamma U}([f]).$$

Not every automorphism θ of D can appear in $\mathsf{\Gamma U}([f])$; a necessary and sufficient condition is that

$$[\theta, \sigma] \in \mathsf{Int}(D);$$

up to inner automorphisms, θ has to centralize σ.

2.4.1 The case when h_f is degenerate

There is one issue which we have to address. In characteristic 2, it is possible that $V^{\perp h_f} \neq 0$, while $q_f^{-1}(0) \cap V^{\perp h_f} = 0$. Let's call such a pseudo-quadratic form *slightly degenerate*. This case can be reduced to the non-degenerate case as follows. Let $V' = V/V^{\perp h_f}$, and let

$$\Lambda' = \{c \in D \mid c + \Lambda \in q_f(V^{\perp h_f})\}.$$

It can be checked that Λ' is a form parameter. Define $(\tilde{q}, \tilde{h})$ on V' by

$$\tilde{h}(u + V^{\perp h_f}, v + V^{\perp h_f}) = h(u, v) \text{ and } \tilde{q}(v + V^{\perp h_f}) = f(v, v) + \Lambda'.$$

One can check that this pair is a pseudo-quadratic form $[\tilde{f}] = (\tilde{q}, \tilde{h})$; there is a canonical bijection

$$\mathsf{Gr}_k^{[f]}(V) \longrightarrow \mathsf{Gr}_k^{[\tilde{f}]}(V').$$

Furthermore, there is a corresponding isomorphism $\mathsf{GL}(V)_{[f]} \cong \mathsf{GL}(V')_{[\tilde{f}]}$. Here is an example. Let D be a perfect field of characteristic 2, let $V = D^5$, and let f denote the bilinear form given by the matrix

$$f \sim \begin{pmatrix} 0 & 1 & & & \\ & 0 & & & \\ & & 0 & 1 & \\ & & & 0 & \\ & & & & 1 \end{pmatrix}$$

Thus $q_f(x) = x_1 x_2 + x_3 x_4 + x_5^2$. The associated bilinear form h_f is symplectic and degenerate; its matrix is

$$h_f \sim \begin{pmatrix} 0 & 1 & & & \\ 1 & 0 & & & \\ & & 0 & 1 & \\ & & 1 & 0 & \\ & & & & 0 \end{pmatrix}$$

The process above gives us an isomorphism between an orthogonal generalized quadrangle and a symplectic generalized quadrangle, and a group isomorphism

$$\mathsf{O}(q_f, D) \cong \mathsf{Sp}_4(D)$$
$$\mathsf{O}(5, 2^k) \cong \mathsf{Sp}(4, 2^k) \text{ for the finite case } D = \mathbb{F}_{2^k}.$$

In Tits [32] and Bruhat-Tits [5], the chosen form parameter is always the minimal one, $\Lambda = D_{\sigma, -\varepsilon}$. Therefore, Tits allows his forms to be slightly degenerate (in our terminology). The resulting theory is the same; the choice of a bigger Λ makes the vector spaces smaller and avoids degenerate hermitian forms, which is certainly more elegant; the expense is that in this way we don't really see groups like $\mathsf{O}(5, 2^k)$, since they are identified with their isomorphic images belonging to non-degenerate forms, $\mathsf{O}(5, 2^k) \cong \mathsf{Sp}(4, 2^k)$ — one should keep that in mind.

2.5 Properties of form parameters

We discuss some properties of pseudo-quadratic forms and their form parameters.

In general, we have

$$\Lambda = D^{\sigma, -\varepsilon} \quad \Longrightarrow \quad \left(q_f(v,v) = 0 \quad \Longleftrightarrow \quad h_f(v,v) = 0 \right)$$

so the theory of trace hermitian forms is included in the pseudo-quadratic forms as a subcase. So the question is:

Why pseudo-quadratic forms?

The answer is given by the Fundamental Theorem of Polar Spaces 2.6.3. But first, we mention a few cases where pseudo-quadratic forms are *not* necessary. As we mentioned above, this is the case if $\Lambda = D^{\sigma, -\varepsilon}$. Now if $\mathrm{char}(D) \neq 2$, then $D_{\sigma, -\varepsilon} = D^{\sigma, -\varepsilon}$, so in characteristic different from 2, (σ, ε)-hermitian forms suffice.

2.5.1 Lemma *If* $\mathrm{char}(D) \neq 2$, *then there is a natural isomorphism*

$$\mathsf{TrHerm}_{\sigma, \varepsilon}(V) \cong \Lambda\text{-}\mathsf{Quad}_{\sigma, \varepsilon}(V).$$

$\square$

We consider some more special cases of form parameters. Note that

$$\Lambda = 0 \quad \Longrightarrow \quad \left(\varepsilon = 1 \text{ and } \sigma = \mathrm{id}_D \text{ and } D \text{ commutative} \right).$$

Suppose now that D is commutative. If $\sigma = \mathrm{id}_D$, then $D_{\mathrm{id}_D, -\varepsilon} = D(1 - \varepsilon)$. So either $\Lambda = D$, or $\varepsilon = 1$. If $D \neq \Lambda \neq 0$, then we have necessarily $\mathrm{char}(D) = 2$, and Λ is a D^2-submodule of D, and D is not perfect.

2.5.2 Lemma *Suppose that D is commutative and that $\sigma = \mathrm{id}_D$. If $0 \neq \Lambda \neq D$, then $\mathrm{char}(D) = 2$, the field D is not perfect, and Λ is a D^2-submodule of D. If $\Lambda = D$, then $\varepsilon = -1$.* $\square$

Suppose now that D is commutative and that $\sigma \neq \mathrm{id}_D$. Then $\sigma^2 = \mathrm{id}_D$, and $D_{\sigma, -\varepsilon} \neq 0$. Let $K \subseteq D$ denote the fixed field of σ. After scaling, we may assume that $1 \in D_{\sigma, -\varepsilon}$, which implies that $\varepsilon = -1$. Then $D^{\sigma, 1} = K$, and if $\mathrm{char}(D) \neq 2$, then $\Lambda = K$. For $\mathrm{char}(D) = 2$ we put $\psi(c) = c + c^{\sigma}$; then we have an exact sequence

$$0 \longrightarrow K \longrightarrow D \overset{\psi}{\longrightarrow} D_{\sigma, 1} \longrightarrow 0$$

of finite dimensional vector spaces over K, so $D_{\sigma, 1} = K$, regardless of the characteristic.

2.5.3 Lemma *Suppose that D is commutative and that $\sigma \neq \mathsf{id}_D$. Then $\sigma^2 = \mathsf{id}_D$, and the form parameters are the left translates sK of the fixed field K of σ, for $s \in D^\times$; in particular, $\Lambda = D^{\sigma,\varepsilon}$.* □

2.5.4 Corollary *Form parameters and pseudo-quadratic forms over perfect fields (in particular, over finite fields) are not important.* □

This explains why form parameters and proper pseudo-quadratic forms ($\sigma \neq \mathsf{id}_D$) are not an issue in finite geometry, and why they don't appear in books on algebraic groups over algebraically closed fields. Note also that if D is algebraically closed and if $\sigma \neq 1$, then the fixed field of σ is a real closed field.

The next result is less obvious and was pointed out to me by Richard Weiss.

2.5.5 Proposition (Finite form parameters) *Suppose that $\Lambda \neq 0$ is a finite form parameter. Then D is finite.*

Proof. If $\sigma = \mathsf{id}_D$, then D is commutative. If $\mathsf{char}(D) = 2$, then $\Lambda \subseteq D$ is a D^2-module. If $\mathsf{char}(D) \neq 2$, then $\Lambda = D$; in any case, Λ has a subset of the same cardinality as D.

If $\sigma \neq \mathsf{id}_D$, then it suffices to consider the minimal case where $\Lambda = D_{\sigma,-\varepsilon}$. We rescale in such a way that $1 \in D_{\sigma,-\varepsilon}$; then σ is an involution and $\varepsilon = -1$. Put $\psi(x) = x + x^\sigma$ and let $\lambda = \psi(x) \in \Lambda$. We claim that $\lambda^k \in \Lambda$, for all $k \geq 1$. Indeed, $(x + x^\sigma)^k$ can be written as a sum of 2^k monomials of the form

$$x^{(\sigma^{\nu_1})}x^{(\sigma^{\nu_2})}\cdots x^{(\sigma^{\nu_k})},$$

where $\nu_i \in \{0,1\}$. Now

$$x^{(\sigma^{\nu_1})}x^{(\sigma^{\nu_2})}\cdots x^{(\sigma^{\nu_k})} + x^{(\sigma^{\nu_k+1})}x^{(\sigma^{\nu_{k-1}+1})}\cdots x^{(\sigma^{\nu_1+1})} = \psi(x^{(\sigma^{\nu_1})}x^{(\sigma^{\nu_2})}\cdots x^{(\sigma^{\nu_k})}).$$

If Λ is finite, then all elements $\lambda \in \Lambda \setminus \{0\}$ have thus finite multiplicative order. This implies that D is commutative, see Herstein [16] Cor. 2 p. 116. From Lemma 2.5.3 above, we see that $\mathsf{card}(\Lambda) + \mathsf{card}(\Lambda) = \mathsf{card}(D)$, so D is finite. □

(It is a consequence of this proposition that there exist no semi-finite spherical irreducible Moufang buildings: if one panel of such a building is finite, then every panel is finite. This is a problem about Moufang polygons, and, by the classification due to Tits and Weiss (as described by Van Maldeghem in these proceedings [35], the only difficult case is presented by the classical Moufang quadrangles associated to hyperbolic spaces of rank 2; there, the line pencils are parametrized by the set Λ.)

This section shows that pseudo-quadratic forms are important only if either D is a finite field of characteristic 2 and if $\sigma = \mathsf{id}_D$ (and then we are dealing with quadratic forms), or if D is a non-commutative skew field of characteristic 2. On the other hand, none of the results about classical groups becomes really simpler if pseudo-quadratic forms are excluded, so we stick with them.

2.6 Polar spaces and pseudo-quadratic forms

We state the classification of polar spaces.

Suppose that $[f]$ is a non-degenerate pseudo-quadratic form of (finite) index $\mathrm{ind}[f] = m \geq 2$. Let $\mathsf{PG}^{[f]}(V) = (\mathsf{Gr}_1^{[f]}(V), \ldots, \mathsf{Gr}_m^{[f]}(V), *)$ and

$$\mathsf{PG}^{[f]}(V)_{1,2} = (\mathsf{Gr}_1^{[f]}(V), \mathsf{Gr}_2^{[f]}(V), *).$$

Then $\mathsf{PG}^{[f]}(V)_{1,2}$ is a (possibly weak) polar space and a subgeometry of the (possibly weak) polar space $\mathsf{PG}^{h_f}(V)_{1,2}$. A polar space isomorphic to such a space is called *embeddable*. Here is the first analogue of the Fundamental Theorem of Projective Geometry.

2.6.1 Theorem (Fundamental Theorem of Polar Spaces, I)
Let

$$\mathsf{PG}^{[f]}(V)_{1,2} \xrightarrow[\cong]{\phi} \mathsf{PG}^{[f']}(V')_{1,2}$$

be an isomorphism of embeddable (weak) polar spaces of finite ranks $m, m' \geq 3$. Then $m = m'$, and there exists an isomorphism of skew fields $D \xrightarrow[\cong]{\theta} D'$ and a θ-semilinear isomorphism $V \xrightarrow[\cong]{\Phi} V'$ such that the pull-back $\Phi^*([f'])$ is proportional to $[f]$.

For a proof see Tits [32] Ch. 8, or Hahn & O'Meara 8.1.5. □

This result and the next one are partly due to Veldkamp; the full results were proved by Tits. Cohen [7] and Scharlau [28] are good references for the classification, and for newer results in this area. The situation is more complicated for embeddable polar spaces of rank 2; we refer to Tits [32] Ch. 8. The theorem above deals with embeddable polar spaces. In fact, all polar spaces of higher rank are embeddable.

2.6.2 Theorem (Fundamental Theorem of Polar spaces, II)
Suppose that $(\mathcal{P}, \mathcal{L}, *)$ is a (weak) polar space of rank $m \geq 4$. Then $(\mathcal{P}, \mathcal{L}, *)$ is embeddable.

For a proof see Tits [32] 8.21, combined with Thm. 2.2.2. See also Scharlau [28] Sec. 7. □

This result is not true for polar spaces of rank 3; there exist polar spaces which have Moufang planes over alternative fields as subspaces, and such a polar space cannot be embeddable. However, this is essentially the only thing which can go wrong.

2.6.3 Theorem (Fundamental Theorem of Polar spaces of rank 3)
Let $(\mathcal{P}, \mathcal{L}, *)$ be a polar space of rank 3. If $(\mathcal{P}, \mathcal{L}, *)$ is not embeddable, then either there exists a proper alternative field A, and the maximal subspace are projective Moufang planes over A, or $(\mathcal{P}, \mathcal{L}, *) \cong A_{3,2}(D)$, for some skew field D.

For a proof see Tits [32] 7.13, p. 176, and 9.1, combined with Thm. 2.2.2. See also Scharlau [28] Sec. 7. □

The polar spaces containing proper Moufang planes are related to exceptional algebraic groups of type E_7; these are the only polar spaces of higher rank which do not come from classical groups. If D is commutative, then $A_{3,2}(D)$ is related to the Klein correspondence, $\mathbf{D}_3 = \mathbf{A}_3$.

Finally, we should mention the following result which is a consequence of Tits' classification.

2.6.4 Proposition (Tits)
Let $(\mathcal{P}, \mathcal{L}, *)$ be a weak polar space of rank $m \geq 3$, such that every subspace of rank $m - 2$ is incident with precisely two subspaces of rank $m - 1$. Then either $(\mathcal{P}, \mathcal{L}, *) \cong A_{3,2}(D)$, or $(\mathcal{P}, \mathcal{L}, *) \cong \mathrm{PG}^h(V)_{1,2}$, where V is a $2m$-dimensional vector space over a field D, and h is a non-degenerate symmetric bilinear form (i.e. $\sigma = \mathrm{id}_D$ and $\varepsilon = 1$) of index m (in other words, V is a hyperbolic module of orthogonal type, see the next section). □

Thick polar spaces of rank 2 are the same as generalized quadrangles. Similarly as projective planes, these geometries can be rather 'wild' and there is no way to classify them. The classification of the Moufang quadrangles due to Tits and Weiss is a major milestone in incidence geometry. For results about generalized quadrangles we refer to Van Maldeghem's article [35] in these proceedings. Polar spaces of possibly infinite rank were considered by Johnson [18].

2.7 Polar frames and hyperbolic modules

> *We show how the classification of pseudo-quadratic forms is reduced to the anisotropic case.*

Let V be an m-dimensional vector space over D, and put

$$H = V \oplus V^\sigma.$$

We define a form $f \in \mathsf{Form}_\sigma(H)$ by

$$f((u, \xi), (v, \eta)) = \xi(v).$$

As a matrix, f is represented as

$$f \sim \begin{pmatrix} 0 & 1_m \\ 0 & 0 \end{pmatrix}$$

where 1_m denotes the $m \times m$ unit matrix. The space H with the pseudo-quadratic form $[f]$ (relative to a form parameter $(\Lambda, \sigma, \varepsilon)$) is called a *hyperbolic module* of rank m; if $m = 1$ then H is 2-dimensional and we call it a *hyperbolic line*. (Hyperbolic lines are often called *hyperbolic planes*; this depends on the viewpoint, linear algebra vs. projective geometry.) The following theorem is crucial.

2.7.1 Theorem *Let $[f]$ be a non-degenerate pseudo-quadratic form of finite Witt index $\mathsf{ind}([f]) = m$ in a vector space V. Then there exists a hyperbolic module H of rank m in V, and V splits as an orthogonal sum*

$$V = H \oplus V_0,$$

where $V_0 = H^{\perp_{h_f}}$. If $H' \subseteq V$ is another hyperbolic module of rank m, then there exists an isometry of V which maps H' onto H.

This follows from Witt's Theorem, see Hahn & O'Meara 6.1.12, 6.2.12 and 6.2.13 — the proofs apply despite the fact that Hahn & O'Meara work always with finite dimensional vector spaces. What is needed in their proof is only that H has finite dimension. □

Thus, the subspace V_0 is unique up to isometry. This subspace (together with the restriction of $[f]$) is called the *anisotropic kernel* of $[f]$. Since the hyperbolic module has a relatively simple structure, the study of pseudo-quadratic forms is reduced to the anisotropic case; a pseudo-quadratic form is determined its Witt index and its anisotropic kernel (and by $\sigma, \varepsilon, \Lambda$, of course).

The *building associated to a polar space* $(\mathcal{P}, \mathcal{L}, *)$ is constructed as follows. If $(\mathcal{P}, \mathcal{L}, *)$ is thick, then the vertices are the subspaces of the polar space, and the simplices are sets of pairwise incident vertices. The resulting building has rank m and type $\mathbf{C}_m$, see Tits [32] Ch. 7. If the polar space is weak, then a new geometry is introduced: the vertices are all subspaces of rank different from $m - 2$, and two vertices are called incident if one contains the other, or if their intersection has rank $m - 2$. This is again an m-sorted structure (there are two classes of subspaces of rank $m-1$), and the resulting simplicial complex is a building of type $\mathbf{D}_m$, see Tits [32] Ch. 6 and 7.12, 8.10. It is easy (but maybe instructive) to check that this makes the weak polar space $A_{3,2}(D)$ into the building $\Delta(D^4)$ obtained from the projective space $\mathsf{PG}(D^4)$. The buildings related to polar spaces are also discussed in Brown [4], Cohen [7], Garrett [12], Ronan [26], Scharlau [28], and Taylor [31].

Finally, we mention the classical groups obtained from non-degenerate pseudo-quadratic forms of index $m \geq 2$. Scaling the form by a suitable constant $s \in D^{\times}$, the following cases appear.

Symplectic groups

This is the situation when $(\sigma, \varepsilon, \Lambda) = (\mathrm{id}_D, -1, D)$. Here D is commutative, $q_f = 0$ and h_f is alternating. The dimension of V is even (and V is hyperbolic), and $2\,\mathsf{ind}(h_f) = \dim(V)$. The corresponding polar space is thick.

Orthogonal groups

This is the situation when $(\sigma, \varepsilon, \Lambda) = (\mathrm{id}_D, 1, 0)$. Here D is commutative and h_f is symmetric, and $2\,\mathsf{ind}[f] \leq \dim(V)$. The corresponding polar space is thick if and only if $2\,\mathsf{ind}[f] < \dim(V)$. If $2\,\mathsf{ind}[f] = \dim(V)$, then the corresponding building is the $\mathbf{D}_m$-building (the *oriflamme geometry*) described above, and V is hyperbolic.

Defective orthogonal groups

This is the situation when $(\sigma, \varepsilon) = (\mathrm{id}_D, 1)$ and $0 \neq \Lambda \neq D$. Here D is commutative and h_f is symmetric. This occurs only in characteristic 2 over non-perfect fields (in the perfect case, $\Lambda = D$ and we are in the symplectic case).

Classical unitary groups

This is the situation when $\sigma \neq \mathrm{id}_D = \sigma^2$, $\varepsilon = 1$ and $\Lambda = D^{\sigma,-1}$. Here D need not be commutative and h_f is $(\sigma, 1)$-hermitian. Since Λ is maximal, the hermitian form h_f describes $[f]$ completely and q_f is not important. The corresponding polar space is thick.

Restricted unitary groups

This is the situation when $\sigma \neq \mathrm{id}_D = \sigma^2$, $\varepsilon = 1$ and $\Lambda < D^{\sigma,-1}$. Here D is of characteristic 2 and not commutative. The corresponding polar space is thick.

2.8 Omissions

By now it should be clear that the theory of pseudo-quadratic forms and the related geometries is rich, interesting, and sometimes difficult. There are many other interesting topics which we just mention without further discussion.

Root elations

Root elations in polar spaces are more complicated than root elations in projective spaces. This is due to the fact that there are two types of half-apartments and, consequently, two types of root groups. One kind is isomorphic to the additive group of D, while the other is related to the anisotropic kernel V_0 of V, and to the form parameter Λ. These root groups are nilpotent of class 1 or 2. We refer to Van Maldeghem [34] for a detailed description of the root groups. The root elations are Eichler transformations (also called Siegel transformations), which are special products of transvections, and the group generated by these maps is the elementary unitary group $\mathsf{EU}([f])$.

K-theory

Starting with the abstract commutator relations for the root groups of a given apartment, one can construct unitary version of the Steinberg groups, and unitary K-groups. Hahn & O'Meara [15] give a comprehensive introduction to the subject (for hyperbolic V). There is a natural map $\mathsf{K}_0(D) \longrightarrow \mathsf{KU}_0(D)$ whose cokernel is the Witt group of D, another important invariant.

Permutation groups

If $\mathrm{ind}[f] \geq 2$, then the action of the unitary group on $\mathsf{Gr}_1^{[f]}(V)$ is not 2-transitive. Instead, one obtains interesting examples of permutation groups of rank 3 (i.e. with 3 orbits in $\mathsf{Gr}_1^{[f]}(V) \times \mathsf{Gr}_1^{[f]}(V)$).

Moufang sets

If $\mathrm{ind}[f] = 1$, then the corresponding unitary group is 2-transitive on $\mathsf{Gr}_1^{[f]}(V)$, and there is a natural Moufang set structure.

Isomorphisms

As in the linear case, one can ask whether two unitary groups can be (abstractly or as permutation groups) isomorphic. Indeed, there are several interesting isomorphisms related to the Klein correspondence and to Cayley algebras. Many results in this direction can be found in Hahn & O'Meara [15].

References

[1] R. Alperin and R. Dennis, K_2 of quaternion algebras. J. Algebra **56** (1979) 262–273.

[2] E. Artin, *Geometric algebra.* Reprint of the 1957 original. John Wiley & Sons, Inc., New York (1988).

[3] A. Bak, *K-theory of forms.* Princeton University Press, Princeton (1981).

[4] K. Brown, *Buildings.* Springer-Verlag, New York (1989).

[5] F. Bruhat and J. Tits, Groupes réductifs sur un corps local. (French) Inst. Hautes Études Sci. Publ. Math. **41** (1972) 5–251.

[6] F. Buekenhout and E. Shult, On the foundations of polar geometry. Geom. Dedicata **3** (1974) 155–170.

[7] A. Cohen, *Point-line spaces related to buildings.* In: *Handbook of incidence geometry*, 647–737. F. Buekenhout ed., North-Holland, Amsterdam (1995).

[8] P. Cohn, *Skew fields.* Cambridge University Press, Cambridge (1995).

[9] J. Dieudonné, *La géométrie des groupes classiques.* (French) Springer-Verlag, Berlin Heidelberg New York (1971) (3rd ed.)

[10] C.-A. Faure and A. Frölicher, Morphisms of projective geometries and semilinear maps. Geom. Dedicata **53** (1994) 237–262.

[11] C.-A. Faure and A. Frölicher, Dualities for infinite-dimensional projective geometries. Geom. Dedicata **56** (1995) 225–236.

[12] P. Garrett, *Buildings and classical groups.* Chapman & Hall, London (1997).

[13] T. Grundhöfer, Commutativity of finite groups according to Wedderburn and Witt. Arch. Math. **70** (1998) 425–426.

[14] T. Grundhöfer, *Basics on buildings.* (these proceedings)

[15] A. Hahn and T. O'Meara, *The classical groups and K-theory.* Springer-Verlag, Berlin (1989).

[16] I. N. Herstein, *Rings with involution.* University of Chicago Press, Chicago (1976).

[17] D. Hughes and F. Piper, *Projective planes.* Springer-Verlag, New York-Berlin (1973).

[18] P. M. Johnson, Polar spaces of arbitray rank. Geom. Dedicata **35** (1990) 229–230.

[19] M.-A. Knus, *Quadratic and Hermitian forms over rings.* Springer-Verlag, Berlin (1991).

[20] H. Lüneburg, Ein neuer Beweis eines Hauptsatzes der projektiven Geometrie. (German) Math. Z. **87** (1965) 32–36.

[21] H. Lüneburg, Über die Struktursätze der projektiven Geometrie. (German) Arch. Math. **17** (1966) 206–209.

[22] J. Milnor, *Introduction to algebraic K-theory.* Princeton University Press, Princeton (1971).

[23] G. Pickert, *Projektive Ebenen.* (German) Zweite Auflage. Springer-Verlag, Berlin-New York (1975).

[24] U. Rehmann, Zentrale Erweiterungen der speziellen linearen Gruppe eines Schiefkörpers. (German) J. Reine Angew. Math. **301** (1978) 77–104.

[25] D. J. S. Robinson, *A course in the theory of groups.* Springer-Verlag, New York (1982).

[26] M. Ronan, *Lectures on buildings.* Academic Press, Inc., Boston (1989).

[27] J. Rosenberg, *Algebraic K-theory and its applications.* Springer-Verlag, New York (1994).

[28] R. Scharlau, *Buildings.* In: *Handbook of incidence geometry,* 477–645. F. Buekenhout ed., North-Holland, Amsterdam (1995).

[29] W. Scharlau, *Quadratic and Hermitian forms.* Springer-Verlag, Berlin (1985).

[30] J. Strooker, The fundamental group of the general linear group. J. Algebra **48** (1977) 477–508.

[31] D. Taylor, *The geometry of the classical groups.* Heldermann Verlag, Berlin (1992).

[32] J. Tits, *Buildings of spherical type and finite BN-pairs.* Lecture Notes in Mathematics, Vol. 386. Springer-Verlag, Berlin-New York (1974).

[33] J. Tits, *Twin buildings and groups of Kac-Moody type*. In: *Groups, combinatorics & geometry (Durham 1990)*, 249–286. M. Liebeck and J. Saxl ed., London Math. Soc. Lecture Notes 165, Cambridge Univ. Press, Cambridge (1992).

[34] H. Van Maldeghem, *Generalized polygons*. Birkhäuser Verlag, Basel (1998).

[35] H. Van Maldeghem, *An introduction to generalized polygons*. (these proceedings)

TWIN BUILDINGS

BERNHARD MÜHLHERR

FACHBEREICH MATHEMATIK
UNIVERSITÄT DORTMUND
44221 DORTMUND, GERMANY
EMAIL: BERNHARD.MUEHLHERR@MATHEMATIK.UNI-DORTMUND.DE

1. INTRODUCTION

The goal of this paper is to give a survey of the main ideas in the classification of 2-spherical twin buildings. A program for such a classification had been outlined by J. Tits in [Ti92]. There are two main conjectures in this program: Conjecture 1 concerns uniqueness, Conjecture 2 concerns existence. Conjecture 1 was proved in [MR95] under a certain condition (co) which is 'almost always' satisfied; this result relies heavily on results proved in [Ti92]. It turned out that Conjecture 2 of [Ti92] has to be slightly modified; as already suspected in loc. cit. it was 'too optimistic'. There is not yet a direct proof of this modified conjecture. Its validity is just a consequence of the classification which has been recently accomplished by the author. The classification is achieved by giving several construction procedures of twin buildings.

The interesting fact about these constructions is that they provide a link with several concepts developed by Tits in the 'prehistory' of buildings and that they shed new light on the classification of spherical buildings achieved in [Ti74] and on the recent classification of Moufang polygons to appear in [TW0*].

We start for this reason with some historical remarks. After introducing some notation and giving some examples, we describe the main ideas in the classification of spherical buildings. In Section 6 we provide information about the origin and the main ideas of the theory of twin buildings. In Sections 7 to 9 we explain how the classification of 2 spherical twin buildings works in principle. In the course of the paper several topics are only touched and others are even not mentioned for sake of clarity. For this reason there is an appended section in which we make further comments on some topics mentioned in the paper and give some information about several other aspects of twin building theory.

2. SOME HISTORY AND MOTIVATION

The purpose of this section is to provide a short historical introduction to buildings. Good places where there reader can find more information about the history of buildings and their motivation are [Ti78] and the introductions of [Ti74], [Ti81] and [Ti92] (see also 10.1 below).

The theory of buildings is a synthesis of projective geometry and Lie theory. Starting with a vector space V over a division ring K there are two important objects one can associate with V: there is the group $\mathrm{Gl}(V)$ of K-linear transformations of V and there is the projective geometry $\mathbf{P}(V)$, the geometry of non-trivial subspaces of V. If we let K be the field of complex numbers and if V has dimension $n + 1$ for a natural number $n \in \mathbf{N}$, then $\mathrm{Sl}(V)$ is a simple Lie group of type $\mathbf{A}_n$ - the Dynkin diagram of $\mathrm{Sl}(V)$.

In the 1950's J.Tits made the crucial observation that the Dynkin diagram $\mathbf{A}_n$ encodes a lot of information about the projective geometry $\mathbf{P}(\mathbf{C}^{n+1})$. Moreover, the properties which one can read off from this diagram are properties of arbitrary projective geometries of rank n, i. e. , of projective geometries over arbitrary division rings. The Dynkin diagram $\mathbf{A}_n$ describes in fact the most important properties of the local structure of projective geometries of rank n. The fact that there is a link between the Dynkin diagrams of type $\mathbf{A}_n$ and the projective geometries of rank n motivates the question about the geometries which correspond to the other Dynkin diagrams (and in particular to the corresponding groups!). In order to find good axioms for this class of geometries J.Tits introduced diagram geometries over Coxeter diagrams (see 10.2).

The main motiviation for Tits' investigations was the idea to use combinatorial techniques in order to study Lie-groups (especially of exceptional type) in a fashion similar to using projective geometry for understanding linear groups. The class of groups for which these ideas turned out to be applicable is now known as the class of groups with a BN-pair. The first important class of BN-pairs are the analogues of Lie groups over arbitrary fields constructed by Chevalley in [Ch55]. Roughly speaking, this construction associates with each Dynkin diagram $\mathbf{D}$ and each field k a group $\mathbf{D}(k)$ and, given such a group, it then provides a diagram geometry (over the diagram $\mathbf{D}$) in a natural way. The second important class which should be mentioned here (and which contains the first) is the class of semi-simple algebraic groups defined over arbitrary fields. A further class of interest in our context is the class of classical groups which has a non-trivial intersection with the class of algebraic groups.

As already mentioned, the combinatorial ideas described above apply to the class of groups with BN-pairs. These are groups which contain two subgroups B and N satisfying certain axioms. As an immediate consequence of the BN-pair axioms one obtains the so-called Bruhat-decomposition of the group in question. The appearance of the Bruhat-decomposition motivated a new axiomatic approach to the combinatorial structures associated with these groups. It turned out to be easier to axiomatize the flag-complex of the desired diagram geometries rather than the geometries themselves. Therefore J.Tits introduced buildings as simplicial complexes endowed with an apartment system. In comparison to the earlier approach the new set of axioms was more powerful because it gave better control over the global structure. One remarkable fact about the list of axioms is that one does not start with a particular Coxeter diagram in advance, but recovers it from the basic theory.

Later, in the 1980's [Ti81], a new set of axioms for buildings was given by introducing chamber systems over Coxeter diagrams. The principal idea is that one can axiomatize the graph of maximal simplices (called chambers) by a short list of axioms which uses the Coxeter group of type M. This approach was pushed further to the most recent point of view, namely to consider buildings as a metric spaces whose points are the chambers and where the metric has its values in the corresponding Coxeter group.

3. SOME NOTATION

As it is clear from the previous section Coxeter diagrams play a crucial role in the theory of buildings. Therefore we shall give a formal definition of them here: Let I be a set. A Coxeter matrix over I is a matrix $M = (m_{ij})_{i,j \in I}$ whose entries are in the set $\mathbf{N} \cup \{\infty\}$ such that $m_{ii} = 1$ for all $i \in I$ and $m_{ij} = m_{ji} \geq 2$ for all $i \neq j \in I$. The Coxeter diagram corresponding to M is the labeled graph $(I, E(M))$ where $E(M) := \{\{i, j\} \mid m_{ij} > 2\}$ and where each edge is labeled by the corresponding m_{ij} (there is the convention that the labels '3' are omitted if one draws such a diagram). As the Coxeter matrix and the corresponding Coxeter diagram carry the same information we do not distinguish between them formally. We call a Coxeter diagram *irreducible* if its underlying graph is connected and we call it *2-spherical* if all entries in the matrix are finite. The *rank* of a Coxeter diagram is the cardinality of the set of its vertices.

Let $M = (m_{ij})$ be a Coxeter matrix over a set I. A Coxeter system of type M is a pair (W, S) such that W is a group, $S = \{s_i \mid i \in I\}$ is a set of generators of W satisfying the relations $(s_i s_j)^{m_{ij}}$ for all $i, j \in I$ and such that the set S together with these relations form a presentation of W. The Coxeter diagram M is called *spherical* if the corresponding Coxeter group is finite. If this is the case, then there exists a unique element $r \in W$ whose length with respect to the generating set S is maximal and this element r is an involution.

Although the formal definition of a building won't be needed in the sequel we give it here for the sake of completeness. We recall the definition of a building as a metric space following [Ti92]: Let I be a set, M a Coxeter matrix over I, let (W, S) be the Coxeter system of type M and let $l : W \to \mathbf{N}$ be the associated length function. A *building* of type M is a pair $\Delta = (\mathbf{C}, \delta)$

where $\mathbf{C}$ is a set (whose elements are called chambers) and $\delta : \mathbf{C} \times \mathbf{C} \to W$ is a *distance function* satisfying the following axioms where $x, y \in \mathbf{C}$ and $w = \delta(x, y)$:

(Bu 1): $w = 1$ if and only if $x = y$;

(Bu 2): if $z \in \mathbf{C}$ is such that $\delta(y, z) = s \in S$, then $\delta(x, z) = w$ or ws, and if, furthermore, $l(ws) = l(w) + 1$, then $\delta(x, z) = ws$;

(Bu 3): if $s \in S$, then there exists $z \in \mathbf{C}$ such that $\delta(y, z) = s$ and $\delta(x, z) = ws$.

4. EXAMPLES

Projective geometries. For each $n \in \mathbf{N}$, let $\mathbf{A}_n$ be the Coxeter diagram with vertex set $\{1, 2, \dots, n\}$ and edge-set $\{\{i, i + 1\} \mid 1 \leq i \leq n - 1\}$ for which all edges have label 3. The following theorem describes all buildings of type $\mathbf{A}_n$:

Theorem: *The buildings of type $\mathbf{A}_n$ - considered as simplicial complexes - are precisely the flag complexes of the projective geometries of rank n.*

Starting from a vector space of dimension $n + 1$ then there is a nice way of associating with any two maximal flags of the corresponding projective geometry a distance in the group $\text{Sym}(n + 1)$ which is precisely the Coxeter group associated with the diagram $\mathbf{A}_n$ (see [Gr01] for details). Hence there is also a natural link between projective geometries and $\mathbf{A}_n$-buildings considered as metric spaces.

Polar geometries. Let n be a natural number. The Coxeter diagram $\mathbf{C}_n$ is the Coxeter diagram whose underlying graph coincides with the underlying graph of $\mathbf{A}_n$ and where all labels are equal to 3 except the label of $\{n - 1, n\}$ which is equal to 4.

Let V be a vector space over a division ring K, let σ be an involution of K and let $h : V \times V \to K$ be a trace-valued σ-hermitian form of Witt index n. The flag complex of the geometry of non-trivial singular subspaces with respect to h is an example of a $\mathbf{C}_n$-building (considered as a simplicial complex). It is known from [Ti74] that for $n \geq 4$, all $\mathbf{C}_n$-buildings arise from constructions which are variations of the construction above.

Another fact about the $\mathbf{C}_n$-buildings described above which will be of relevance in the sequel is the following. If we assume that V has dimension $m + 1$, then we can interpret the hermitian form as an involutory automorphism of the corresponding $\mathbf{A}_m$-building (a polarity!) whose set of fixed simplices is precisely the building associated with the singular subspaces. Thus we have an example of a group acting on a building Δ whose set of fixed points is again a building $\tilde{\Delta}$ which is however of a different type.

BN**-Pairs.** Let G be a group. A BN-pair for G is a pair of subgroups of G satisfying certain axioms. These axioms involve a Coxeter system whose associated Coxeter diagram is called the *type* of the BN-pair in question.

From what is said in Section 2 about BN-pairs it is not surprising that one can associate in a canonical way with each group G with a BN-pair of type M a building of the same type.

Here the converse is definitely not true in general. An arbitrary building need not come from a BN-pair construction. Nevertheless, given a building Δ and a group G acting on Δ such that the action satisfies certain properties of transitivity, then one can define subgroups B and N of G which constitute a BN-pair for G having the same type as Δ. The building which one obtains from this BN-pair is then canonically isomorphic to Δ.

5. THE CLASSIFICATION OF SPHERICAL BUILDINGS

Given a natural number $n > 2$ and a division ring K, $\mathbf{P}(K^{n+1})$ gives rise to a building of type $\mathbf{A}_n$ in a natural way. By the fundamental theorem of projective geometry and the fact that projective geometries of rank n are in one-to-one correspondence with the buildings of type $\mathbf{A}_n$ it follows that there is a one-to-one correspondence between the $\mathbf{A}_n$-buildings and the division rings K. As the full group of automorphisms contains the group $\mathrm{PSl}(n + 1, K)$ there is hence a natural bijective correspondence between the projective geometries of rank n and the groups $\mathrm{PSl}(n + 1, K)$ for $n \geq 3$.

Now we argue on the level of groups: if K is commutative, then $\mathrm{PSl}(n + 1, K)$ is a Chevalley-group, if K is finite-dimensional over its center k, then $\mathrm{PSl}(n + 1, K)$ is an algebraic group (defined over k) and $\mathrm{PSl}(n + 1, K)$ is by definition a classical group for any K. It is a fact that Chevalley-groups, isotropic algebraic groups and classical groups all have a (canonical) BN-pair whose associated Coxeter diagram is spherical. Hence one can associate a spherical building with each group in either class. The spherical buildings obtained from such a group are hence 'of algebraic origin'. Now the classification of spherical buildings can be stated as follows:

Theorem (Tits [Ti74]**):***Each irreducible spherical building of rank at least 3 is of algebraic origin.*

Here a building is said to be of algebraic origin if it comes from a BN-pair associated with a classical group, an algebraic group or a so-called mixed group. Mixed groups are a third kind of groups which have to be included in order to deal with certain phenomena which occur in small characteristics (see 10.3).

Uniqueness and existence. As is clear from the discussion in the previous subsection, Tits' classification of the irreducible spherical buildings of rank at least 3 is a far-reaching generalization of the fundamental theorem of projective geometry. So let's first look at the example of projective geometries. Suppose we are given two projective geometries Γ, Γ' of rank $n \geq 3$ and that we can find a plane π in Γ and a plane π' in Γ' which are isomorphic as projective planes. Then we already know that Γ and Γ' are isomorphic because the planes π and π' determine the underlying fields. Even more is true: each isomorphism from π onto π' can be extended to an isomorphism from Γ onto Γ'. This fact can be proved by purely combinatorial arguments without knowing in advance that Γ, Γ' come from vector spaces. As a result one has that a projective plane in a projective geometry Γ of rank at least 3 determines Γ up to isomorphism. A similar local-global principle holds for spherical buildings. In this more general context the projective planes have to be replaced by the so-called foundations which describe the local structure at a chamber of the building in question. Tits proved in [Ti74] the following important local-global-theorem:

Theorem: *Let Δ, Δ' be two buildings of spherical type M, then each isomorphism of a foundation of Δ onto a foundation of Δ' extends to an isomorphism from Δ onto Δ'.*

This theorem provides a strong tool for settling the uniqueness part of the classification. But it yields a lot more. It can be used to prove the fact that spherical buildings of rank at least 3 have a big group of automorphisms and that they satisfy, in particular, the so-called Moufang property. This reduces the possible local data of a spherical building of rank at least 3 drastically. The reduction generalizes the well known fact that planes in a projective geometry of rank at least 3 have to be Desarguesian.

Thus, there remains a short list of possible local structures for spherical buildings. This list was shown by Tits to be precisely the list of local data of the buildings which are of algebraic origin in the sense above. This of course settled the problem of existence because the existence of the corresponding groups was already established (see however 10.3).

There is another important fact about the local-global theorem stated above: Its proof uses the *opposition relation* on the set of chambers in a spherical building. That such a relation exists is due to the fact that there is a longest element r in the corresponding finite Coxeter group which is an involution. Thus, two chambers of a spherical building are called *opposite* if their distance is equal to r. As there is no opposition relation available for non-spherical buildings one cannot prove a similar theorem in this case; in fact, one can show that it becomes false in the non-spherical case and - even worse - that there is no hope for classification results in general (see for instance [RT87]).

6. Kac-Moody-groups and twin buildings

In Chevalley's construction of the analogues of semi-simple Lie groups over arbitrary fields one uses the corresponding semi-simple Lie algebra. The Lie algebra in question is already determined by its Cartan matrix. The data of all possible Cartan matrices are encoded in the corresponding Dynkin diagrams mentioned in Section 2. There is a natural generalization of Cartan matrices and they yield infinite-dimensional generalizations of semi-simple Lie algebras - the Kac-Moody algebras. In [Ti87] it is proved that one can associate with each generalized Cartan matrix A and each field k a group (more precisely: a class of groups). The groups one obtains for a given field k by this construction are called the Kac-Moody groups over k. If the generalized Cartan matrix is a Cartan matrix in the classical sense, then the Kac-Moody groups for this matrix over k are precisely the groups constructed by Chevalley. Hence, Kac-Moody groups over fields are natural generalizations of Chevalley groups over fields; in fact, they can be considered as the infinite-dimensional versions of Chevalley-groups.

Similar to the case of Chevalley groups one can find BN-pairs in Kac-Moody groups; but there is a new phenomenon. In Kac-Moody groups there are two different BN-pairs of the same type which have the subgroup N in common. Thus one has a triple of subgroups (B_+, B_-, N) in each Kac-Moody group G such that (B_+, N) and (B_-, N) are BN-pairs of the same type for G. This yields two different buildings Δ_+, Δ_- on which the group G acts in a natural way. Moreover, the fact that these two BN-pairs have the same 'N' provides a natural relation $\mathbf{O}$ between the chambers of Δ_+ and the chambers of Δ_-. The relation $\mathbf{O}$ has properties which are very similar to the properties of the opposition relation on the set of chambers of a spherical building. Based on this observation, M. Ronan and J. Tits introduced twin buildings as triples $(\Delta_+, \Delta_-, \mathbf{O})$ consisting of two buildings Δ_+, Δ_- of the same type M and an opposition relation $\mathbf{O}$ between the chambers of the two buildings. Formally, this was done by requiring the existence of a group-valued codistance function between the chambers of the two buildings. It turns out that each spherical building can be considered as a twin building in a canonical way through the existence of the ordinary opposition relation on the set of its chambers; hence twin buildings generalize spherical buildings in a natural way.

Here is the precise definition of a twin building (see [Ti92]): As in Section 3 let M be a Coxeter diagram, let (W, S) be the Coxeter system of type M and $l : W \to \mathbf{N}$ the associated length function. A twin building of type M is a triple $\Delta = (\Delta_+, \Delta_-, \delta_*)$ where $\Delta_\epsilon = (\mathbf{C}_\epsilon, \delta_\epsilon)$ is a building of type M for $\epsilon \in \{+, -\}$ and where $\delta_* : \mathbf{C}_+ \times \mathbf{C}_- \cup \mathbf{C}_- \times \mathbf{C}_+ \to W$ is a mapping (called the codistance) satisfying the following axioms for all $\epsilon \in \{+, -\}$, $x \in \mathbf{C}_\epsilon, y \in \mathbf{C}_{-\epsilon}$ and $w = \delta_*(x, y)$:

(Tw 1): $\delta_*(y, x) = w^{-1}$;

(Tw 2): if $z \in \mathbf{C}_{-\epsilon}$ is such that $\delta_{-\epsilon}(y, z) = s \in S$ and $l(ws) = l(w) - 1$, then $\delta_*(x, z) = ws$;

(Tw 3): if $s \in S$, there exists $z \in \mathbf{C}_{-\epsilon}$ such that $\delta_{-\epsilon}(y, z) = s$ and $\delta_*(x, z) = ws$.

7. The classification of twin buildings

In his fundamental paper on twin buildings [Ti92] J. Tits gives a survey of the basic theory of twin buildings and Kac-Moody groups. As there is an opposition relation in twin buildings there arises the natural question whether the arguments used in the classification of spherical buildings can be generalized to the twin case. A large part of [Ti92] deals with this question. As a first observation it turns out that there is no hope for a classification if there are infinities in the Coxeter diagram; hence there is hope for such a classification in the 2-spherical case only. For the subclass of 2-spherical twin buildings it is proved in [Ti92], however, that there is a local-global-theorem for the two halves of the buildings. More precisely, it is proved that the local structure of one of the two buildings determines the structure of Δ_+ and the structure of Δ_-; but it is not clear at this point whether the opposition relation of the twinning is determined as well. That this is indeed the case is proved in [MR95] for twin buildings satisfying a certain condition (co); this condition is always satisfied if there are at least 5 chambers per panel. Thus there is a local-global theorem for almost all 2-spherical twin buildings. As in the spherical case, this settles the uniqueness part of the classification and it restricts the possible local structures. It follows from the local-global theorem that the local structure of a 2-spherical twin building has to be a so-called *Moufang foundation*. Using the classification of Moufang polygons of J. Tits and R. Weiss, it is possible to classify all Moufang foundations for a given 2-spherical Coxeter diagram. Hence the classification of the twin buildings of that type is reduced to the question of which of these foundations are indeed the local structure of a 2-spherical building. It turns out that 'most' of them are in fact local structures of a twin building. Therefore one has to prove the existence of twin buildings with a prescribed local structure in order to finish the classification.

As described above, J.Tits used the existence of the corresponding (classical, algebraic or mixed) groups in order to settle the existence part in the classification of spherical buildings. In the twin situation the groups are not available because there is not yet a theory of forms of Kac-Moody groups which would guarantee the existence of all desired groups (see however 10.4). In [Mu99/2] the author developed geometric existence criterions for twin buildings which are sufficient to prove the existence of all 2-spherical twin buildings and thus finish the classification. The classification is carried out in the case where the local structure of the building is 'sufficiently nice' in loc. cit. (a special case of this was done in [Mu99/1]); in the remaining cases it is reduced to the validity of certain conjectures concerning Moufang polygons and Moufang foundations. These have been proved in the meantime (partly) in joint work with H. Van Maldeghem and V. Vermeulen.

In the next two sections we describe the main ideas in the classification of 2-spherical twin buildings as it was outlined in [Mu99/2].

8. THREE CONSTRUCTIONS - ANOTHER VIEW ON SPHERICAL BUILDINGS

As it is clear from the discussion above the main problem left in the classification is to prove the existence of 2-spherical twin buildings with prescribed local structures. The main idea is to use the existence of certain buildings in order to produce new ones. There are basically 3 constructions which will be of interest.

(CS) The first construction is based on the observation that a convex (where the notion of convexity is defined in a suitable way) subset of the set of chambers in a twin building (and hence also in a spherical building) containing 2 opposite chambers is a twin building of the same type as the orginal one. A typical example of such a situation is the following: if k is a subfield of a field k' then the building of type $\mathbf{A}_n$ over the field k can be obtained as a convex subbuilding of the building of the same type over k' (because we are dealing with projective geometries in this case). An instance which is of more relevance in the present context is that we can consider spherical buildings associated with mixed groups (and also some very exotic Moufang quadrangles of indifferent type) as a convex subbuildings of spherical buildings associated with Chevalley groups of the same type.

(L) The second construction is standard. Given an inductive system of buildings of the same type M, then its limit is again a building of type M. Examples which are of interest in the present context are the $\mathbf{C}_n$-buildings associated with hermitian forms of Witt index n which may be defined over vector spaces of infinite dimension. These buildings can be considered as limits of buildings associated with hermitian forms on finite-dimensional vector spaces.

(FP) The third construction procedure to obtain new twin buildings from old ones is the fixed point construction: given a group G acting on a twin building Δ, one can show (under certain conditions) that the set of fixed points of G in Δ becomes a twin building $\tilde{\Delta}$ in a natural way. One example of such a situation is that of a polarity of an $\mathbf{A}_n$-building which fixes a $\mathbf{C}_k$-building as described in Section 4. This particular example shows that the fixed point building need not have the same type as the original one (as was the case in the other constructions).

Now we can restate Tits' classification theorem for spherical buildings as follows.

Theorem: *Let Δ be an irreducible spherical building of rank at least 3. Then one of the following holds:*

1. *Δ is associated with a classical group over a division ring which is infinite-dimensional over its center.*

2. *Δ can be obtained from the class of spherical buildings associated with Chevalley groups over algebraically closed fields by a sequence of constructions (CS), (FP), (L).*

9. 2-SPHERICAL TWIN BUILDINGS

The form in which Tits' classification of spherical buildings is stated in the previous section generalizes (with slight changes) to the twin situation, as follows:

Theorem:*Let* Δ *be a 2-spherical irreducible twin building of rank at least 3 which satisfies Condition (co). Then one of the following holds:*

1. Δ *is asscociated with a classical group over a division ring which is infinite-dimensional over its center. In this case* Δ *is of spherical or of affine type.*
2. Δ *can be obtained from the class of twin buildings associated with Kac-Moody groups over algebraically closed fields by a sequence of constructions (CS), (FP), (L), (FP), (FP).*

We add the following remarks concerning this result and its proof:

1. One may dream of a short statement of the classification by giving a short list of classes of groups which have a twin BN-pair as in the spherical case. In a narrow sense, such a list does not exist because there is a certain freeness in constructing 2-spherical buildings. So, in a sense, the statement above is the best possible.

2. The most important procedure to obtain twin buildings from other ones is the fixed point construction. It was already emphasized that the type of the fixed point building may be different from the original type. This happens already in the spherical case. Nevertheless there is a completely new phenomenon. In the constructions (FP) of the theorem there are instances where Δ has infinite rank and $\tilde{\Delta}$ has finite rank. There is a general fixed point theorem which is applicable in all these situations. Its proof uses the fact that buildings are incidence geometries, simplicial complexes and chamber systems at the same time. One considers a group G acting on a building Δ considered as simplicial complex. The group acts on the associated flag-complex and, using arguments from diagram geometry, one can show that the chamber system associated with a certain subcomplex of the fixed point complex is a building $\tilde{\Delta}$. It is a remarkable fact that the different ideas which had been developed in the history of buildings in order to find good axioms for them are also useful in the proof of this theorem.

3. An irreducible Coxeter diagram M can be considered as a quotient of a group G acting on its universal cover, which is a Coxeter diagram $\bar{M}$ whose underlying graph is a tree. Given a 2-spherical twin building of type M, one can find in almost all cases a twin building of type $\bar{M}$ on which the group G acts in such a way that Δ is precisely the fixed point building of this action.

This essentially reduces the existence problem to the problem of proving existence for twin buildings over 'tree-diagrams'. At the same time it explains the last '(FP)' in Case 2 of the theorem. Roughly speaking, the constructions (CS),(FP),(L),(FP) are used to get all 2-spherical twin buildings over tree diagrams; the remaining buildings are obtained from these by the process described above.

4. The classification of all possible local structures uses the classification of Moufang polygons. There is also a 'constructive version' of the latter classification in which one has the sequence (CS),(FP),(L),(FP). Here the last '(FP)' is due to the appearance of the generalized octagons which are obtained by Suzuki-Ree twists. That there is no such (FP) in the case of spherical buildings of rank at least 3 is due to the fact that these octagons cannot be part of the local structure of a spherical building. But of course this is possible in the twin case which explains the second last (FP) in Case 2 of the theorem above.

5. As already mentioned, the possible local structures of buildings can be abstractly axiomatized and are called foundations. Foundations are amalgams of generalized polygons glued along the panels of one chamber. From the local-global theorem one knows that the local structure of a 2-spherical building is a Moufang foundation. This means that all polygons involved are Moufang polygons and the glueings are compatible with the induced Moufang structures. As mentioned in the previous remark there is also a 'constructive version' of the classification of Moufang polygons. This version provides a kind of construction rule for each Moufang polygon which starts from a building associated with a Chevalley group over an algebraically closed field.

Let **F** be a Moufang foundation. As it is an amalgam of Moufang polygons the constructive version of the classification gives a construction rule for each of the polygons involved. This yields a construction rule for the desired twin building by amalgamating the constructions of the polygons in a way which is compatible with the stucture of the given foundation. In practice the following happens. In most cases the construction according to this rule is feasible. In the remaining cases it doesn't work; but by recognizing that the construction is not feasible one obtains valuable information for non-existence arguments.

10. Notes and Comments

10.1. References on (twin) buildings. References for the general theory of buildings are [Ti74],[Br89], [Ro89], [Sc95] and [Gr01] in this volume. The earliest references for the general theory of twin buildings are [Ti88/89], [Ti89/90] and [Ti92]; in these references several proofs for the basics are only sketched or even omitted. In the first chapter of [Ab96/1] one can find a more detailed treatment of the foundations of the theory of twin buildings and twin BN-pairs. The references which are most relevant for the classification of 2-spherical twin buildings are [Ti88/89], [Ti89/90], [Ti92], [MR95], [Ti97/98], [Mu99/1] and [Mu99/2]. Apart from those there is work in progress (partly joint work with H. Van Maldeghem and V. Vermeulen) aimed at making the complete classification available.

10.2. Diagram geometries. The concept of a diagram geometry is due to J. Tits; we emphasized that he introduced geometries over Coxeter diagrams because the notion of diagram geometry today is more general. The generalization is due to F. Buekenhout who used Tits' original ideas for his investigations of the sporadic simple groups [Bu79]. The theory of diagram geometries became a big research area for people interested in finite (simple) groups. We refer to [Ti78] and the introduction of [Ti81] for information about buildings and diagram geometries and to [BP95] for a survey on the subject.

10.3. Mixed groups. In a Chevalley group the commutation relations between certain subgroups (called the root groups) play a significant role. These commutation relations do not depend on the field over which the group is defined - they only depend on the underlying Dykin diagram. In the formulae describing them there occur integers which are smaller than or equal to 3. It is due to this fact that in small characteristics several root groups commute which are not meant to commute. This has the effect that in those characteristics one can define certain 'sub-Chevalley groups' of the original Chevalley group. It turns out that the subgroups in question can be interpreted as Chevalley groups which are defined over two different fields at the same time. These groups have been introduced by Tits in [Ti74] 10.3.2. as mixed groups of the corresponding type. The construction above can be generalized further in a certain special case. This yields the indifferent Moufang quadrangles introduced in [Ti76] and their corresponding groups (which do not have - to my knowledge - a special name yet).

We mentioned that the existence part of the classification of spherical buildings was not a problem when the classification was finished because the corresponding groups where then available. This is not completely true because the mixed groups had been introduced in [Ti74] 10.3.2. Obviously, this was done in order to have appropriate algebraic models for the corresponding buildings.

10.4. Forms of Kac-Moody groups. It is true that there is not yet a theory of forms of Kac-Moody groups which would be strong enough to settle the existence problem. Nevertheless there is a theory of Galois-descent in Kac-Moody theory. We refer to the work of G. Rousseau (see [BBBR95], [Ro90]), which was pushed further by B. Rémy in his thesis [Ré99/1] . In [Hée93] (see also [Hée90]) there are related results on Steinberg-Ree twists in Kac-Moody theory.

Nevertheless, as it was already divined in [Ti92], twin-buildings yield a wide class of groups which appear as some sort of 'non-Galois twists' of Kac-Moody groups. These groups are not covered by the existing algebraic approaches.

10.5. Further topics. The main purpose of this paper is to report on the classification of 2-spherical twin buildings. Therefore much of what is said about twin buildings restricts to what is relevant for the classification and to the 2-spherical case. Nevertheless there are several other things which I want to mention here.

1 There is not much known about the combinatorics of general twin buildings. For instance, it would be interesting to know how restrictive the twin building axioms are for arbitrary diagrams. Except for the 2-spherical case (where there is a classification) there is only one further case for which non-trivial results are available - namely the case of twin trees. These are precisely the twin buildings of rank two where the diagram consists of one edge which is labeled by infinity. Here one cannot hope for a classification and probably the same is true for arbitrary twin buildings whose Coxeter diagram involves an infinity.

For a survey on the theory of twin trees we refer to M. Ronan's paper [Ro01] in this volume.

2 In the previous item we mentioned that not much is known about general twin buildings as far as classification results are concerned. Nevertheless there are general results on geometric properties of twin buildings and character- izations of them. We refer to [AR98], [AVM99], [AVM00], [Mu96], [Mu97], and [Ro00].

3 Twin buildings have their origin in the appearance of Kac-Moody groups and twin BN-pairs. It is therefore not surprising that the theory of twin buildings can be applied to studying these groups and to proving that they have interesting properties. The earliest reference to applications of twin buildings in group theory is the work of P. Abramenko [Ab96/1] where they are used to prove finiteness properties of S-arithmetic groups. Another in- teresting application of twin buildings is that they can be used to produce lattices in Kac-Moody groups over finite field (see [Ré99/2]). Further refer- ences to group theoretical applications of twin buildings are [Ab96/2], [AM97], [Ré99/1], [Ré00] and [Ro90].

REFERENCES

[Ab96/1] P. Abramenko *Twin Buildings and Applications to S-Arithmetic Groups.* LNM 1641, Springer, Berlin, 1996.

[Ab96/2] P. Abramenko *Group actions on twin buildings.* Bull. Belg. Math. Soc. - Simon Stevin **4**,No. 4, (1996), 391–406.

[AM97] P. Abramenko, B. Mühlherr *Présentations de certaines BN-paires jumelées comme sommes amalgamées.* C. R. Acad. Sci. Paris, t. **325**, (1997), Série I, p. 701–706.

[AM01] P. Abramenko, B. Mühlherr *On the Levi-decomposition in twin-BN-pairs.* Preprint Dortmund/Bielefeld, 2001.

[AR98] P. Abramenko, M. Ronan *A characterization of twin buildings by twin apart- ments.* Geom. Dedicata **73**, No1, 1–9.

[AVM00] P. Abramenko, H. Van Maldeghem *On opposition in spherical buildings and twin buildings.* Ann. Comb. **4**, No.2, (2000),125–137.

[AVM99] P. Abramenko, H. Van Maldeghem *1-Twinnings of Buildings.* SFB-Preprint 99– 132, Bielefeld, 1999; to appear in Math. Z..

[BBBR95] V. Back-Valente, N. Bardy-Panse, H. Ben Messoud, G. Rousseau *Formes presque déployées d'algèbres de Kac-Moody, Classification et racines relatives.* J. Alg. **171**, (1995), 43–96.

[Bo68] N. Bourbaki *Groupes et Algèbres de Lie, Chapitres 4,5 et 6.* Hermann, Paris, 1968.

[Br89] K. Brown *Buildings*. Springer, Berlin, 1989.

[Bu79] F. Buekenhout *Diagrams for geometries and groups*. J. Comb. Theory (A) **27**, (1979), 121–151.

[BP95] F. Buekenhout, A. Pasini *Finite diagram geometries extending buildings*. in *Handbook of Incidence Geometry*. F. Buekenhout (ed.), Elsevier, Amsterdam, 1995, 1143–1254.

[Ch55] C. Chevalley *Sur certains groupes simples*. Tôhoku Math. J.(2) **7**, (1955), 14–66.

[Gr01] T. Grundhöfer *Basics on buildings*. this volume.

[Hée90] J.-Y. Hée *Construction de groupes tordus en théorie de Kac-Moody*. C. R. Acad. Sc. Paris **310** (1990), 77–80.

[Hée93] J.-Y. Hée *Sur la torsion de Steinberg-Ree des groupes de Chevalley et de Kac-Moody*. Thèse d'Etat, University Paris 11 (1999).

[Mu96] B. Mühlherr *A Rank 2 Characterization of Twinnings*. Europ. J. Combin. **19** (1998), 603–612.

[Mu97] B. Mühlherr *On the simple connectedness of a chamber system associated to a twin building*. Preprint, Dortmund, 1997.

[Mu99/1] B. Mühlherr *Locally split and locally finite twin buildings of 2-spherical type*. J. Reine Angew. Math. **511**, (1999), 119–143.

[Mu99/2] B. Mühlherr *On the Existence of 2-Spherical Twin Buildings*. Habilitationsschrift, Dortmund, 1999.

[MR95] B. Mühlherr, M. Ronan *Local to Global Structure in Twin Buildings*. Invent. Math. **122** (1995), 71–81.

[Ré99/1] B. Rémy *Formes presque déployées des groupes de Kac-Moody sur un corps quelconque*. PhD-thesis, Nancy, 1999.

[Ré99/2] B. Rémy *Construction de réseaux en théorie de Kac-Moody*. C. R. Acad. Sc. Paris **329**, 475–478 (1999).

[Ré00] B. Rémy *Immeubles de Kac-Moody hyperboliques groupes non isomorphes de même immeuble*. Preprint, Nancy, 2000.

[Ro89] M. Ronan *Lectures on Buildings*. Academic Press, San Diego, 1989.

[Ro00] M. Ronan *Local isometries of twin buildings*. Math. Z. **234**, (2000), 435–455.

[Ro01] M. Ronan *Twin trees and twin buildings*. this volume.

[RT87] M. Ronan, J. Tits *Building buildings*. Math. Annalen **278** (1987), 291–306.

[Ro90] G. Rousseau *L'immeuble jumelé d'une forme presque déployée d'une algèbre de Kac-Moody*. Bull. Soc. Math. Belg. **42**, (1990), 673–694.

[Sc95] R. Scharlau *Buildings*. in *Handbook of Incidence Geometry*, ed. F. Buekenhout, Elsevier, Amsterdam 1995.

[Ti68] J. Tits *Les problème de mots dans les groupes de Coxeter*. Symposia Mathematica (INDAM, Rome, 1967/68), Academic Press, London, 1969, vol. 1, 175–185.

[Ti74] J. Tits *Buildings of spherical type and finite BN-pairs*. LNM **386**, Springer, Berlin, 1974, (2nd edition 1986).

[Ti76] J. Tits *Quadrangles de Moufang*. Preprint, Paris, 1976 (unpublished).

[Ti78] J. Tits *Buildings and Buekenhout Geometries*. Proceedings, Durham, 1978.

[Ti81] J. Tits *A local approach to buildings*. In C. Davis et al.(eds.), *The Geometric Vein - the Coxeter Festschrift*. Springer, Berlin - Heidelberg - New York 1981.

[Ti87] J. Tits *Uniqueness and presentation of Kac-Moody groups over fields*. J. Alg. **105** (1987), 542–573.

[Ti88/89] J. Tits *Résumé des cours et travaux*. Annuaire du Collège de France, 89e année, 1988–1989, 81–95.

[Ti89/90] J. Tits *Résumé des cours et travaux*. Annuaire du Collège de France, 90e année, 1989–1990, 87–103.

[Ti92]   J. Tits *Twin Buildings and groups of Kac-Moody type*. In M.W.Liebeck and
          J. Saxl(eds.), *Groups, Combinatorics and Geometry* (Durham 1990), London
          Math. Soc. Lecture Note Ser. **165**, Cambridge University Press, 249–286.
[Ti97/98] J. Tits *Résumé des cours et travaux*. Annuaire du Collège de France, 98e année,
          1997–1998, 97–112.
[TW0*]   J. Tits, R. Weiss *Moufang Polygons*. Book in preperation, 2001.

Twin Trees and Twin Buildings

Mark Ronan

1 Introduction

A twin tree comprises two trees, both without end points, and an integer-valued "codistance" between vertices in one tree and vertices in the other. This codistance, defined in section 2, satisfies some straightforward properties reminiscent of the distance between vertices in a single tree, and it compels the two trees to be isomorphic to one another.

Single trees have been used extensively in mathematics, mainly in connection with groups that act on them. For example the group GL_n over a field having a discrete valuation (a field such as the p-adic numbers) acts very naturally on a Bruhat-Tits building, and when $n = 2$ this building is a tree — see [S] for details on the action of GL_2 on trees. The theory of buildings has been extended by J. Tits and the author (see [T6]) to include twin buildings, which arise from Kac-Moody groups. A twin building of rank two is a twin tree, so Kac-Moody groups of rank two act on twin trees.

The theory of twin trees has been developed jointly by Tits and the author (see [RT1] and [RT2]), and part of the purpose of this paper is to adumbrate the principal results published so far, and to indicate the direction of current research. I shall also explain the connection with twin buildings and describe how twin trees can be used to obtain interesting groups acting on certain hyperbolic buildings.

The present paper is laid out as follows. Section 2 presents the definition and basic, non-group theoretical results on twin trees from [RT1], and section 3 draws a comparison with the more familiar theory of generalised polygons. Groups appear in section 4 where it will be explained that although the two trees in a twin each admit the action of an enormous group, the automorphism group of a twin tree is very much more restricted.

Twin buildings, which are a higher dimensional analogue of twin trees, make their appearance in section 5, in connection with Kac-Moody groups. The rank 2 residues of a twin building are either generalised polygons or twin trees. In the former case there is a Local-to-Global theorem, stated in sect. 5. In the latter case there has been little research to date, but a recent paper by B. Rémy and the author creates hyperbolic twin buildings using twin tree residues. These buildings admit interesting lattices, as explained in section 6.

The one new part of this paper is section 7 where a notion of "defect" is introduced for a "double apartment" in a twin tree, those of defect 0 being the twin apartments that appear earlier in sect. 2. In order to define this defect we need the horoballs that arise in a natural way in any twin tree, and these are then used in section 8 which summarizes the construction of twin trees from [RT2].

2   The definition and elementary properties of a twin tree

This section gives a definition of twin trees, and summarizes most of the combinatorial results from [RT1]. The group theoretic results of that paper will appear in sect. 4. Before defining a twin tree, let us first define a single tree in the following way. Take a set T, whose elements are called vertices, and suppose that for any two vertices x and y in T we are given a non-negative integer $\text{dist}(x, y) = \text{dist}(y, x)$, called the *distance* between them. We suppose $\text{dist}(x, y) = 0$ if and only if $x = y$, and when $\text{dist}(x, y) = 1$ we say that x and y are *adjacent* or are *neighbours*. With this distance function, T is a tree precisely when the following condition is satisfied:

> *if* $\text{dist}(x, y) = n$ *and if* y' *is adjacent to* y, *then* $\text{dist}(x, y') = n\pm1$; *furthermore if* $n > 0$, *then* -1 *occurs for a unique such neighbour* y'.

A twin tree can now be defined as a pair of trees together with a non-negative integer $\text{codist}(x, y) = \text{codist}(y, x)$, called the *codistance*, between vertices of one and vertices of the other. This codistance satisfies the following condition:

> *if* $\text{codist}(x, y) = n$ *and if* y' *is adjacent to* y, *then* $\text{codist}(x, y') = n \pm 1$; *furthermore if* $n > 0$, *then* $+1$ *occurs for a unique such neighbour* y'.

The two trees in a twin will be called its *components*, and two vertices x and y in different components will be called *opposite* when $\text{codist}(x, y) = 0$. Opposite vertices necessarily have the same valency, and when the trees are thick (meaning that every vertex has valency at least three), then vertices in the same tree at even distance from one another, or vertices in opposite trees at even codistance from one another, have the same valency. This elementary result is Prop. 1 in [RT], and I repeat it here.

2.1 Proposition *Opposite vertices have the same valency. Furthermore if the trees are thick, then vertices at even distance, or even codistance, from one another have the same valency. In particular when the two trees are thick they are isomorphic to one another.*

Sketch of Proof: If x and y are opposite, then the relation of being at codistance 2 sets up a bijection between the neighbours of x and the neighbours of y, proving the first assertion. For the second assertion it suffices to prove that if x and w are two vertices in the same tree at distance 2 from one another, then there is a vertex y in the other tree that is opposite both. Let z be the vertex adjacent to x and w, and take any vertex v opposite z. Being at codistance 2 sets up a bijection between the neighbours of v and those of z. Therefore there are precisely two neighbours of v that are not opposite x and w. Since v has valency at least three, it has a neighbour opposite both x and w. *QED*

Trees in which vertices at even distance from one another have the same valency are called *semi-homogeneous*. In [RT1] we gave the example of a twin tree associated to the group GL_2 over a ring $k[t, t^{-1}]$ of Laurent polynomials, in which case the valency of each vertex is one greater than the cardinality of k. When k is a finite field these trees are homogeneous of valency $1 + q$, where q is a prime power. However, semi-homogeneous examples of any desired valencies exist, and this question will be more fully addressed in section 8.

Ends and Apartments

In a tree in which each vertex has valency at least two, an *apartment* is a path that is infinite in both directions (and therefore has no end points). A *half-apartment* is a path that is infinite in one direction and has one end point. Two half-apartments are said to have the same *end* if their intersection is a half-apartment. Having the same end is an equivalence relation on the set of half-apartments, and the equivalence classes are called the *ends* of the tree. Each apartment has two ends, and any two ends a and b span a unique apartment, denoted (ab). A single end e plus a vertex x span a half-apartment (xe).

Given a tree in which each vertex has finite valency the number of ends is uncountable, but for a *twin tree* of similar valency the twinning picks out a countable set of ends, in the following way.

Take two vertices x and y in different components of the twin, and let $(y = y_0, y_1, y_2, \dots)$ be a path along which the codistance from x increases monotonically. If $\mathrm{codist}(x, y) > 0$, then this path is unique because there is only one direction in which codistance can increase, but in any case it is unique after the first step. The path determines an end of one tree, namely the tree containing y. The ends obtained in this way are called the *ends of the twinning*. Let us state this as a lemma.

2.2 Lemma *Any two non-opposite vertices in different components of the twin determine a unique end of the twinning in each tree of the twin.*

If the valency of each vertex is finite, then the number of vertices is countable and hence there are only countably many ends of the twinning. In particular the twinning picks out a proper subset of all ends.

Now let T_+ and T_- denote the two trees of the twin, let x be a vertex in T_+ and y a vertex in T_- that is not opposite x. By (2.2) the pair (x, y) determines an end e_+ in T_+, and an end e_- in T_-. Starting with two other vertices x' and y' gives two ends e'_+ and e'_-. Now here is a striking result.

2.3 Lemma *([RT1];3.4) With the notation just given, if $e'_+ = e_+$, then $e'_- = e_-$.*

By definition every end of the twinning arises as in (2.2), so (2.3) gives a natural bijection between the ends of the twinning for T_+ and for T_-. More precisely:

2.4 Proposition *Let $E_\pm$ denote the set of ends of the twinning for $T_\pm$. Then the previous lemma gives a natural bijection between E_+ and E_-.*

We may therefore refer to an end e of the twin tree without specifying to which of the two trees it belongs.

Twin Apartments

A *twin apartment* of (T_+, T_-) is a pair of apartments A_+ in T_+ and A_- in T_- having the property that every vertex of one is opposite a unique vertex of the other.

Twin apartments play an important role for twin trees (and more generally for twin buildings), similar to the role apartments play for buildings of spherical type. In sect. 4 they will be used in discussing roots and root groups. Our first observation about twin apartments is the following.

2.5 Lemma *(see [RT1];3.5) For any twin apartment (A_+, A_-) the ends of A_+ correspond, under the natural bijection of (2.4), to the ends of A_-.*

This lemma implies that a twin apartment (A_+, A_-) has two ends a and b, in the sense that the apartment (ab) spanned by a and b is A_+ in one tree and A_- in the other. Given two ends of the twinning, one can ask whether they necessarily span a twin apartment, but the answer is no. Only some of them have this property, and knowing those that do is equivalent to knowing the codistance function and hence knowing the twinning.

2.6 Proposition *([RT1];3.6) In a thick twin tree the set of twin apartments uniquely determines the twinning.*

Sketch of proof: Given two vertices x and y in different components of a twin tree, it is straightforward to show that $\text{codist}(x, y)$ equals the distance from x to the set of vertices opposite y. It therefore suffices to prove that the set of twin apartments determines all pairs of opposite vertices. This is done as follows: three ends of a tree determine a vertex in that tree, so if a, b and c are ends of a twinning, they determine vertices x and y in each tree. When each pair of these ends spans a twin apartment, then it is simple to show that x and y are opposite. Conversely given two opposite vertices x and y, take three vertices adjacent to x; and hence at codistance 1 from y; these vertices, along with y, determine three ends of the twinning a, b, and c, as in (2.2), and one shows that any two of these ends span a twin apartment. In this way the set of twin apartments determines all opposite pairs of vertices and hence determines the codistance; see [RT1;3.6] for more details. □

Example: Although the set of twin apartments determines the twinning, the set of ends does not. For example, the group GL_2 over a ring $k[t, t^{-1}]$ of Laurent polynomials acts on a twin tree, as mentioned earlier. One component is the tree for GL_2 over the local field $k((t))$, and the other is for GL_2 over $k((t^{-1}))$ — see [RT1] for details. The ends of the twin tree are in bijective correspondence with the set of 1-spaces in a 2-dimensional vector space V over the field of rational functions $k(t)$. The group $G = \text{GL}_2(k[t, t^{-1}])$ is transitive, but not 2-transitive, on this set of 1-spaces; in other words it is transitive on the set of ends of the twinning, but not on the pairs of ends, only some of which span twin apartments. This subgroup G of $\text{GL}_2(k(t))$ arises by choosing a basis β for V and taking all 2-by-2 matrices, with basis β and with entries in $k[t, t^{-1}]$. Every basis gives a pair of 1-spaces in V, and hence a pair of ends. As g ranges over G, the pairs of 1-spaces of V generated by the bases $g(\beta)$ span the twin apartments. If β' is a basis not conjugate to β by any element of G, then it gives a different subgroup G of $\text{GL}_2(k(t))$. Each such $\text{GL}_2(k[t, t^{-1}])$ subgroup gives a different twinning of the same two trees. These twinnings all have the same set of ends, but different sets of twin apartments. For further details, see [RT1].

Remark 1. There is a very simple criterion to determine whether an end belongs to a given twinning. Take any half-apartment h leading to that end, and take any vertex x in the other component of the twin. Then the end belongs to the twinning precisely when h contains only finitely many vertices opposite x. In this case the codistance from x to the vertices of h increases monotonically after finitely many steps — see [RT1;p.472].

Remark 2. A tree has a natural topology on its set of ends and the ends of a twinning are dense in this topology. Conversely given a semi-homogeneous tree whose two valencies are at most countable, any dense subset of the ends must be the set of ends of a twinning. This result is due to Bennett [Be] who also deals with uncountable valencies by defining "separating well-orderable" subsets of ends.

3   Comparison with Generalised Polygons

Spherical buildings of rank 2 — often called generalised polygons — are bipartite graphs whose girth (length of a shortest circuit) is twice their diameter. They can also be defined in terms of distances between vertices. The distance between two vertices is the length of a shortest path joining them, and when they are at distance 1 they are said to be adjacent. Here is an alternative definition in the spirit of twin trees.

A generalised m-gon is a graph of diameter m satisfying the following property for any two vertices x and y:

> *if* dist$(x, y) = n$ *and if* y' *is adjacent to* y, *then* dist$(x, y') = n \pm 1$; *furthermore if* $m > n > 0$, *then* -1 *occurs for a unique such neighbour* y'.

When m is infinity this is the definition of a tree, as given above.

A generalised m-gon can be naturally twinned with itself (this is true of all spherical buildings), and the codistance between vertices is given by codist$(x, y) = m - $ dist(x, y). This codistance satisfies the following property which is immediate from the distance property above:

> *if* codist$(x, y) = n$ *and if* y' *is adjacent to* y, *then* codist$(x, y') = n \pm 1$; *furthermore if* $m > n > 0$, *then* $+1$ *occurs for a unique such neighbour* y'.

When codist$(x, y) = 0$ (which is the same as dist$(x, y) = m$) the two vertices x and y are said to be opposite. When m is infinity this is the definition of codistance in a twin tree, as given earlier.

Generalised polygons are more like twin trees, than single trees, because they contain circuits, called apartments, in which every vertex is opposite exactly one other vertex, and these play an analogous role to the twin apartments in a twin tree. This analogy will recur in the next section when we define roots and root groups. In some sense twin trees are generalised polygons of "infinite diameter".

Our first proposition (2.1), on vertex valencies in twin trees, is also true for generalised polygons. If two vertices are opposite they must have the same valency, and if the generalised polygon is thick then two vertices have the same valency when the distance between them is even. In this case (when the distance is even) the two vertices are said to have the same *type*.

Generalised polygons originally arose from groups of Lie type, in much the same way that twin trees arise from Kac-Moody groups (see the next section). For a thick generalised m-gon arising in this way $m = 2, 3, 4, 6,$ or 8. It is an obvious question to ask whether this restriction on the possible values of m can be relaxed if we do not require that it arises from a group of Lie type. The first results in this direction were a free construction by Tits (see [T4]) showing the existence of thick generalised m-gons for any m. This construction resulted in infinite valency. Restricting to finite valency, and hence to finite generalised m-gons, however, W. Feit and G. Higman used representation theory to show the following.

3.1 Theorem *(Feit-Higman [FH]) For a thick finite generalised m-gon, m equals $2, 3, 4, 6$ or 8.*

Their proof used an algebra that operates on the set of edges in a generalised m-gon. When the number of edges is finite one has a finite dimensional representation of this algebra. In addition to proving non-existence results, they obtained restrictions on the possible valencies. For example if $s + 1$ and $t + 1$ are the valencies of the two types of vertices then for $m = 4$, $s + t$ must divide $st(st + 1)$, and for $m = 6$ or 8 there are stronger restrictions. Later results [Hi] and [Ha] showed that for $m = 4$ or 8, $s \leq t^2$ and $t \leq s^2$, and for $m = 6$, $s \leq t^3$ and $t \leq s^3$. For twin trees there are no such restrictions, as I mentioned earlier (see sect. 8).

4 Moufang Twin Trees

We have already mentioned that the automorphism group of a twin tree is far more restricted than that of a single tree, and the following theorem demonstrates this by exhibiting a rigidity that will be used below when we define root groups. It is the same as the well-known theorem for spherical buildings that says an automorphism fixing two opposite chambers and all chambers adjacent to one of them must be the identity. The chambers of a tree, or a twin tree, are its edges. An edge c, having vertices x and y, is said to be *opposite* an edge d if x is opposite one vertex of d and y is opposite the other.

4.1 Theorem *([RT1];4.1) Let T be a thick twin tree and let c and d be opposite edges of T. The only automorphism fixing c and d, and fixing all edges that meet c in a vertex, is the identity.*

The proof, which is essentially the same as for spherical buildings, goes as follows: if c and d are opposite edges then there is a natural bijective correspondence between the set of edges meeting c and those meeting d. If an automorphism fixes the edges meeting c then it fixes those meeting d. One then works by induction outwards from c, and from d, using pairs of opposite edges.

Each pair of opposite edges lies in a twin apartment, by [RT1;3.5], so the theorem above has the following corollary.

4.2 Corollary *Given an edge c in a thick twin tree, the group of automorphisms fixing every edge meeting c acts freely on the set of twin apartments containing c.*

When the twin tree is Moufang (see the definition below) this action is transitive, in which case the group fixing all edges meeting a given edge is the full "unipotent subgroup" of an edge stabilizer. To define the Moufang property we need the concept of roots and root groups.

Roots and root groups

A *root* is half a twin apartment in the following sense. Let e and f be ends spanning a twin apartment A. Every pair of opposite vertices x and y in A determines two roots: $\alpha = (xe) \cup (ey)$ and $\beta = (xf) \cup (fy)$. We call α and β *opposite* roots of A, and write $\beta = -\alpha$. The *boundary* of α, denoted $\partial\alpha$, is $\{x, y\}$.

Given a root α, the group of all automorphisms fixing α and every edge containing a vertex of $\alpha - \partial\alpha$ is denoted U_α. Every apartment containing α has edges that are opposite edges in $\alpha - \partial\alpha$, so by (4.1) U_α acts freely on the set of apartments containing α. When this action is transitive we call U_α a *root group*. If U_α is a root group for all roots α we call the twin tree *Moufang*. The following theorem shows that it is only necessary to consider roots in a single twin apartment.

4.3 Theorem *([RT1];4.5) If U_α is a root group for all roots α in a given twin apartment A, then the group generated by these U_α is transitive on the set of twin apartments, and the twin tree is Moufang.*

The set of roots in a twin apartment splits naturally into two halves, one for each end. Roots in the same half contain the same end and can be naturally indexed by the integers: $\ldots, \alpha_{n-1}, \alpha_n, \alpha_{n+1}, \ldots$ where the boundary vertices of α_n are adjacent to those of α_{n+1} for each n. Given α_m and α_n, the roots α_i as i ranges between m and n are precisely those roots containing $\alpha_m \cap \alpha_n$. As in the case of spherical buildings, there are commutator relations between root groups. In the following theorem U_n denotes the root group for the root α_n.

4.4 Theorem *([RT1];4.7) Given $m < n$, the group $[U_m, U_n]$ generated by the commutators $ghg^{-1}h^{-1}$, for g in U_m and h in U_n, is contained in the product $U_{m+1} \cdots U_{n-1}$, this product being defined as $\{1\}$ if $n = m + 1$.*

Root groups are defined in the same way for generalized polygons, and if U_α is a root group for all roots α in a given apartment then, as in (4.3), the generalised polygon is Moufang. In this case an analogue of the Feit-Higman Theorem (3.1) was proved by Tits [T3] and Weiss [W].

4.5 Theorem *Thick Moufang polygons exist only for $m = 2, 3, 4, 6$ and 8.*

Action of root groups on vertex stars

Given a root α and a vertex x on its boundary, each edge containing x and not contained in α lies in a unique twin apartment containing α (see [RT1;4.4]). In a Moufang twin tree therefore the root group U_α acts freely transitively on this set of edges. To examine this action in more detail let $\text{St}(x)$ denote the set of edges containing x.

4.6 Theorem *([RT1];4.9) Fix a vertex x in a Moufang twin tree, and an edge c containing x. Then for roots α containing c and having x as a boundary vertex, all root groups U_α induce the same group of permutations on $\text{St}(x)$.*

If $U(x, c)$ denotes this permutation group, then as c ranges over $\text{St}(x)$, the groups $U(x, c)$ form a Moufang set in the sense of Tits [T6;4.4]. The same result holds for generalized polygons, in which case $\text{St}(x)$ typically has the structure of a projective line over a field k and the groups above are root groups of $\text{SL}_2(k)$. In any case they are root groups of a rank 1 group of Lie type; this follows from the classification of Moufang polygons achieved by Tits and Weiss [TW].

The study of Moufang twin trees is not nearly as advanced as that of Moufang polygons, but it is already clear that a classification along the same lines is not possible. For example, Tits [T7] has shown that in the trivalent case (three edges per vertex) there are uncountably many isomorphism classes of Moufang twin trees, because the root groups all have order two and one can create uncountably many different commutator relations.

5 Kac-Moody Groups and Twin Buildings

A Kac-Moody group G over a field k is an infinite dimensional analogue of a Chevalley group over k. It contains a set of root groups U_α parametrized by the roots α of a Coxeter complex W, and these root groups generate G. Each root group is isomorphic to the additive group of k, and the set of roots can be split (in many different ways) into two parts: a positive part and a

negative part. The positive root groups generate a group U_+ and the negative
root groups generate U_-. Let N denote the subgroup of G normalizing the
set of root groups, and let T be the subgroup normalizing each individual
root group. As in the case of Chevalley groups, N/T is isomorphic to W;
T is called a torus, and is isomorphic to a direct product of copies of the
multiplicative group of k. Let B_+ and B_- denote the groups generated by T
along with U_+ and U_- respectively. Then the pairs (B_+, N) and (B_-, N) are
BN-pairs for G — see for example [R1] for basic information on root groups
and BN-pairs.

These BN-pairs generate two buildings Δ_+ and Δ_-. The chambers of
Δ_+ and Δ_- can be regarded as the cosets in G of B_+ and B_- respectively.
The usual Bruhat decompositions of G as B_+WB_+ and B_-WB_- give the
W-valued distance between chambers in the same building. A further decomposition B_+WB_- gives a W-valued "codistance" between chambers of
Δ_+ and Δ_-. This provides (Δ_+, Δ_-) with the structure of a twin building
— see [T6].

When W is finite (this is the spherical case) it has a unique longest word,
and this element of W conjugates B_+ to B_-. When W is infinite there is no
longest word and the subgroups B_+ and B_- are not conjugate in G. Twin
buildings are, however, similar to spherical buildings in the sense that two
chambers can be opposite one another. A chamber in one component of
the twin is *opposite* a chamber in the other component when the codistance
between them is the identity element of W. A spherical building is canonically
twinned with itself, so spherical buildings are a special case of twin buildings.

The rank 2 residues of a building are either generalised polygons or trees.
When they are generalised polygons the building is called 2-*spherical*. When
a twin building is 2-spherical, then Tits' methods [T2] of using opposite rank
2 residues can be adapted to study its local structure. In particular one can
prove that the rank 2 residues must be Moufang (see [R2] Theorem 4 for
a complete proof). Moreover one can show that the local structure usually
determines the global structure, as the following theorem makes precise. The
notation $E_2(c)$ means the union of the rank 2 residues containing the chamber
c, and is referred to as the *local structure*. An *isometry* is a map preserving
distances and codistances.

5.1 Theorem *([MR]) Let (c, d) and (c', d') be pairs of opposite chambers in
twin buildings Δ and Δ' respectively, and assume Δ has no residues of type
$Sp_4(2)$, $G_2(2)$, $G_2(3)$ or $^2F_4(2)$. Then any isometry from $E_2(c) \cup \{d\}$ onto
$E_2(c') \cup \{d'\}$ extends uniquely to an isometry from Δ to Δ'. In particular the
local structure of Δ determines its global structure.*

The reason for excluding certain residues over the fields of 2 or 3 elements
is that in these cases the set of chambers opposite a given chamber is not

connected — see [MR] for more details.

Using this Local-to-Global theorem the classification of 2-spherical twin buildings can be achieved by classifying the local data. This difficult problem has been dealt with by Mühlherr — see [M1] and [M2].

Kac-Moody data

The data that determines a Kac-Moody group is a generalised Cartan matrix, along with a ground field k. In the rank 2 case a generalised Cartan matrix has the form $\begin{pmatrix} 2 & a \\ b & 2 \end{pmatrix}$, where a and b are negative integers. If $ab = 1, 2$ or 3, then the group is $\mathrm{SL}_3(k)$, $\mathrm{Sp}_4(k)$ or $\mathrm{G}_2(k)$, and the building is a generalised m-gon for $m = 3, 4$ or 6 respectively. If $ab \geq 4$ the building is a twin tree.

When the field k is algebraically closed the only simple rank 2 algebraic groups are the three listed in the previous paragraph. These are the "split" cases, but when k is not algebraically closed there are non-split cases. For example when k is finite the non-split cases are unitary groups, whose buildings are generalised 4-gons, and $^3D_4(k)$ whose building is a generalised 6-gon; there is also the Ree group $^2F_4(k)$ (which is not an algebraic group) whose building is a generalised 8-gon. A classification of non-split algebraic groups is given by Tits [T1], but as yet there is no classification of non-split Kac-Moody groups having rank 2, still less of Moufang twin trees.

Lifting the 2-spherical restriction in (5.1) allows some rank 2 residues to be trees. Two opposite residues of this kind form a twin tree. Here now are some examples of twin buildings involving twin trees at the rank 2 level.

6   Some twin buildings involving twin trees

In this section I shall explain some recent work, by B. Rémy and the author, using Moufang twin trees to construct Moufang hyperbolic buildings. The starting point is a tiling of the hyperbolic plane $\mathbf{H}^2$ by equilateral right-angled m-gons. Pick one such m-gon c and label its edges π_i in a natural cyclic order as i ranges over the integers mod m. Each π_i lies in a unique geodesic, and we let s_i denote the reflection of $\mathbf{H}^2$ across this geodesic. The reflections s_i generate a Coxeter group W, and the tiling of $\mathbf{H}^2$ is a metric realization of the associated Coxeter complex $(W, \{s_i\}_{i \in \mathbf{Z}/m})$.

In W the reflections s_i and s_{i+1} commute, and the corresponding geodesics of $\mathbf{H}^2$ intersect at right angles. On the other hand, if $j \neq i, i \pm 1$, then s_i and s_j generate an infinite dihedral group, and the corresponding geodesics of $\mathbf{H}^2$ do not intersect. The panels $w\pi_i$ are said to have *type* i, and in any geodesic containing such a panel, all panels have type i. These geodesics are the *walls* of *type i*. Each wall of type i bounds two half-spaces of $\mathbf{H}^2$, called *roots*, again of *type i*. The two roots for a given wall are said to be *opposite* one another. The roots containing c are called the *positive roots*; those not containing c are the *negative roots*.

For each i in $\mathbf{Z}/m$ choose a field k_i, and to each root α of type i attach a group U_α isomorphic to the additive group of k_i. For each w in W, let Φ_w denote the set of roots containing c but not wc, and define U_w as the direct product of the groups U_α for α in Φ_w. Treating W as a partially ordered set, via the Bruhat ordering, U_w is a subgroup of U_v whenever $w \le v$. We then define U_+ as the inductive limit of the U_w for W in W.

For each i in $\mathbf{Z}/m$ let H_i denote the multiplicative group of the field k_i, and define the *torus H* to be the direct product of the H_i. The group H_i acts in a natural way on each root group of type i, and centralizes those of type $\ne i$. Thus the torus H acts on U_+, and we define the (positive) Borel subgroup B_+ as a semi-direct product of U_+ and H.

Let $\alpha = \alpha_i$ be the root containing c and having π_i on its boundary, and let $-\alpha$ be the opposite of α. The root groups U_α and $U_{-\alpha}$ can be identified with root groups generating $\mathrm{SL}_2(k_i)$, and for each i in $\mathbf{Z}/m$ the *Levi factor* L_i of type i is the direct product of $\mathrm{SL}_2(k_i)$ and the groups H_j for $j \ne i$. One can then define a minimal parabolic subgroup P_i of type i using L_i and U_+. These minimal parabolics contain B_+ and can be amalgamated to form a group Λ — see [RR] for details. This group Λ acts on a twin building of type W, but first let us quote an existence and uniqueness theorem about buildings of type W.

6.1 Theorem *([GP] and [Bo]) Given a set of cardinals q_i for each $i \in \mathbf{Z}/m$, there exists a unique hyperbolic building whose apartments are tilings of $\mathbf{H}^2$, as above, such that the cardinality of the set of chambers on a panel of type i is $q_i + 1$. This building carries a natural $CAT(-1)$ metric.*

Remark. When the q_i are all equal to a given prime power q, the building in (6.1) comes from a (non-unique) Kac-Moody group over the field of order q.

Definition. Let $\Delta(m, \{q_i + 1\})$ denote the building of (6.1).

6.2 Theorem *([RR]) When each q_i is the cardinality of a field k_i, then the building $\Delta(m, \{q_i + 1\})$ above belongs to a Moufang twinning having the group Λ, defined above, as an automorphism group.*

Remark 1. The rank 2 residues of $\Delta(m, \{q_i + 1\})$ are of two kinds. Those of type $\{i, i+1\}$ are complete bipartite graphs, but those of type $\{i, j\}$, where $j \ne i \pm 1$, are trees. These trees are semi-homogeneous with valencies $q_i + 1$ and $q_j + 1$. In a twinning, as in (6.2), two such residues that are opposite form a twin tree whose automorphism group induces $SL_2(k_i)$ on the star of vertices of one type, and $SL_2(k_j)$ on the star of vertices of the other type.

Remark 2. A semi-homogeneous tree has many different twinnings (see the last section), and the same is presumably true for the building $\Delta(m, \{q_i + 1\})$. Some of these will be Moufang, but I expect that "most" will not.

Finally, I want to end this section with a further result from [RR] showing that the use of fields of different characteristics leads to non-linearity of the group Λ.

6.3 Theorem *([RR])* *When the q_i are not all powers of the same prime number, then the group Λ is not linear over any field.*

7 Horoballs and the defect of an apartment in a twin tree

This is the only section of the paper that contains a result not given elsewhere in the literature. This involves an integer, associated to each pair of twin ends, that is zero precisely when the pair spans a twin apartment. It will be defined in terms of horoballs and horospheres, which are used extensively in the construction of twin trees given in [RT2], and will be needed in the next section.

Horoballs and horospheres

Recall that an end e and a vertex x in a tree T span a half-apartment (xe) having vertex x and end e. Two vertices x and y will be said to be *equidistant* from e if they are equidistant from some (hence any) vertex in the half-apartment $(xe) \cap (ye)$. Being equidistant from e is an equivalence relation on the set of vertices, and the equivalence classes are called *horospheres of T centred at e*. If S is such a horosphere, the union H of the (xe) for x in S is called a *horoball centred at e*, and S is called its *boundary*, denoted ∂H. The *interior* of H is $H - \partial H$. Given two vertices x and y, one says that x is *closer to e than y* if x is closer than y to some (hence any) vertex of $(xe) \cap (ye)$. In other words x lies in the interior of the horoball centred at e and having y as a boundary vertex.

Given a vertex v in one tree, and an end e of the twinning, let $H_0(v, e)$ denote the set of vertices x in the other tree for which the codistance from v increases monotonically along (xe). It is a horoball with centre e — see [RT2;2.1].

Now let a and b be any two ends of the twinning. They belong to both trees and therefore span an apartment (ab) in each tree of the twin. We

call (ab) a *double apartment*, and we are going to define its *defect*. Choose
a vertex v in the double apartment (ab), and define the defect of (ab) to be
the distance between the horoballs $H_0(v, a)$ and $H_0(v, b)$. We show this is
independent of v.

7.1 Lemma *With the notation above, the defect of (ab) is well-defined.*

Proof: Choose a vertex v in (ab) and let v' be a vertex of (ab) adjacent to
v. We will show that the distance between $H_0(v, a)$ and $H_0(v, b)$ is the same
as the distance between $H_0(v', a)$ and $H_0(v', b)$. For simplicity of notation
regard the end a as being on the left, and b as being on the right. Given two
vertices of (ab) in the same tree, the one closer to a will be said to be *on the
left* and the one closer to b will be *on the right*. Without loss of generality
suppose v' is on the right of v.

Let y and z be the vertices of (ab) that are boundary vertices of $H_0(v, a)$
and $H_0(v, b)$ respectively. Both y and z are opposite v and the distance
between them is the distance between $H_0(v, a)$ and $H_0(v, b)$. Let y' and
z' be the vertices of (ab) adjacent on the left to y and z respectively. The
codistance from v to y' is 1 and increases monotonically along $(y'a)$. Therefore
the codistance from v' to y' is 0 and increases monotonically along $(y'a)$; this
shows that y' is the boundary vertex of $H_0(v', a)$ in (ab). A similar argument,
reversing the roles of v and v', shows that z' is the boundary vertex of $H_0(v', b)$
in (ab). Since $\operatorname{dist}(y', z') = \operatorname{dist}(y, z)$, this shows that using v' in place of v
gives the same defect.

By an obvious induction, any vertex x of (ab) in the same component as
v gives the same defect. To complete the proof it only remains to replace v
by a vertex of (ab) in the other component. We shall replace it by z. First
notice that v is the boundary vertex of $H_0(z, b)$ in (ab). We must therefore
show that the boundary vertex x of $H_0(z, a)$ in (ab) is at distance $\operatorname{dist}(y, z)$
from v. To do this, notice that v is the boundary vertex of $H_0(y, a)$ in (ab),
and as one moves from y to the vertex of (ab) on its right, this boundary
vertex moves one step to the left. Since z is at distance $\operatorname{dist}(y, z)$ to the right
of y, an obvious induction shows that the boundary vertex x of $H_0(z, a)$ is at
distance $\operatorname{dist}(y, z)$ to the left of v. Thus $\operatorname{dist}(v, x) = \operatorname{dist}(y, z)$ proving that v
and z define the same defect of (ab), as required. $\square$

7.2 Proposition *Given ends a and b of the twinning, the double apartment
(ab) has defect 0 if and only if it is a twin apartment.*

Proof: By (7.1) the defect of (ab) is zero if and only if for any vertex v
of (ab) the horoballs $H_0(v, a)$ and $H_0(v, b)$ have a vertex of (ab) in common.
This is equivalent to saying that each vertex v of (ab) is opposite a unique
vertex of (ab) in the other component.

8 Construction of twin trees

The main purpose of this section is to explain the construction of twin trees given in [RT2]. As usual T will denote a semi-homogeneous tree. A vertex v in a tree twinned with T determines a function on the set of vertices in T by assigning codist(v, x) to each vertex x of T. Such a function, taking non-negative integer values and satisfying the codistance axiom given in the second section, can be created by starting at one vertex and working outwards. As in [RT2] let Σ denote the set of such functions, and if $\sigma \in \Sigma$, then let codist(σ, x) denote the integer σ assigns to a vertex x of T. The elements of Σ are candidates for vertices in a tree twinned with T.

Define a graph T^*, called the *universal twin* of T, as follows. Its vertices are the elements of Σ, and two vertices σ and τ are joined by an edge in T^* if and only if codist$(\sigma, x) = $ codist$(\tau, x) \pm 1$ for all vertices x in T.

8.1 Proposition *([RT2];3.2) Any tree twinned with T embeds as a full subgraph of T^*, preserving its codistance with T.*

The problem then is to extract subgraphs of T^* that are trees twinned with T. This is done in [RT2] by starting with a single vertex and working outwards. The neighbourhood that one creates for each vertex must be "uniform" in the following sense. Given a vertex σ of T^* a set U of neighbours of σ in T^* is called *uniform* if the following condition is satisfied. For each vertex v in T such that codist$(\sigma, v) = 0$, the property of being at codistance 2 sets up a bijection between U and the neighbours of v.

8.2 Proposition *([RT2];3.3) A connected subgraph S of T^* is a tree twinned with T if and only if the neighbourhood of each vertex in S is uniform.*

Uniform neighbourhoods are then constructed using ends. Each vertex σ of T^* determines a set of ends of T, namely those ends e such that as x approaches e, codist(σ, x) increases monotonically after at most a finite number of steps. Let $E(\sigma)$ denote this set of ends. If σ is adjacent to τ in T^* then $E(\sigma) = E(\tau)$, so in each connected component of T^* all vertices determine the same set of ends.

Constructing subgraphs, as in (8.2), we work within a connected component T° of T^*. The choice of component makes no difference because every automorphism of T induces an automorphism of T^*, and the automorphism group of T is transitive on the set of vertices of T^* having the same type (meaning they are opposite vertices of the same type in T) — see [RT2;3.1]. In particular Aut T is transitive on the set of connected components of T^*.

Having chosen a connected component T° of T^*, let E denote the set of ends $E = E(\sigma)$ for any vertex σ of T°.

Each vertex σ assigns to each end e in E a horoball of T, namely $H_0(\sigma, e)$ in the notation of the last section. This is the set of vertices x in T for which the codistance from σ increases monotonically along (xe). Let $H_1(\sigma, e)$ denote the subset of $H_0(\sigma, e)$ comprising vertices at codistance ≥ 1 from σ. If τ is a vertex of T° adjacent to σ then either $H_0(\tau, e) = H_1(\sigma, e)$ or $H_1(\tau, e) = H_0(\sigma, e)$. In the latter case $H_0(\sigma, e)$ is a subset of $H_0(\tau, e)$, and the set of ends e for which this occurs forms a subset of E that we denote $E_{\sigma\tau}$.

8.3 Lemma *([RT2];sect.4]) If U is a uniform set of neighbours of σ then the subsets $E_{\sigma\tau}$, as τ ranges over U, partition E.*

This partition satisfies a condition best explained by defining a new tree, denoted T/σ, which has two types of vertices: the set E of ends, and the set V_σ of vertices v in T for which $\mathrm{codist}(\sigma, v) = 0$. The first type are called *end-vertices*, and the second type are called *0-vertices*. An end-vertex e is joined in T/σ to a 0-vertex v precisely when v is a boundary vertex of $H_0(\sigma, e)$. Thus the neighbourhood of e in T/σ is the horosphere $\partial H_0(\sigma, e)$. On the other hand the neighbourhood of a 0-vertex v in T/σ is in a natural bijective correspondence with its neighbourhood in T. This allows the following characterization of uniform sets of neighbours.

8.4 Lemma *([RT2];4.2) A partition of E is determined by a uniform subset of neighbours of σ, as in (8.3), if and only if every 0-vertex is adjacent in T/σ to a unique end-vertex in each part of the partition.*

Since T/σ is a tree it is straightforward to construct suitable partitions of E. In [RT2;sect.4] these partitions are called σ-colourings. The set of "colours" has cardinality equal to the valency in T of a vertex of V_σ. Every neighbour of a 0-vertex in T/σ is assigned one of these colours, no two neighbours receiving the same colour. To produce such a colouring, one simply starts at one vertex of V_σ and works outwards. This construction works for any semi-homogeneous tree T and leads to the following result.

8.5 Theorem *([RT2];5.2) Every semi-homogeneous tree admits a twinning.*

This summarizes the first half of [RT2]; the second half is devoted to a careful examination of the local structure of a tree twinned with T.

A vertex of T° along with a uniform set of neighbours is called a *uniform 1-ball*. I have already mentioned that the automorphism group of T is transitive on the set of vertices of a given type in T^*. The following proposition strengthens this result.

8.6 Proposition *([RT2];6.2) Given any vertex σ in T°, the automorphism group of T is transitive on the set of uniform 1-balls centred at σ. Moreover* *the stabilizer of any uniform 1-ball induces the full symmetric group on its set of boundary vertices.*

This makes the situation seem rather loose, but things change dramatically when one moves to vertices at distance 2 from σ in a tree twinned with T. Define a *uniform n-ball* with *centre* σ to be a subgraph of T° having the following properties:

(i) all its vertices are at distance $\leq n$ from σ;
(ii) all vertices on the boundary are at distance n from σ;
(iii) the neighbourhood of each vertex not on the boundary is uniform.

Theorem (8.5) appears in [RT2] as a particular case of the following result.

8.7 Theorem *([RT2];5.2) Every uniform n-ball in T° lies in a tree twinned with T, and the vertices at distance n from a vertex in any tree twinned with T form a uniform n-ball.*

A careful, and rather technical, analysis of uniform 2-balls in [RT2] leads to a very different result from (8.6). Let us call two subsets of T° *conjugate* if there is an automorphism of T carrying one to the other.

8.8 Theorem *([RT2];8.1) Let T be a thick semi-homogeneous tree whose set of vertices has cardinality α. Then there are 2^α conjugacy classes of uniform 2-balls of either type in T°.*

8.9 Corollary *([RT2];8.2) A thick semi-homogeneous tree whose set of vertices has cardinality α admits 2^α isomorphism classes of twinnings, and among these 2^α have trivial automorphism group.*

The methods of proof show that the uniform 2-balls at different vertices of a tree twinned with T can be independent of one another, so twin trees are rather wild. On the other hand although a uniform 2-ball can have a wild structure, the Moufang case imposes strong restrictions, and it would be interesting to analyse the possible structures in that case.

Before leaving the construction of twin trees, I want to mention two related papers. One is by Fon-der-Flaass [F] who gives an alternative construction of twin trees. His methods amount to constructing twin trees along with an isomorphism between the two trees, sending each vertex to one opposite (i.e. one at codistance 0). The other is by Abramson and Bennett [AB] who extend a "partial twinning" of two trees, showing for example how one twin tree can be embedded in another.

9   References

[AB] M. Abramson and C.D. Bennett, Embeddings of Twin Trees, Geom. Ded. 75 (1999) 209-215.

[Be] C.D. Bennett, A Topological Characterization of End Sets of a Twinning of a Tree, Europ. J. Combin. 22 (2001) 27-35.

[Bo] M. Bourdon, Sur les immeubles fuchsiens et leur type de quasi- isometrie, Erg. Thy. and Dyn. Sys. (to appear).

[FW] W. Feit and G. Higman, The Nonexistence of Certain Generalized Polygons, J. Algebra 1 (1964) 114-131.

[F] D G. Fon-der-Flaass, A combinatorial Construction for Twin Trees, Europ. J. Combin. 117 (1996) 177-189.

[GP] D. Gaboriau and F. Paulin, Sur les immeubles hyperboliques, preprint, Univ. Paris 11, 1998.

[Ha] D. Haemers, Eigenvalue Techniques in Design and Graph Theory, Proefschrift, Mathematisch Centrum, Amsterdam 1979.

[Hi] D. Higman, Invariant Relations, Coherent Configurations and Generalized Polygons, pp. 247-263 in Combinatorics part 3 (ed. M. Hall and J. Van Lint), Reidel, Dordrecht 1975.

[M1] B. Mühlherr, Locally split and locally finite twin buildings of 2- spherical type, J. reine angew. Math. 511 (1999) 119-143

[M2] B. Mühlherr, Twin Buildings, these proceedings.

[MR] B. Mühlherr and M. Ronan, Local to global structure in twin buildings, Inventiones math. 122 (1995) 71-81.

[RR] B. Rémy and M. Ronan, Topological groups of Kac-Moody type, Fuchsian twinnings and their lattices, preprint.

[R1] M. Ronan, Lectures on Buildings, Academic Press, 1989

[R2] M. Ronan, Local isometries of twin buildings, Math. Z. 234 (2000) 435-455.

[S] J.-P. Serre, Arbres, Amalgames, SL_2; Astérisque 46 (1977). English version: Trees, Springer-Verlag 1980.

[T1] J. Tits, Classification of Algebraic Semisimple Groups, in Proc. Symp. Pure Math vol.9 (Algebraic Groups and Discontinuous Subgroups, Boulder 1965), 33-62, Am. Math. Soc. 1966.

[T2] J. Tits, Buildings of spherical type and finite BN-pairs, Lecture Notes in Mathematics 386, Springer Verlag 1974.

[T3] J. Tits, Non-existence de certains polygones généralisés, I, II, Inventiones math. 36 (1976) 275-284; 51 (1979) 267-269.

[T4] J. Tits, Endliche Spiegelungsgruppen, die als Weylgruppen auftreten, Inventiones math. 45 (1977) 283-295.

[T5] J. Tits, Uniqueness and Presentation of Kac-Moody Groups over Fields, J. Algebra 105 (1987) 542-573.

[T6] J. Tits, Twin Buildings and Groups of Kac-Moody Type, in Groups, Combinatorics and Geometry. Durham 1990 (ed. M.Liebeck and J.Saxl), London Maths. Soc. Lect. Notes 165, 249-286, Cambridge U. Press 1992.

[T7] J. Tits, private communication

[TW] J. Tits and R. Weiss, Moufang Polygons, Springer Monographs in Mathematics (to appear).

[W] R. Weiss, The Nonexistence of Certain Moufang Polygons, Inventiones math. 51 (1979) 261-266.

Simple groups of finite Morley rank of even type

Tuna Altınel

Institut Girard Desargues

UPRES-A 5028 Mathématiques

Batiment 101 (mathématiques)

Université Claude Bernard Lyon-1

43 blvd du 11 novembre 1918

69622 Villeurbanne CEDEX, France

e-mail: altinel@desargues.univ-lyon1.fr

1 Introduction

A fundamental problem in the study of groups of finite Morley rank is the classification of the infinite simple ones. It was conjectured independently by Gregory Cherlin and Boris Zil'ber that they are simple algebraic groups over algebraically closed fields. This conjecture, which is not an ordinary conjecture in the sense that the classification of the finite simple groups is not an ordinary theorem, remains open. Nevertheless, in recent years there has been considerable progress in the study of some subclasses of the infinite simple groups of finite Morley rank. This progress, which uses a large number of ideas from the classification of the finite simple groups, has culminated in the following theorem:

Theorem 1.1 *A simple K^*-group of finite Morley rank of even type is an algebraic group over an algebraically closed field of characteristic* 2.

In this survey an outline of the arguments used in the proof of this result will be given.

2 Background

In this survey general definitions and results about groups of finite Morley rank will be mentioned only if they are needed. The reader is referred to [22] for a good introduction to the algebraic theory, to [49] and [57] for discussions with emphasis on model theoretic aspects.

Nevertheless it seems useful for the reader's convenience to recall the definition of a *connected component of a subgroup* of a group of finite Morley rank. Definable subgroups of groups of finite Morley rank satisfy the *descending chain condition*. As a result the connected component of a definable subgroup H of a group G of finite Morley rank, which is defined as the intersection of the definable subgroups of finite index in H, exists and is a definable subgroup. It is denoted $H°$. A group of finite Morley rank is said to be connected if it is equal to its connected component, or equivalently if it has no proper definable subgroup of finite index.

In a somewhat dual way, if X is an arbitrary subset of a group G of finite Morley rank, it is possible to consider the intersection of the definable subgroups of G which contain X. This subgroup, the *definable closure* of X, denoted by $d(X)$, is a definable subgroup, again by the descending chain condition on definable subgroups. If X is a subgroup of G then its connected component is defined as $X \cap d(X)°$. In general, the connected component of an arbitrary subgroup X of G is denoted $X°$.

2.1 Types of groups

The following theorem of Alexandre Borovik and Bruno Poizat is the most general known result about Sylow theory in groups of finite Morley rank:

Theorem 2.1 ([23]) *Let G be a group of finite Morley rank.*

1. *The Sylow 2-subgroups of G are conjugate.*

2. *If S is a Sylow 2-subgroup of G, then $S°$ is the central product of a connected, definable (in G), nilpotent subgroup B of bounded exponent, and of a divisible abelian subgroup T. In general T is not definable in G.*

The following definitions will be handy in the sequel:

Definition 2.2 *1. A definable connected subgroup of bounded exponent of a group of finite Morley rank is called a* unipotent *subgroup.*

2. *A divisible abelian subgroup of a group of finite Morley rank is called a* torus.

It is worth noting that it is not known whether a unipotent subgroup of a group of finite Morley rank is nilpotent. On the other hand by Theorem 2.1, unipotent 2-subgroups of groups of finite Morley rank are nilpotent.

Theorem 2.1 can be seen as a rough approximation to the Sylow 2-structure in simple algebraic groups over algebraically closed fields whose Sylow 2-subgroups are infinite and either of bounded exponent or finite extensions of 2-tori according to the characteristic of the field over which they are defined (respectively 2 or different from 2). In view of this the following definition is made:

Definition 2.3 *A group of finite Morley rank is said to be*

- *of* even type *if its Sylow 2-subgroups are infinite of bounded exponent;*

- *of* odd type *if its Sylow 2-subgroups are finite extensions of nontrivial 2-tori;*

- *of* mixed type *if the connected components of its Sylow 2-subgroups are a central product of a nontrivial unipotent 2-subgroup and a nontrivial 2-torus;*

- *of* degenerate type *it the connected components of its Sylow 2-subgroups are trivial, i.e. if its Sylow 2-subgroups are finite.*

These four types provide an indispensable case division in the analysis of simple groups of finite Morley rank. Results of varying levels of generality some of which will be mentioned in the sequel exist for each type. For the time being it is worth mentioning that there should not exist simple groups of finite Morley rank of the last two types if simple groups of finite Morley rank are all algebraic, and that the least understood and by far the most difficult category of simple groups of finite Morley rank are those of degenerate type. The elimination of this type would correspond to a strong version of the Feit-Thompson theorem.

2.2   An inductive setting

As mentioned in the introduction, ideas and methods from the classification of the finite simple groups have been crucial for recent progress in the classification of simple groups finite Morley rank. The context in which these have been used also corresponds to an approach developed by finite group theorists in the last twenty years: *the revisionist approach.* The revisionist approach is the study of a minimal counterexample to a classification result, minimality being measured according to the context. In the context of groups of finite Morley rank this measure is the Morley rank. Since simple groups of finite Morley rank have not been classified the following notions have been introduced to provide firm grounds for a classification project of an inductive nature:

Definition 2.4   *1. A group of finite Morley rank all of whose infinite definable simple sections are algebraic groups over algebraically closed fields is said to be a K-group.*

   2. A group of finite Morley rank all of whose proper *infinite definable simple sections are algebraic groups over algebraically closed fields is said to be a K^*-group.*

In these definitions, the word *section* is used in the ordinary sense of group theory: if $K \triangleleft H \leq G$ where G is any group then H/K is a section of G. A section is definable if H and K are definable subgroups of G.

   The notions of K- and K^*-group provide a good inductive setting in the light of the Cherlin-Zil'ber conjecture since a minimal counterexample to this conjecture is a K^*-group. They turn out to be especially useful when "large" simple groups are studied, large meaning that a group has proper infinite definable simple sections. On the other hand the additional information obtained by assuming that a group of finite Morley rank is a K^*-group decreases when sections (at least those that govern the structure of a group) become solvable, which can be seen as a property of "smallness". Indeed the K^*-assumption is vacuously true for a simple group of finite Morley rank whose proper definable connected subgroups are solvable. It should be noted that these "small configurations", as in finite group theory, are the most difficult ones. Strikingly enough they are the configurations whose analyses deviate the most from finite group theory. Moreover, they are the only simple K^*-groups which can be of degenerate type because of the following easy K-group fact:

Fact 2.5 ([1]) *Let G be a connected nonsolvable K-group of finite Morley rank. Then $G/\sigma(G)$ is isomorphic to a direct sum of simple algebraic groups over algebraically closed fields. In particular, the definable connected $2^\perp$-sections are solvable.*

Here $\sigma(G)$ denotes the *solvable radical*, the largest normal solvable subgroup of G. By results of Nesin ([47]), in a group of finite Morley rank $\sigma(G)$ exists and is definable.

   A consequence of this discussion is that the K^* assumption not only provides a good inductive setting but also to some extent measures what can be done using known tools. It also underlines that the heart of the classification problem is the analysis of groups of degenerate type. In fact, the following statement is equivalent to Theorem 1.1:

Theorem 2.6 *Let G be a simple group of finite Morley rank of even type, with no infinite definable simple section of degenerate type. Then G is an algebraic group over an algebraically closed field of characteristic 2.*

Before finishing this section it is worthwhile making some comments on the proof of Fact 2.5 as well as on types of arguments which are frequently done to analyze K-groups. Fact 2.5 depends on the following two important theorems on groups of automorphisms definable in structures of finite Morley rank. Theorem 2.8 is recurrently used in K-group arguments.

Theorem 2.7 ([22], Theorem 8.3) *Let $\langle K, +, \cdot, 0, 1, G \rangle$ be a structure of finite Morley rank where $\langle K, +, \cdot, 0, 1 \rangle$ is an infinite field and G is a group that acts on K as field automorphisms. Then the action of G is trivial.*

Theorem 2.8 ([22], Theorem 8.4) *Let $\mathcal{G} = G \rtimes H$ be a group of finite Morley rank where G and H are definable, G is an infinite simple algebraic group over an algebraically closed field, and $C_H(G) = 1$. Then viewing H as a subgroup of $\mathrm{Aut}(G)$, we have $H \leq \mathrm{Inn}(G)\Gamma$ where $\mathrm{Inn}(G)$ is the group of inner automorphisms of G and Γ are the graph automorphisms of G.*

Although not used in proving Fact 2.5, there is another type of result which is frequently needed in the analysis of K-groups. This is the understanding of central extensions of simple algebraic groups. In the context of groups of finite Morley rank the following result is the main tool for this understanding:

Theorem 2.9 ([7], Corollary 1) *Let G be a perfect group of finite Morley rank such that $G/Z(G)$ is a simple algebraic group. Then G is an algebraic group. In particular, it follows from general properties of algebraic groups that $Z(G)$ is finite.*

2.3 Solvable groups

A highly developed theory of solvable groups was indispensable to the classification of the finite simple groups. This is also the case for the classification of the simple groups of finite Morley rank. It suffices to recall that in small configurations a large proportion of the information about the structure of the group under analysis comes from its solvable subgroups.

Historically, the study of solvable groups of finite Morley rank can be divided into three periods. In the first one structural results on similarities between connected solvable groups of finite Morley rank and connected solvable algebraic groups were obtained ([44, 45], work by Zil'ber). The following result by Nesin summarizes the most important results proven in this first period and is very useful:

Theorem 2.10 ([45]) *If G is a connected solvable group of finite Morley rank then $G/F^\circ(G)$ (and hence $G/F(G)$) is divisible abelian.*

For a group G of finite Morley rank, $F(G)$ denotes its largest normal nilpotent subgroup, the *Fitting subgroup*. By results of Nesin and Belegradek ([46],[12]) $F(G)$ exists, is definable, and generated by the normal nilpotent subgroups of G.

It became clear that theorems with a more finitistic spirit were necessary and this started the second period which involves such results as various versions of the Schur-Zassenhaus theorem and the Hall theorem ([20, 21, 8]). From this period it seems appropriate for the purposes of this article to recall two theorems. The first one is a version of the Shur-Zassenhaus theorem while the second is a connectedness result for maximal π-subgroups (*Hall π-subgroups*) of a connected solvable group of finite Morley rank where π is a set of prime numbers.

Theorem 2.11 ([21], Proposition C) *Let G be a solvable group of finite Morley rank, and H a normal Hall π-subgroup of G of bounded exponent. Then any subgroup K of G with $K \cap H = 1$ is contained in a complement to H in G, and the complements of H in G are definable and conjugate to each other.*

Theorem 2.12 ([22], Theorem 9.29; [32], Corollaire 7.15) *The Hall π-subgroups of a connected solvable group of finite Morley rank are connected.*

Despite their effectiveness in handling many situations the results proven in the second period rely heavily on torsion elements. But groups of finite Morley rank have many elements of infinite order and solvable groups of finite Morley rank are no exception. Moreover, as opposed to algebraic groups there are no such notions available in a group of finite Morley rank as *semisimple* or *unipotent* elements which would refine the understanding elements of infinite order. As a result methods independent from torsion become necessary and this is one of the basic features of the third period. The first important work in this direction was proven by Frank Wagner and is on the *Carter subgroups* of various classes of solvable stable groups with a certain connectedness assumption ([56]). More recently Olivier Frécon developed a new approach to the results of Wagner based on the study of the *abnormal subgroups* of connected solvable groups of finite Morley rank ([32]). This allowed him to obtain deep structural information about these groups as well as to develop a completely new and substantial analysis of solvable groups of finite Morley rank in general with no connectedness assumption ([33, 30, 31]). As will be mentioned later some results proven during this third period have been crucial in the analysis of some small configurations key to the proof of Theorem 1.1.

It still remains to see whether and how the results obtained in this third period will be applied in tackling the simple groups whose proper definable connected subgroups are solvable, and in particular those of degenerate type. There is however a point where all this solvable group theory information reaches its limits and becomes vacuous. This is when the proper definable connected subgroups of a simple group of finite Morley rank are nilpotent. Such simple groups of finite Morley rank, also known as *bad groups*, still form the most difficult part of the classification of the simple groups of finite Morley rank as their analysis lies beyond the limits of the universe where finite group theoretic ideas are effective. Indeed, even the Carter subgroup theory originated in finite groups and its subsequent generalizations frequently inherited some traits from finite group theory. On the other hand, it is fairly easy to show that there exists no bad finite groups.

2.4 A quick survey of results on simple K^*-groups not of even type

To put this paper in a well-defined context it seems appropriate to give a quick overview of the classification results in the context of K^*-groups which are not of even type. With the exception of simple K^*-groups of mixed type, there remain a large number of open problems of varying levels of difficulty. In mixed type the following theorem by Eric Jaligot gives a definite answer:

Theorem 2.13 ([40]) *There exist no simple K^*-groups of mixed type.*

An important work on simple K^*-groups of odd type was [17] where Borovik proved a trichotomy theorem which clearly described the different paths to be followed to achieve a complete classification theorem of the simple K^*-groups of odd type. This was followed by the work of Ayşe Berkman which treated the generic cases ([15]). The remaining configurations however are still waiting to be handled except some recent work by Jaligot ([36]). Here one should also mention Christine Altseimer's work on identification theorems ([9]) and applications of the *Thompson rank formula* to groups of odd type ([10]).

In the K^*-context, the proper definable connected subgroups of groups of degenerate type are all solvable as it has been explained above. This reduction does not make their analysis any easier and degenerate configurations remain intact. Recently however, Jaligot offered a new look to the main problems involved (Chapter 5 of [39] and [37]).

Another survey by Borovik ([18]) is a very useful source where the reader can find out about how the classification of simple K^*-groups of even type evolved.

2.5 Outline of the paper and some notions

The rest of the paper will be devoted to an outline of the classification of simple K^*-groups of even type. This is a project with several phases reminiscent of the classification of the finite simple groups. The first phase is the analysis of small groups. In the context of K^*-groups of even type this analysis consists of two characterizations of $SL_2(K)$ with K algebraically closed of characteristic 2. These characterizations are used as tools in the second phase where intermediate configurations are analyzed. These configurations are related to the structure of parabolic subgroups and their interactions. They provide crucial information for the third phase where a faithful adaptation of the *third-generation approach* to the classification of the finite simple groups using the theory of amalgams allows the construction of BN-pairs and eventually an identification with simple algebraic groups over algebraically closed fields of characteristic 2.

Before going any further it seems appropriate to introduce some more notation and go over a few properties of groups of finite Morley rank. For any subset X of a group G, $I(X)$ will denote the involutions in X. For a group G of finite Morley rank $\mathfrak{B}(G)$ will denote the subgroup of G generated by its unipotent 2-subgroups, $O(G)$ the largest definable connected normal subgroup without involutions.

Evidently, any step of the classification of simple K^*-groups of even type involves arguments about Sylow 2-subgroups. Most of the time however, in the context of groups of even type one is more interested in the connected component of a Sylow 2-subgroup than in the entire subgroup itself. As a result the notation *Sylow° 2-subgroup* will be used to denote the connected component of a Sylow 2-subgroup. An illustration of the usefulness of focusing on Sylow° 2-subgroups is an immediate corollary of Theorems 2.10 and 2.12: a connected solvable group G of finite Morley rank of even type has a unique Sylow 2-subgroup which is connected and contained in $F(G)$.

3 Characterizations of SL_2: weakly embedded subgroups

The classification of the finite simple groups with *strongly embedded subgroups* was a very important step in the classification of the finite simple groups. This result of Helmut Bender ([13]) and Michio Suzuki ([53, 54]) not only was a characterization theorem for a family of small finite simple groups but also an effective tool for further analysis. A natural question was whether an analogous concept would be as useful for groups of finite Morley rank. The answer is affirmative for simple K^*-groups of even type and the work which was done to achieve this answer forms the most fundamental part of the entire classification of simple K^*-groups of even type. This section is devoted to an outline of this work.

The following is the definition of a strongly embedded subgroup in the context of groups of finite Morley rank:

Definition 3.1 *Let G be a group of finite Morley rank and M a proper definable subgroup of G. M is said to be a* strongly embedded subgroup *of G if it satisfies the following conditions:*

1. *M contains involutions.*

2. *For any $g \in G \setminus M$, $M \cap M^g$ does not contain involutions.*

Apart from the definability assumption this is exactly the same definition as in finite group theory. As expected some of the elementary properties of finite groups with strongly embedded subgroups are true in the context of groups of finite Morley rank as well:

Fact 3.2 ([35] Theorem 9.2.1; [22], Theorem 10.19) *Let G be a group of finite Morley rank with a strongly embedded subgroup M. Then the following hold:*

1. *A Sylow 2-subgroup of M is a Sylow 2-subgroup of G.*

2. *$I(G)$ is a single conjugacy class.*

3. *$I(M)$ is a single conjugacy class in M.*

One major ingredient which makes such a fact provable in the context of groups finite Morley rank is the following analogue of an elementary but crucial property of involutions in finite groups:

Fact 3.3 ([22], Proposition 10.2) *Let G be a group of finite Morley rank and i, j involutions in G. Then either i and j are $d(ij)$-conjugate or they commute with another involution.*

A well-known characterization of finite groups with strongly embedded subgroups also has an analogue of finite Morley rank:

Fact 3.4 *Let G be a group of finite Morley rank with a proper definable subgroup M. Then the following are equivalent:*

1. *M is a strongly embedded subgroup.*

2. *M has involutions, $C_G(i) \leq M$ when $i \in I(M)$, and $N_G(S) \leq M$ whenever S is a Sylow 2-subgroup of M.*

3. M *has involutions, and* $N_G(S) \leq M$ *whenever* S *is a nontrivial 2-subgroup of* M.

Given the nature of the Cherlin-Zil'ber conjecture and the fact the only examples of simple algebraic groups over algebraically closed fields with strongly embedded subgroups are $SL_2(K)$ where K is of characteristic 2 with the Borel subgroups as strongly embedded subgroups, the natural conjecture about simple groups of finite Morley rank with strongly embedded subgroups is the following:

Conjecture 1 *A simple group of finite Morley rank with a strongly embedded subgroup is isomorphic to* $SL_2(K)$ *where* K *is algebraically closed of charateristic 2.*

In its full generality this conjecture is far from being proven.

In the context of simple K^*-groups of even type it quickly became clear that a weaker notion than that of a strongly embedded subgroup is more frequently encountered in practice ([4]). This is the content of the following definition:

Definition 3.5 *Let* G *be a group of finite Morley rank and* M *a proper definable subgroup of* G. M *is said to be a* weakly embedded subgroup *if it satisfies the following conditions:*

1. M *has infinite Sylow 2-subgroups.*

2. *For any* $g \in G \setminus M$, $M \cap M^g$ *has finite Sylow 2-subgroups.*

This is an infinite notion and in general it is not even a weakening of the notion of strongly embedded subgroup. Only the first property in Fact 3.2 is true in general for groups of finite Morley rank with weakly embedded subgroups. It is not hard to construct counterexamples to the conjugacy statements. However there exists a general characterization in the spirit of Fact 3.4:

Fact 3.6 ([4], Proposition 3.3) *Let* G *be a group of finite Morley rank. A proper definable subgroup* M *of* G *is a weakly embedded subgroup if and only if the following hold:*

1. M *has infinite Sylow 2-subgroups.*

2. *For any unipotent 2-subgroup* U *and 2-torus* T *in* M, $N_G(U) \leq M$, *and* $N_G(T) \leq M$.

In the context of groups of even type this characterization becomes much more compact:

Fact 3.7 ([5], Corollary 2.29) *Let G be a group of finite Morley rank of even type and M a proper definable subgroup of G containing a Sylow 2-subgroup S of G. Then M is a weakly embedded subgroup if and only if for any unipotent 2-subgroup U of S, $N_G(U) \leq M$.*

At the risk of overwhelming the reader with an inflation of characterizations of weak embedding, one last such statement, more precisely a detection principle whose proof contains ideas which will turn out to be very important later (Theorem 5.2), will be stated:

Fact 3.8 ([5], Proposition 2.33; [6], Fact 2.27) *Let G be a simple group of finite Morley rank of even type and H a proper definable subgroup with infinite Sylow 2-subgroups. Assume that if U is a nontrivial unipotent 2-subgroup of H then $N_G^\circ(U) \leq H$. Then G has a weakly embedded subgroup. More precisely, if S is a Sylow$^\circ$ 2-subgroup of H then there is a subgroup H_1 of H containing S such that $N_G(H_1)$ is weakly embedded in G.*

This fact is proven by constructing a graph whose vertices are nontrivial unipotent 2-subgroups. Two vertices are connected if they have infinite intersection. Under the given assumptions this graph is disconnected and this disconnectedness implies the existence of a weakly embedded subgroup: the stabilizer of a connected component of the graph, or equivalently the normalizer of the subgroup generated by the vertices of a connected component of the graph.

One should also note that in a group of finite Morley rank which is not of degenerate type the notion of weakly embedded subgroup becomes a weakening of the notion of strongly embedded subgroup. Because of this and the fact that it is much easier to prove the existence of a weakly embedded subgroup in a simple group of even type, in the rest of the paper the notion of strong embedding will not be visible although weak embedding is omnipresent.

The main result about simple K^*-groups of even type with weakly embedded subgroups is the following theorem by Jaligot:

Theorem 3.9 ([38]; [39], Théorème 4.1) *Every simple K^*-group of even type with a weakly embedded subgroup is isomorphic to $\mathrm{SL}_2(K)$ with K algebraically closed of characteristic 2.*

This is a fundamental result which is the culmination of work spread over three papers ([1], [5], [38]). As it will become clearer in the sequel it is the key to the entire classification project of the simple K^*-groups of even type. In the rest of this section a summary of the arguments will be given. Until the end of this section, unless stated otherwise, G will denote a simple K^*-group of even type with a weakly embedded subgroup M.

In retrospect, the contribution of [1] and [5] to the analysis of simple K^*-groups of even type with a weakly embedded subgroup can be seen as a reduction to a particular case. [1] considers a simple K^*-group of even type with a strongly embedded subgroup and most of the time uses additional hypotheses. In [5], M is a weakly embedded subgroup which is not strongly embedded. A comparison of Facts 3.4 and 3.6 shows that the difference is witnessed by an "offending involution" $\alpha \in M$ such that $C_G^\circ(\alpha) \not\leq M$. K-group arguments show that $C_G^\circ(\alpha) = \mathfrak{B}(C_G^\circ(\alpha)) \times O(C_G^\circ(\alpha))$ with $\mathfrak{B}(C_G^\circ(\alpha)) \cong SL_2(K)$ where K is algebraically closed of characteristic 2. Moreover $\mathfrak{B}(C_G^\circ(\alpha)) \cap M^\circ$ is a Borel subgroup of $\mathfrak{B}(C_G^\circ(\alpha))$. A consequence of this is that $\alpha \in I(M \setminus M^\circ)$.

Despite these differences between the strongly and weakly embedded cases, the first step in both situations is to prove that M° is solvable. When M° is strongly embedded, this is achieved fairly smoothly using a particular factorization: $M^\circ = C_G^\circ(i)K^\circ$ where $i \in I(M^\circ)$, and K is a definable $2^\perp$-subgroup. When M is weakly embedded, the argument is more involved and requires the analysis of all possible isomorphism types of the Sylow$^\circ$ 2-subgroups of G. This is obtained by proving the following analogue of a result by Landrock and Solomon ([42]):

Theorem 3.10 ([5], Theorem 4.1) *Let $H = S \rtimes T$ be a group of finite Morley rank, where S is a definable connected 2-group of bounded exponent, and T is definable. Assume that S has a definable subgroup A such that $A \rtimes T \cong K_+ \rtimes K^\times$ for some algebraically closed field K of characteristic 2, with the multiplicative group acting naturally on the additive group. Assume also that α is a definable involutory automorphism of H such that $C_H^\circ(\alpha) = A \rtimes T$. Under these assumptions S is isomorphic to one of the following groups:*

1. *If S is abelian then either S is homocyclic with $I(S) = A^\times$, or $S = E \oplus E^\alpha$, where E is an elementary abelian group isomorphic to K_+. In the latter case, $A = \{xx^\alpha : x \in E\}$.*

2. *If S is nonabelian then S is an algebraic group over K whose underlying set is $K \times K \times K$ and the group multiplication is as follows:*

   For $a_1, b_1, c_1, a_2, b_2, c_2 \in K$,

   $$(a_1, b_1, c_1)(a_2, b_2, c_2) = (a_1+a_2, b_1+b_2, c_1+c_2+\epsilon\sqrt{a_1 a_2}+\sqrt{b_1 b_2}+\sqrt{b_1 a_2}),$$

   where ϵ is either 0 or 1.

   In this case α acts by $(a, b, c)^\alpha = (a, a+b, a+b+c+\sqrt{ab})$ and $[\alpha, S] = \{(0, b, c) : b, c \in K\}$.

In particular, if S is nonabelian then S has exponent 4.

The discussion of the structure of the centralizer of the offending involution α shows that the hypotheses of Theorem 3.10 are satisfied in a simple K^*-group of even type with a weakly but not strongly embedded subgroup. It is worth noting that the statement of Theorem 3.10 is an exact analogue of the theorem by Landrock and Solomon but the methods used to prove it are fairly different.

Once Thereom 3.10 is proven, an analysis of the possibilites shows that M° cannot be nonsolvable. This solvability result has strong implications on the structure of M and on the control of fusion in M°. To start, as was explained at the end of Section 2.5 M has a unique Sylow$^\circ$ 2-subgroup. Although M is not strongly embedded M° behaves very much like a strongly embedded sugbroup in that if $i \in I(M^\circ)$ then $C_G(i) \leq M$, and also if $g \in G \setminus M$ then $M^g \cap I(M^\circ) = \emptyset$. It follows from this that G has at least two conjugacy classes, the involutions conjugate to an involution in $M \setminus M^\circ$ and those conjugate to an involution in M°, and M controls fusion of involutions in M°.

Theorem 3.10 is also crucial to carry out the rest of the analysis when M is weakly embedded. A close analysis of each possibility which arises in the statement of Theorem 3.10 reduces the discussion to the case where G has homocyclic Sylow$^\circ$ 2-subgroups. This is the point where Jaligot's analysis starts.

Before discussing Jaligot's contribution, it is worth mentioning that when M is weakly embedded the nonabelian possibilities for the Sylow$^\circ$ 2-subgroups are eliminated using the Thompson rank formula, a finite Morley rank analogue of the Thompson order formula in finite group theory. As has been mentioned above G has at least two conjugacy classes of involutions. This allows the use of the Thompson rank formula (after some preparation in order to prove the definability of the *Thompson map*), which severely restricts the rank of G.

When M is weakly embedded, the remaining case is the one where the Sylow$^\circ$ 2-subgroups of G are homocyclic. On the other hand, a strongly embedded M has not been discussed above beyond the solvability of M°. There is a good reason for this: in both cases M° has the same structure except for one property. When M is strongly embedded it is not known whether the Sylow$^\circ$ 2-subgroups of G are abelian. This, however, does not cause problems as the missing piece of information is provided by the following theorem of Mark Davis and Ali Nesin:

Theorem 3.11 ([26]) *A free Suzuki 2-group of finite Morley rank is abelian.*

A *Suzuki 2-group* is a pair (S, T) where S is a nilpotent 2-group of bounded exponent and T is an abelian group that acts on S by group automorphisms and which is transitive on $I(S)$. A Suzuki 2-group is said to be *free* if T acts on S *freely*, i.e. for any $g \in S$ and $t \in T$, $g^t = g$ implies either $g = 1$ or $t = 1$.

A Suzuki 2-group (S, T) is said to be abelian if S is abelian. Whenever the commutativity of the Sylow° 2-subgroups is needed, a modest preparation, which certainly uses the solvability of $M°$, shows that the assumptions of Theorem 3.11 are satisfied in G.

As a result, Jaligot is able to avoid the use of separate treatments for strong and weak embedding as well as any recourse to $I(M \setminus M°)$. In the rest of this section his arguments will be summarized.

To recapitulate, G is a simple K^*-group of even type with a weakly embedded subgroup M. $M°$ is solvable and $A = \langle I(M°) \rangle$ is an elementary abelian 2-subgroup on the involutions of which $M°$ acts transitively. The proof of Theorem 3.9 is an argument divided into two parts. In the first part it is assumed that if H is an infinite definable subroup of $C_G^°(A)$ then $C_G(H) \leq M$. With this assumption it is proven that $G \cong SL_2(K)$ with K algebraically closed of characteristic 2. The second and much more involved part of the proof considers the case when there exists an infinite definable subgroup $H \leq C_G^°(A)$ such that $C_G(H) \not\leq M$. This eventually yields a contradiction.

The main effort in the proof of the first part is devoted to a computation of the rank of G using *strongly real elements* (an element is strongly real if it is the product of two involutions). After obtaining a factorization of $M°$ as a semidirect product of $F°(M)$ and a definable $2^\perp$-subgroup T, Jaligot investigates how two distinct conjugates of M intersect. It turns out this intersection meets the Fitting subgroups trivially which is an important piece of information for the double transitivity argument later. Afterwards he introduces the following notation which goes back to Suzuki and was first introduced to the realm of groups of finite Morley rank by Nesin ([27]):

$$T(w) = \{m \in M° : m^w = m^{-1}\};$$

$$X_1 = \{w \in i^G \setminus M : T(w) = 1\};$$

$$X_2 = \{w \in i^G \setminus M : T(w) \neq 1\}.$$

Here $i \in I(M°)$. A faithful reproduction of computations from [27] yields the following equality:

$$\mathrm{rk}\,(G) = \mathrm{rk}\,(T) + 2\mathrm{rk}\,(F(M)).$$

This information immediately yields the conclusion that M is connected and the action of G on the cosets of M is doubly transitive. A quick argument shows that G is actually a *split Zassenhaus group*, a doubly transitive group where the stabilizer of a 3-point set is the identity, and the stabilizer of a 1-point-set is the semidirect product of a 2-point-set stabilizer and of a normal subgroup $(M = F°(M) \rtimes T)$. The following fundamental result by DeBonis and Nesin then gives the desired conclusion:

Theorem 3.12 ([28]) *Let G be an infinite nonsolvable split Zassenhaus group of finite Morley rank. Let B denote a 1-point stabilizer and assume $B = U \rtimes T$ where T is a 2-point stabilizer. If U has a central involution and $T \neq 1$ then $G \cong \mathrm{SL}_2(K)$ with K algebraically closed of characteristic 2.*

The second part of the proof starts by fixing a nontrivial definable connected subgroup H of M such that $C_G(H) \not\leq M$ and which is maximal with respect to this property. K-group arguments imply that $C_G^\circ(H) = \mathfrak{B}(C_G(H)) \times O(C_G(H))$ and $\mathfrak{B}(C_G(H)) \cong \mathrm{SL}_2(K)$ with K algebraically closed of characteristic 2. Moreover $\mathfrak{B}(C_G(H)) \cap M$ is a Borel subgroup $A \rtimes T$ of $\mathfrak{B}(C_G(H))$. From this it is possible to obtain useful structural information about M° which already hints at the tightness of the situation:

1. $M^\circ = F^\circ(M) \rtimes (H \times T)$.

2. For every $t \in T^\times$, $C_{F^\circ(M)}(t) = 1$.

3. For every $h \in H^\times$, $C_{F^\circ(M)}(h) = A$.

4. $F^\circ(M) = O(M) \times A$ with $O(M)$ torsion-free.

5. $O(M) = O(F) \rtimes H$.

It is worth emphasizing that the solvable nonnilpotent cores as in point 5 above are not algebraic groups and their complete elimination is achieved only with the final contradiction in Jaligot's proof. This is the point where results from the third period of the theory of solvable groups of finite Morley rank enter the scene.

The next step in the proof is, as in the first part, the computation of the rank of G. However this time the arguments are much lengthier. The general strategy is similar to that used in the first part and Jaligot starts by introducing the following notation:

$$T(w) = \{m \in M^\circ : m^w = m^{-1}\};$$

$$X_1 = \{w \in i^G \setminus M : \mathrm{rk}\,(T(w)) < \mathrm{rk}\,(T)\};$$

$$X_2 = \{w \in i^G \setminus M : \mathrm{rk}\,(T(w)) \geq \mathrm{rk}\,(T)\};$$

$$W = \{w \in i^G \setminus M : T^f \leq T(w) \text{ for some } f \in F^\circ(M).$$

The arguments that lead to the computation of $\mathrm{rk}\,(G)$ again require a close analysis of strongly real elements but the implementation of each step of the general strategy requires new ideas and tools, notably the conjugacy of Carter subgroups of connected solvable groups of finite Morley rank. Eventually the following equality is obtained:

$$\mathrm{rk}\,(G) = \mathrm{rk}\,(HT) + 2\mathrm{rk}\,(F(M)).$$

The passage from this point to the double transitivity is not easy either. It again necessitates the study of how two distinct conjugates of M intersect. Jaligot proves that this intersection meets $F(M)$ trivially. It is striking that the argument uses a particular class of abnormal subgroups of M° whose properties were closely investigated by Frécon in [32], namely the *generalized centralizers*.

Once the investigation of intersections of conjugates of M is over, the double transitivity of G is immediate. However, G is not a Zassenhaus group and Theorem 3.12 cannot be used. This turns out to be a major difficulty because at this point all sources of information are exhausted. As a result Jaligot starts computational arguments using elements of order 3 which are present in M as $T \cong K^\times$ and K is algebraically closed of characteristic 2. The main computational tools are the functions obtained through a parametrization of the group using its rank 1 BN-pair structure: for a fixed $w \in I(G \setminus M)$ which inverts T (a Weyl group element in $\mathfrak{B}(C_G(H))$), and $f \in F^\circ(M)^\times$, one has a unique presentation

$$f^w = \phi(f)w\alpha(f)\beta(f)$$

where $\phi(f), \alpha(f) \in F^\circ(M)^\times$ and $\beta(f) \in HT$. These computational ideas had already been used by Nesin in his analysis of permutation groups of finite Morley rank but Jaligot takes them much further and achieves a rather surprising contradiction.

Theorem 3.9 is of central importance to the classification of the simple K^*-groups of even type as it will become clearer in the following sections. The following immediate application, the *elimination of cores in 2-local subgroups*, is a recurrent tool:

Theorem 3.13 ([5], Theorem 10.13; [39], Fait 4.69) *Let G be a simple K^*-group of even type. Then $O(N_G(P)) = 1$ for any definable 2-subgroup P of G.*

4 Characterizations of SL_2: strongly closed abelian 2-subgroups

The second characterization of SL_2 is the study of simple K^*-groups of even type with *strongly closed abelian 2-subgroups*. This notion is strongly related to fusion analysis and key to the treatment of components. It is defined as follows:

Definition 4.1 *Let A, G be groups.*

1. If $A \leq H \leq G$ then A is strongly closed in H *relative to G if whenever* $a \in A$, $g \in G$, and $a^g \in H$, $a^g \in A$.

2. *When G is of finite Morley rank, A is said to be a* strongly closed 2-subgroup *of G if A is a 2-subgroup, is contained in a Sylow° 2-subgroup of G, and is strongly closed in at least one such Sylow° 2-subgroup.*

The following theorem has been proven in the context of simple K^*-groups of even type:

Theorem 4.2 ([6], Theorem 1.1) *Let G be a simple K^*-group of even type with an infinite definable strongly closed abelian 2-subgroup. Then G contains a weakly embedded subgroup.*

This is not the statement given in [6] but it describes better how the characterization is obtained: the entire body of [6] is devoted to proving that under the assumptions of Theorem 4.2, a simple K^*-group of even type contains a weakly embedded subgroup. Once this is done Theorem 3.9 implies that the only possibility is $SL_2(K)$ with K algebraically closed of characteristic 2.

Special cases of Theorem 4.2 will be used several times in the following sections in the treatment of components before starting the application of the amalgam method. The proof of Theorem 4.2 itself involves several component elimination arguments as will be seen shortly. However in order to carry out any argument one needs to render the very notion itself somewhat more robust (notice that the definition of a strongly closed 2-subgroup is made with respect to one fixed Sylow° 2-subgroup) and obtain basic fusion information which eventually allows to consider a "minimal" strongly closed abelian 2-subgroup. This is the content of the following basic fact (part 4):

Fact 4.3 ([6], Lemma 3.2) *Let G be a group of finite Morley rank and of even type, and let A be a definable abelian 2-subgroup of G such that A is strongly closed in a Sylow° 2-subgroup S of G.*

1. *A is strongly closed in any Sylow° 2-subgroup that contains A.*

2. *A is a normal subgroup of any Sylow° 2-subgroup that contains it.*

3. *$N(A°)$ controls fusion in A.*

4. *Any definable $N(A°)$-invariant subgroup of A is also strongly closed.*

5. *If the Sylow 2-subgroups of G are connected and N is a definable normal subgroup of G then AN/N is strongly closed in G/N.*

It follows that A can be assumed to be connected.

In the rest of this section, unless otherwise stated, G will denote a simple K^*-group of even type with a strongly closed abelian 2-subgroup A which is also infinite definable connected and $N_G(A)$-minimal.

The starting point of the arguments is the following important K-group fact:

Fact 4.4 ([6], Lemma 3.3) *Let H be a connected K-group of even type with an infinite definable definable connected strongly closed abelian 2-subgroup A. If A is not normal in H then H has a normal subgroup $L \cong \mathrm{SL}_2(K)$ with K algebraically closed of characteristic 2, and $L \cap A$ is Sylow 2-subgroup of L; thus $H = L \times C_H(L)$ (Theorem 2.8).*

This K-group fact, together with Theorem 2.8 yields a direct product decomposition of A: $A = (L \cap A) \times C_A(L)$. Whether $C_A(L) = 1$ or not plays a crucial role in the proof of Theorem 4.2 and gives the first component analysis in this proof:

Definition 4.5 *Let G be a group of finite Morley rank of even type with a strongly closed abelian 2-subgroup A. Suppose G has a definable subgroup $L \cong \mathrm{SL}_2(K)$ normalized by A such that $L \cap A$ is a Sylow 2-subgroup of L. In particular, $A = (L \cap A) \times C_A(L)$. If $C_A(L) \neq 1$ then L is called an A-special component.*

The first major step in the proof of Theorem 4.2 is the elimination of A-special components in G. *Pseudoreflections* turn out to be a very powerful tool in this discussion:

Definition 4.6 *If A is an elementary abelian 2-group then a nontrivial torus T acting on A is called a group of pseudoreflections on A if $A = C_A(T) \times [A, T]$, and T acts faithfully on the second factor and transitively on its nonzero elements.*

The following result about K-groups containing pseudoreflections is proven:

Theorem 4.7 ([6], Theorem 4.4) *Let $A \rtimes H$ be a connected K-group of even type, in which A is a definable elementary abelian 2-subgroup and H acts irreducibly and faithfully on A. Assume that H contains a group T of pseudoreflections on A. Then A can be given a vector space structure over an algebraically closed field K in such a way that $H \cong \mathrm{GL}(A)$ acting naturally.*

The proof of Theorem 4.7 relies crucially on results by Jack McLaughlin on groups generated by transvections ([43]).

It is worth noting that in eliminating the A-special components of G, arguments involving weakly embedded subgroups are also used although such weak embedding configurations can be eliminated in these particular cases without using the full force of Theorem 3.9.

The elimination of the A-special components sets the stage for the second major step in the proof of Theorem 4.2. This is a dichotomy theorem which states that either G has a weakly embedded subgroup or G has a definable subgroup $L \cong \mathrm{SL}_2(K)$ (K algebraically closed of characteristic 2) such that A is a Sylow 2-subgroup of L and $C_G(L)$ contains a nontrivial unipotent 2-subgroup. This is the point where Theorem 3.9 becomes indispensable to arrive at a characterization of $\mathrm{SL}_2(K)$.

The second possibility which arises in the dichotomy is eliminated by a lengthy computational argument which uses variants of the Thompson rank formula as well as a close analysis of the structure of the Sylow° 2-subgroups of G. It turns out that it is actually possible to prove a more general result which proves useful in the subsequent stages of the classification of simple K^*-groups of even type. Before stating this theorem, it is necessary to give the definition of a *standard component* in this context:

Definition 4.8 *Let G be a group of finite Morley rank of even type and L a definable connected quasisimple subgroup of G. Then L is called a* standard component *for G if $C_G(L)$ contains an involution, and $L \lhd C_G^\circ(i)$ whenever i is such an involution.*

As such, this definition does not include the following condition which is part of the definition of a standard component in a finite group:

L does not commute with any of its conjugates.

This condition can be proven on the basis of Definition 4.8 but this would require that Theorem 4.2 be already proven. However this is not harmful to what follows and indeed the following theorem turns out to be very useful for later stages of the classification of simple K^*-groups of even type:

Theorem 4.9 ([6], Theorem 1.3) *Let G be a simple K^*-group of even type. Suppose that G has a standard component L of the form $\mathrm{SL}_2(K)$ with K algebraically closed of characteristic 2. Let A be a Sylow 2-subgroup of L and U be the connected component of a Sylow 2-subgroup of $C_G(L)$. If U is nontrivial then AU is a Sylow° 2-subgroup of G.*

As to how this is used to finish the proof Theorem 4.2, it turns out that at this stage of the analysis if U denotes a nontrivial maximal unipotent 2-subgroup in $C_G(L)$ as given by the dichotomy step then AU is not a Sylow° 2-subgroup of G (Lemma 6.4 of [6]). On the other hand, a quick argument shows that L is standard component in G. This, together with Theorem 4.9, yields a contradiction.

5   Intermediate configurations

The last two sections contained the analysis of what corresponds to small configurations in the context of simple K^*-groups of even type. In this section, the analysis of "larger" configurations will be explained. Like the word "small", "large" also has a somewhat vague meaning. It roughly corresponds to groups where normalizers of some definable 2-subgroups are nonsolvable. This was already observed in previous sections but now it is more a phenomenon to be analyzed than a consequence of contradictory assumptions.

Before going into details some notions crucial in the sequel will be reviewed. The first set of definitions will be made under the sole assumption that G is a group of finite Morley rank. A *standard Borel subgroup* of G is a maximal definable connected solvable subgroup of G which contains a Sylow° 2-subgroup. It should be noted that in the context of groups of finite Morley rank, as opposed to that of algebraic groups, one can encounter "nonstandard" Borel subgroups, e.g. definable connected nonsolvable groups without involutions. A *component* of G is a subnormal quasisimple subgroup of G. $E(G)$ will denote (as in finite group theory) the subgroup generated by the components of G. In [46] Nesin analyzed components in groups of finite Morley rank and proved that they are definable subgroups, and in a group of finite Morley rank there are finitely many components. He also showed that if G is connected, then its components are connected and normal in G. Another result of Nesin along the lines of finite group theory states that $E(G)$ is the central product of the components of G. Since in a group of finite Morley rank G, $F(G)$ and $E(G)$ behave as in finite group theory, it is natural to define the *generalized Fitting subgroup* $F^*(G)$ in an analogous way to finite group theory:

$$F^*(G) = F(G)E(G).$$

In the context of groups of finite Morley rank of even type all these subgroups are very useful tools for local analysis as in finite group theory.

Now let G denote a group of finite Morley rank of even type. A 2-local subgroup of G is the normalizer of a definable 2-subgroup. A 2-*local°* subgroup is the connected component of the normalizer of a definable *connected* 2-subgroup. A 2-local° subgroup which contains a standard Borel subgroup is called a *parabolic subgroup*. A related notion is $O_2(G)$, which is the largest normal 2-subgroup of G. In the even type context this subgroup is definable and nilpotent.

In groups of finite Morley rank different types of groups which are expected to correspond to different types of simple algebraic groups according to the characteristic of the field were defined using Theorem 2.1. In finite group theory, a definition of "even type" or more correctly *characteristic 2-type* has been the condition that $F^*(H) = O_2(H)$ for every 2-local subgroup H of a finite group. Given that this is also a property of the normalizers of unipotent subgroups of a simple algebraic group over an algebraically closed field of characteristic 2, it is a natural question whether in the abstract context of a group of finite Morley rank of even type one has $F^*(H) = O_2(H)$ for a 2-local subgroup H. Another question is for which subgroups H this is needed in order to classify simple K^*-groups of even type. It has turned out that in order to carry out the third and final phase of the classification of simple K^*-groups of even type it suffices to prove $F^*(H) = O_2(H)$ only when H is a parabolic subgroup. The results of this section culminate in a proof of this equality as well as the 2-*constraint* property which generally goes with it.

As in finite group theory, an important question is the number of maximal 2-local° subgroups which contain a given 2-local° subgroup. This question brings one quite naturally to consider an analogue of a finite group theoretic concept introduced by Aschbacher:

Definition 5.1 *Let G be a group of finite Morley rank of even type and T a Sylow° 2-subgroup of G. Then $C(G,T)$ is the subgroup of G generated by all subgroups of the form $N_G^\circ(X)$ where X is a definable connected subgroup of T invariant under the action $N_G^\circ(T)$.*

If the answer to the question about the number of maximal 2-local° subgroups is 1 then $C(G,T) < G$ for any Sylow° 2-subgroup of G. As a result one is naturally led to proving the following analogue of *Aschbacher's global $C(G,T)$-theorem*:

Theorem 5.2 ([3]) *Let G be a simple K^*-group of even type with T a Sylow° 2-subgroup. If $C(G,T) < G$ then $G \cong \mathrm{SL}_2(K)$ with K algebraically closed of characteristic 2.*

The general strategy of the proof of this theorem will be explained below. It should be emphasized that what is proven is the existence of a weakly embedded subgroup under the stated hypotheses. Afterwards the conclusion is again an application of Theorem 3.9. It should also be noted that in stages which follow the proof of Theorem 5.2 the notion of $C(G,T)$ replaces that of a weakly embedded subgroup.

The proof of the $C(G,T)$-theorem depends on two other theorems which are important in their own right. The first one is the following analogue of a theorem of Baumann ([11])

Theorem 5.3 ([3]) *Let G be a group of finite Morley rank of even type. Let M be a definable connected subgroup of G such that $M/O_2(M) \cong \mathrm{SL}_2(K)$ with K algebraically closed of characteristic 2 and $F^*(M) = O_2(M)$. Assume that for a Sylow 2-subgroup S of M:*

(P)   no nontrivial definable connected subgroup of S is normalized by both M and $N_G(S)$

Then the following hold:

1. *S contains a nontrivial definable connected subgroup which is normalized by both M and $N_G^\circ(S)$.*

2. *If in addition*

   (P')   no nontrivial definable subgroup of S is normalized by both M and $N_G(S)$

   then there exists an automorphism α definable in G such that $S = Z(O_2(M))^\alpha O_2(M)$.

Before making any comments on this theorem, it is worth taking a detour to clarify a point which might look ambiguous to the careful reader. In the statement of the theorem S is assumed to be a Sylow 2-subgroup of M. This might look potentially dangerous as most of the time in the context of groups of even type only Sylow$^\circ$ 2-subgroups are considered. It can however be easily proven that in connected K-group of even type Sylow 2-subgroups are connected, and M is such a group. Similarly, if H is a connected K-group of even type then $O_2(H)$ is connected.

It is the first part of Theorem 5.3 that will be used in this section but the second part is kept as well since it is relevant to subsequent sections. The original proof of Baumann's theorem is technical and difficult. There is however an amalgam theoretic approach developed by Stellmacher in [52] which is more conceptual and happens to provide the main tool to prove Baumann's theorem. It also goes through with very minor changes in the finite Morley rank context.

The second ingredient in the proof of the $C(G, T)$-theorem is the following result on "pushing up":

Theorem 5.4 ([3]) *Let G be a simple K^*-group of even type, Q a unipotent 2-subgroup of G such that $Q = O_2(N^\circ(Q))$ with $\mathfrak{B}(N^\circ(Q)/Q) \cong \mathrm{SL}_2(K)$ with K algebraically closed of characteristic 2. Then $N^\circ(Q)$ contains a Sylow$^\circ$ 2-subgroup of G.*

This result is a description of the minimal parabolics in a simple K^*-group of even type. Its relatively short proof has two parts. One first assumes that Theorem 5.3 (more precisely its first part) applies with $M = \mathfrak{B}(N_G^\circ(Q))$. Then the conclusion is immediate. When this condition is negated, $M = F^*(M) = L * Q$ with $L \cong \mathrm{SL}_2(K)$. In this situation it is shown that L is a standard component in the sense given above and Theorem 4.9 proves that a Sylow 2-subgroup of M is a Sylow$^\circ$ 2-subgroup of G.

At this point one is ready to prove Theorem 5.2. Let G be a simple K^*-group of even type which satisfies the hypotheses of Theorem 5.2. The strategy is a frequently used one when the presence of a weakly embedded subgroup is being looked for in a group of even type (see Fact 3.8 and the discussion which follows): the construction of a graph whose vertices are the Sylow$^\circ$ 2-subgroups of G. Two vertices are connected if and only if they have infinite intersection. It is then proven that G has a weakly embedded subgroup if this graph is disconnected. In order to achieve the disconnectedness it suffices to prove the following: for S and T two Sylow$^\circ$ 2-subgroups of G, $C(G, S) = C(G, T)$ whenever S and T are connected by an edge. Note that if, with this last condition satisfied, the graph were connected then $C(G, T)$ (T an arbitrary Sylow$^\circ$ 2-subgroup of G) would be normal in G.

The proof of the argument is carried out by analyzing a counterexample, i.e. an edge between two Sylow$^\circ$ 2-subgroups S and T such that $C(G, S) \neq C(G, T)$ where $Q = (S \cap T)^\circ$ has maximal rank. More precisely the structure of $H = N^\circ(Q)$ is investigated. After some readjustment one can prove that H satisfies the hypotheses of Theorem 5.4 and consequently H contains S and T. Then follows a case discussion again based on whether Theorem 5.3 applies or not, this time with $M = \mathfrak{B}(H)$. If it does, then there is an immediate contradiction. If it does not, then G turns out to have abelian Sylow$^\circ$ 2-subgroups where Theorem 4.2 applies to conclude that $G \cong \mathrm{SL}_2(K)$.

The remaining part of this section will show how all these tools culminate in the proof of the elimination of components from parabolic subgroups in simple K^*-groups of even type. More precisely the following theorem is the final target before undertaking the application of the amalgam method and eventually arriving at comprehensive identification results:

Theorem 5.5 ([19]) *Let G be a simple K^*-group of even type, and P a parabolic subgroup of G. Then $F^*(P) = O_2(P)$.*

Theorem 5.5 is proven in two steps both of which depend on Theorem 5.2 in order to arrive at a contradiction. Let G and P be as in the statement of Theorem 5.5. The first and more involved step is the proof of the fact that $E(P) = 1$. The strategy is to suppose that G is a counterexample and prove the existence of a proper $C(G, T)$ in G. Then the inspection of the structure of $\mathrm{SL}_2(K)$ over an algebraically closed field of characteristic 2 immediately yields a contradiction. One therefore considers a component X in P, and argues to show that if T is a Sylow$^\circ$ 2-subgroup of G which is contained in P (remember that P is a parabolic and hence contains a Sylow$^\circ$ 2-subgroup of G) then $C(G, T) \leq N(X)$. This proves that $C(G, T) < G$.

The second step extends Theorem 3.13 in the context of a parabolic subgroup. More precisely, it is shown that if G is a simple K^*-group of even type with a parabolic subgroup P then $O_{2'}(P) = 1$, where $O_{2'}(P)$ is defined as the largest normal definable $2^\perp$-subgroup of P. This notion which no longer assumes connectedness as opposed to $O(P)$ is well-defined in the context of groups of finite Morley rank of even type. Note that in the context of Theorem 5.5 $O_{2'}(P)$ is already finite by Theorem 3.13. The argument then is to show that if S is a Sylow$^\circ$ 2-subgroup of P, then $C(G, S) \le C_G(O_{2'}(P))$. This forces $O_{2'}(P) = 1$.

Now Theorem 5.5 is an immediate consequence of these two steps. Moreover an immediate corollary is the important 2-constraint property for the parabolic subgroups: in the notation of the theorem, $C_P(O_2(P)) \le O_2(P)$.

6   The final phase

For expository purposes it seems more appropriate to break the logical order of things and to sumarize the main argument which proves Theorem 1.1 ([2]). The situation can be described as follows. G is a fixed simple K^*-group of even type. B is a fixed standard Borel subgroup of G and $S = O_2(B)$ is a Sylow$^\circ$ 2-subgroup of G. $\mathcal{M}$ is defined as the set of 2-local$^\circ$ subgroups of G which contain B, and $\mathcal{M}_0 = \{P \in \mathcal{M} : \mathfrak{B}(P/O_{2'}(P)) \cong \mathrm{SL}_2(K), K$ algebraically closed of characteristic 2$\}$. The subgroup G^* is defined as $\langle P : P \in \mathcal{M}_0 \cup \{B\} \rangle$ ($\mathcal{M}_0$ can be empty). G^* is a definable subgroup of G. A relatively simple K-group argument shows that G^* contains $C(G, S)$. As a consequence Theorem 5.2 implies that either G^* is G or $G \cong \mathrm{SL}_2(K)$ with K algebraically closed of characteristic 2. After this conclusion is obtained, one can assume that $G = G^*$. As G is assumed to be simple, $\mathcal{M}_0$ has at least two elements. A case division follows:

1. $\mathcal{M}_0$ has two elements P_1 and P_2 which generate G.

2. No 2-element subset of $\mathcal{M}_0$ generates G.

These two cases correspond to the following two subsections which eventually prove that in both situations G is an algebraic group over an algebraically closed field of characteristic 2.

The analysis of the first case brings into the realm of groups of finite Morley rank methods known as "a third-generation approach" to the classification of the finite simple groups, notably the theory of amalgams. The information obtained using this approach together with the knowledge about BN-pairs of finite Morley rank ([41]) and recent identification theorems yield the desired answer.

The second case is more direct and uses an adaptation by Berkman and Borovik ([16]) of a theorem of Niles ([48]) to arrive at a BN-pair construction which allows the use of results from [41] to prove Theorem 1.1 in the "generic" case.

6.1 A third-generation approach to simple K^*-groups of even type: amalgams

In this subsection, unless otherwise specified, G will continue to denote the simple K^*-group of even type to be identified with the additional assumption that $\mathcal{M}_0$ has two elements P_1 and P_2 such that $G = \langle P_1, P_2 \rangle$. From this one can deduce the following properties of the setting:

1. P_1 and P_2 are connected groups of finite Morley rank of even type such that $P_1 \cap P_2 = B$, where B is a standard Borel subgroup of both P_1 and P_2.

2. $\mathfrak{B}(P_i/O_2(P_i)) \cong \mathrm{SL}_2(K)$ with K algebraically closed of characteristic 2, for $i = 1, 2$.

3. $F^*(P_i) = O_2(P_i)$, for $i = 1, 2$.

4. For any definable subgroup N of P_1 and P_2, $N \lhd P_1$ and $N \lhd P_2$ implies that $N = 1$.

The coset graph Γ is the graph constructed using the information about P_1, P_2, B in the following way: the edges are $\{P_1 x, P_2 y : x, y \in G\}$, and two distinct edges are connected if they intersect nontrivially. G acts on Γ by right multiplication. The following basic properties are proven in [34], where the amalgam method was introduced, as well as in [29] and [51], which are the main models in order to apply the amalgam method to K^*-groups of even type:

1. Γ is connected.

2. G operates faithfully on Γ.

3. The vertex stabilizers are conjugate to P_1 or P_2, and the edge stabilizers are conjugate to B.

4. A vertex stabilizer acts transitively on the neighbors of the vertex which it stabilizes.

This picture in its entirety can be formulated using a graph-theoretic language:

1. Γ is a connected graph and G is a group of automorphisms of Γ which operates on Γ edge but not vertex-transitively.

2. The vertex stabilizers are connected groups of finite Morley rank.

3. If G_α is the stabilizer of the vertex α, then $F^*(G_\alpha) = O_2(G_\alpha)$.

4. The Sylow 2-subgroups of $G_\alpha \cap G_\beta$ are Sylow 2-subgroups of G_α and G_β whenever α and β are adjacent nodes in Γ.

In general the group G can be taken to be a homomorphic image of the amalgamated product $\tilde{G}$ of P_1 and P_2 over B. In the case where $G = \tilde{G}$, the following result by Jean-Pierre Serre describes Γ:

Theorem 6.1 ([50], Chapter I §4, Theorem 6) *The coset graph Γ is a tree if it is obtained from the amalgamated product as described above.*

At this level of generality there is no reason why (G, Γ) should be of finite Morley rank. However, a remark by Cherlin ([25]) shows that (G, Γ) is "locally" of finite Morley rank. In fact, if α is a vertex in Γ, $\Delta_k(\alpha)$ denotes the vertices of Γ at distance at most k from α, and $G(\alpha, k)$ is the set of elements of G which can be expressed as a product of at most k elements taken from stabilizers of vertices in $\Delta_k(\alpha)$, then the structure $(G(\alpha, k), \Delta_k(\alpha))$ is of finite Morley rank. It has a partially defined multiplication and a partially defined action on $\Delta_k(\alpha)$. This is a crucial observation as otherwise one cannot control definability problems. Moreover one eventually has to pass to the amalgamated product $P_1 *_B P_2$ to carry out the necessary arguments as the following fact shows:

Fact 6.2 ([25]; [29], (3.6)) *Let Γ be a tree and K a Cartan subgroup of G with apartment $T = T(K)$. Suppose that T satisfies both the uniqueness and the exchange conditions, and that $r = s - 1 \geq 2$. Then there is a G-invariant equivalence relation $\sim$ on Γ such that;*

1. *$\Gamma / \sim$ is a generalized r-gon.*

2. *G_α acts faithfully on $G_{\Gamma/\sim}$ for each vertex α in Γ.*

3. *If (G, Γ) is locally of finite Morley rank then $(\overline{G}, \Gamma / \sim)$ is of finite Morley rank, where $\overline{G}$ is the image of G in $\mathrm{Aut}(\Gamma / \sim)$.*

Space limitations make it impossible to give the definition of the terminology used in this fact but it should convince the reader that if the above mentioned conditions are satisfied in the amalgamated product then Theorem 6.1 and Fact 6.2 imply that there exists a group of finite Morley rank with a precise BN-pair structure and which has the same parabolic type as (*parabolic isomorphic to*) the simple K^*-group G of even type which is to be identified. As a result, a lengthy analysis ([25, 24]) following a close combination of those in [29] and [51] is carried out in order to check that a BN-pair of rank 2 can be constructed in $\tilde{G}$. This analysis involves graph theoretic arguments from [34] and [58], both parts of Theorem 5.3 and information about representations of SL_2 as important background material. It is worth emphasizing that in the context of groups of finite Morley rank there exists no representation theory of SL_2 comparable to the finite case. Moreover, representations are infinite dimensional over the prime field. As a result, one is led to replacing dimensions by the Morley rank and considering actions on elementary abelian 2-groups. This preparation is in [3], where it was used to prove the main ingredient for Theorem 5.3 which itself uses an amalgam-theoretic approach. A result of Timmesfeld ([55], Proposition 2.7) on quadratic modules is very helpful in this context. It should also be noted that despite the length of the arguments which are mostly faithful reproductions of those in the finite context, the complications (e.g. sporadics) which arise in the finite case disappear in the context of simple K^*-groups of even type.

When the hypotheses of Fact 6.2 are satisfied, one can apply the following theorem by Linus Kramer, Katrin Tent and Hendrik Van Maldeghem to the quotient $\overline{G}$ (same notation as in Fact 6.2) of the amalgam $\tilde{G}$ of P_1 and P_2 to obtain precise structural information:

Theorem 6.3 ([41], Theorem 3.14) *An infinite simple group of finite Morley rank with a spherical Moufang BN-pair of Tits rank 2 is either* $PSL_2(K)$, $Sp_4(K)$ *or* $G_2(K)$.

The Moufang condition required by Theorem 6.3 is satisfied in this context ([25]).

What remains to be done is to show that the identification of parabolics of G with those of $\overline{G}$ in fact yields an identification of G with the corresponding simple algebraic groups. After an analysis of the Cartan subgroups in G, one can use a recent result by Curtis Bennett and Sergey Shpectorov ([14]) to conclude. In this context the Bennett-Shpectorov theorem can be stated as follows:

Theorem 6.4 ([14]) *Let G be the simple K^*-group of even type to be identified. Let N be the normalizer of a Cartan subgroup C in the fixed standard Borel subgroup B of G and $N_i = N \cap P_i$. Suppose $\langle N_1, N_2 \rangle / C$ is isomorphic to the dihedral group D_{2k} with $k \leq n$, where n is the dihedral group parameter from the corresponding algebraic group (in this case $\overline{G}$). Then $k = n$ and $G \cong \overline{G}$.*

Hence, Theorem 1.1 is proven in this case.

6.2 Niles' theorem

This subsection contains the "generic" identification argument and covers the second case mentioned at the beginning of the section, namely the one where $\mathcal{M}_0$ has at least three elements. Unless otherwise specified G will continue to denote the simple K^*-group of even type to be identified.

The following theorem was proven by Berkman and Borovik.

Theorem 6.5 ([16], Theorem 1.1) *Let G be a group of finite Morley rank and even type. Also assume that G contains definable connected subgroups $P_1, \ldots, P_r$ which satisfy the following:*

1. *$G = \langle P_1, \ldots, P_r \rangle$.*

2. *All the P_i contain one fixed standard Borel subgroup B_G of G.*

3. *If G_i is defined as $\mathfrak{B}(P_i)$ then $\overline{G}_i = G_i/O_2^\circ(G_i) \cong \mathrm{SL}_2(F_i)$ with F_i algebraically closed of characteristic 2.*

4. *If $G_{ij} = \langle G_i, G_j \rangle$ then $\overline{G}_{ij} = O_2^\circ(G_{ij})$ is one of the following groups: $(P)\mathrm{SL}_3(F_{ij})$, $\mathrm{SL}_2(F_{ij}) \times \mathrm{SL}_2(F'_{ij})$, $\mathrm{Sp}_4(F_{ij})$, $\mathrm{G}_2(F_{ij})$ with F_{ij} and F'_{ij} algebraically closed of characteristic 2.*

Then $G_0 = \langle G_1, \ldots, G_r \rangle$ is normal in G and has a definable spherical BN-pair of Tits rank r.

An immediate corollary is that the simple K^*-group G of even type satisfies the hypotheses of Theorem 6.5 and contains a BN-pair of rank at least 3 when $|\mathcal{M}_0| \geq 3$. Then the following theorem completes the proof of Theorem 1.1 in the generic case.

Theorem 6.6 ([41], Theorem 5.1) *Let G be an infinite simple group of finite Morley rank with a definable spherical BN-pair of Tits rank at least 3. Then G is definably isomorphic to an algebraic group over an algebraically closed field.*

A few comments on Theorem 6.5 are in order. Theorem 6.5 uses the following general result by Niles:

Theorem 6.7 ([48], Theorem A) *Let G be an arbitrary group and assume that G satisfies the following conditions:*

1. *There exists a triple (B, N, S) which forms a pairwise BN-pair for G.*

2. *Set $H = B \cap N$. If X is an H-invariant subgroup of B such that $B = XB_s$ and $B = XB_t$, then $B = XB_{s,t}$.*

   Then (B, N, S) forms a BN-pair for G.

Here the notation is the standard one for BN-pairs ($B = B \cap B^s$, $B_t = B \cap B^t$ and $B_{s,t} = \bigcap_{x \in \langle s,t \rangle} B_x$). In a pairwise BN-pair, the BN-pair condition on double cosets is weakened to include only the inclusions involving 2-element subsets of S and the elements contained the subgroups of $N/B \cap N$ generated by these subsets.

Berkman and Borovik prove that (in the notation of Theorem 6.5), G_0 is normal in G and that it has a BN-pair structure in the following way: the Schur-Zassenhaus theorem allows to write $B_G = U \rtimes H$ where U is a Sylow$^\circ$ 2-subgroup of G. Then N_i is defined as $N_{G_i}(H)$ and B_i as $N_{G_i}(U)$ for $1 \leq i \leq r$. These definitions allow to define B and N: $B = \langle B_i : 1 \leq i \leq r \rangle$ and $N = \langle N_i : 1 \leq i \leq r \rangle$. The assumptions on G allow to prove first that $G_0 = \langle B, N \rangle$ and $G_0 \triangleleft G$, and then the pairwise BN-pair condition on G_0. Afterwards the condition 4 of Theorem 6.5 is used to check that the second condition of Theorem 6.7 is satisfied.

7 Acknowledgements

The author thanks Gregory Cherlin and Alexandre Borovik for their company without which most of the results discussed in this survey would not exist.

References

[1] T. Altınel. Groups of finite Morley rank with strongly embedded subgroups. *J. Algebra*, 180:778–807, 1996.

[2] T. Altınel, A. Borovik, and G. Cherlin. K^*-groups of finite Morley rank and of even type. preprint.

[3] T. Altınel, A. Borovik, and G. Cherlin. Pushing up and $C(G,T)$ in groups of finite Morley rank of even type. submitted.

[4] T. Altınel, A. Borovik, and G. Cherlin. Groups of mixed type. *J. Algebra*, 192:524–571, 1997.

[5] T. Altınel, A. Borovik, and G. Cherlin. On groups of finite Morley rank with weakly embedded subgroups. *J. Algebra*, 211:409–456, 1999.

[6] T. Altınel, A. Borovik, and G. Cherlin. Groups of finite Morley rank and even type with strongly closed abelian subgroups. *J. Algebra*, 2000. 420–461.

[7] T. Altınel and G. Cherlin. On central extensions of algebraic groups. *J. Symbolic Logic*, 67:68–74, 1999.

[8] T. Altınel, G. Cherlin, L.-J. Corredor, and A. Nesin. A Hall theorem for ω-stable groups. *J. London Math. Soc.*, (2) 57:385–397, 1998.

[9] C. Altseimer. A characterization of PSp(4,K). *Comm. Algebra*, 27:1879–1888, 1999.

[10] C. Altseimer. The thompson rank formula. Preprint of Manchester Centre for Pure Mathematics, 1999/4. 9pp.

[11] B. Baumann. Über endliche Gruppen mit einer zu $L_2(2^n)$ isomorphen Faktorgruppe. *Proc. Amer. Math. Soc.*, 74:215–222, 1979.

[12] O. V. Belegradek. On groups of finite Morley rank. In *Abstracts of the Eighth International Congress of Logic, Methodology and Philosophy of Science*, pages 100–102, Moscow, USSR, 17-22 August 1987. LMPS '87.

[13] H. Bender. Transitive Gruppe gerader Ordnung, in denen jede Involution genau einen Punkt festlasst. *J. Algebra*, 17:175–204, 1971.

[14] C. Bennett and S. Shpectorov. A remark on a theorem of J. Tits, 2000. preprint.

[15] A. Berkman. The classical involution theorem for groups of finite morley rank. submitted.

[16] A. Berkman and A. Borovik. An identification theorem for groups of finite Morley rank and even type. submitted.

[17] A. Borovik. Simple locally finite groups of finite Morley rank and odd type. In *Proceedings of NATO ASI on Finite and Locally Finite Groups*, pages 247–284, Istanbul, Turkey, 1994. NATO ASI.

[18] A. Borovik. Tame groups of odd and even type. In *Algebraic groups and their representations*, volume 517 of *NATO ASI Series C: Mathematical and Physical Sciences*, pages 341–366. Kluwer Academic Publisher, 1998.

[19] A. Borovik, G. Cherlin, and L.-J. Corredor. Parabolic 2-local subgroups in groups of finite morley rank of even type. Preprint.

[20] A. Borovik and A. Nesin. On the Schur-Zasssenhaus theorem for groups of finite Morley rank. *J. Symbolic Logic*, 57:1469–1477, 1992.

[21] A. Borovik and A. Nesin. Schur-Zassenhaus theorem revisited. *J. Symbolic Logic*, 59:283–291, 1994.

[22] A. V. Borovik and A. Nesin. *Groups of Finite Morley Rank*. Oxford University Press, 1994.

[23] A. V. Borovik and B. Poizat. Tores et *p*-groupes. *J. Symbolic Logic*, 55:565–583, 1990.

[24] The Consortium. Notes on a paper by Stellmacher, 2000. unpublished notes.

[25] The Consortium. Rank 2 amalgams of finite Morley rank. unpublished notes, 2000.

[26] M. Davis and A. Nesin. Suzuki 2-groups of finite Morley rank (or over quadratically closed fields). 1995. preprint.

[27] M. DeBonis and A. Nesin. On CN-groups of finite Morley rank. *J. London Math. Soc.*, (2) 50:532–546, 1994.

[28] M. DeBonis and A. Nesin. On split Zassenhaus groups of mixed characteristic and of finite Morley rank. *J. London Math. Soc.*, (2) 50:430–439, 1994.

[29] A. L. Delgado and B. Stellmacher. Weak (B, N)-pairs of rank 2. In *Groups and graphs: results and methods*, volume 6 of *DMV Seminar*, pages 58–244. Birkhauser, 1985.

[30] O. Frécon. Propriétés locales et sous-groupes de Carter dans les groupes de rang de Morley fini. submitted.

[31] O. Frécon. *Etude des groupes résolubles de rang de Morley fini*. PhD thesis, Université Claude Bernard-Lyon I, 2000.

[32] O. Frécon. Sous-groupes anormaux dans les groupes de rang de Morley fini résolubles. *J. Algebra*, 229:118–152, 2000.

[33] O. Frécon. Sous-groupes de Hall généralisés dans les groupes résolubles de rang de Morley fini. *J. Algebra*, 233:253–286, 2000.

[34] D. M. Goldschmidt. Automorphisms of trivalent graphs. *Ann. Math.*, 111:377–406, 1980.

[35] D. Gorenstein. *Finite Groups*. Chelsea Publishing Company, New York, 1980.

[36] E. Jaligot. FT-groupes. Prépublications de l'Institut Girard Desargues UPRES-A 5028, N°33. Lyon, Janvier 2000.

[37] E. Jaligot. Full Frobenius groups of finite Morley rank and the Feit-Thompson theorem. submitted.

[38] E. Jaligot. Groupes de type pair avec un sous-groupe faiblement inclus. to be published in the Journal of Algebra.

[39] E. Jaligot. *Contributions à la classification des groupes simples de rang de Morley fini.* PhD thesis, Université Claude Bernard-Lyon I, 1999.

[40] E. Jaligot. Groupes de type mixte. *J. Algebra*, 212:753–768, 1999.

[41] L. Kramer, K. Tent, and H. van Maldeghem. Simple groups of finite Morley rank and Tits buildings. *Israel J. Math*, 109:189–224, 1999.

[42] P. Landrock and R. Solomon. A characterization of the Sylow 2-subgroups of $PSU(3,2^n)$ and $PSL(3,2^n)$. Technical Report 13, Aarhus Universitet, Matematisk Institut, January 1975.

[43] J. McLaughlin. Some groups generated by transvections. *Arch. Math.*, 18:364–368, 1967.

[44] A. Nesin. Solvable groups of finite Morley rank. *J. Algebra*, 121:26–39, 1989.

[45] A. Nesin. On solvable groups of finite Morley rank. *Trans. Amer. Math. Soc.*, 321:659–690, 1990.

[46] A. Nesin. Generalized Fitting subgroup of a group of finite Morley rank. *J. Symbolic Logic*, 56:1391–1399, 1991.

[47] A. Nesin. Poly-separated and ω-stable nilpotent groups. *J. Symbolic Logic*, 56:694–699, 1991.

[48] R. Niles. BN-pairs and finite groups with parabolic-type subgroups. *J. Algebra*, 75:484–494, 1982.

[49] B. Poizat. *Groupes Stables.* Nur Al-Mantiq Wal-Ma'rifah, Villeurbanne, France, 1987.

[50] J.-P. Serre. *Trees.* Springer Verlag, Berlin-New York, 1980.

[51] B. Stellmacher. On graphs with edge-transitive automorphism groups. *Ill. J. Math*, 28:211–266, 1984.

[52] B. Stellmacher. Pushing up. *Arch. Math.*, 46:8–17, 1986.

[53] M. Suzuki. On a class of doubly transitive groups I. *Ann. Math.*, (2) 75:105–145, 1962.

[54] M. Suzuki. On a class of doubly transitive groups II. *Ann. Math.*, (2) 79:514–589, 1964.

[55] F. G. Timmesfeld. Groups generated by k-transvections. *Inventiones Math.*, 100:167–206, 1990.

[56] F. Wagner. Nilpotent complements and Carter subgroups in stable $\mathfrak{R}$-groups. *Arch. Math. Logic*, 33:23–34, 1994.

[57] F. Wagner. *Stable Groups*. Number 240 in London Mathematical Society Lecture Note Series. Cambridge University Press, Cambridge, UK, 1997.

[58] R. Weiss. Über lokal s-reguläre Graphen. *J. Comb. Th. Series B*, 20:124–127, 1976.

BN-pairs and
groups of finite Morley rank

Katrin Tent

Mathematisches Institut

Universität Würzburg

Am Hubland,

97074 Würzburg,

Germany

Abstract

We describe how the geometry of Tits buildings and generalized polygons enters the study of groups of finite Morley rank.

1 Introduction

Model theory is concerned with the study and classification of first order structures and their theories. Theories which were hoped to be particularly easy to classify are those having very few models up to isomorphism in 'most' infinite cardinalities. In the 1970's, Zil'ber started the program to classify all uncountably categorical theories, i.e. all theories having (up to isomorphims) exactly one model in every cardinality $\kappa \geq \aleph_1$ - an example being the theory of algebraically closed fields of some fixed characteristic.

It quickly turned out that in such theories, definable, and thus themselves uncountably categorical, groups entered the picture almost inevitably as so-called 'binding groups' between definable sets and hence it became particularly important to classify the definably simple uncountably categorical groups, i.e. groups without definable proper normal subgroups. These are exactly the simple groups of finite Morley rank.

2 Finite Morley Rank

The *Morley rank* is a model theoretic dimension on definable sets, which to some extent behaves like the algebraic dimension in the sense of algebraic geometry.

2.1 Definition Let L be a countable language, $\mathcal{M}$ a (saturated) L-structure.

I. The set $X \subseteq M^n$ is *definable*, if there is some L-formula $\varphi(x_1, \ldots, x_n, \bar{b})$ with $\bar{b} \subseteq M$ such that $X = \{\bar{a}; \ \mathcal{M} \models \varphi(\bar{a}, \bar{b})\}$.

II. The *Morley rank* $RM(X)$ of a definable set X is inductively defined as follows:

(i) $RM(X) \geq 0$ if and only if $X \neq \emptyset$;

(ii) $RM(X) \geq \alpha + 1$ for an ordinal α if and only if there are definable pairwise disjoint sets $X_i \subseteq X$ for $i < \omega$ with $RM(X_i) \geq \alpha$;

(iii) $RM(X) \geq \gamma$ for γ a limit ordinal if and only if $RM(X) \geq \alpha$ for all $\alpha < \gamma$.

We say that $RM(X) = \alpha$ if $RM(X) \geq \alpha$ and $RM(X) \not\geq \alpha + 1$. If $RM(X) = \alpha$, the Morley degree of X is the maximal (finite) number of pairwise disjoint sets $X_i \subseteq X$ with $RM(X_i) = \alpha$ We say that the structure $\mathcal{M}$ has finite Morley rank if *every* definable subset of M^n for each n has this property.

Clearly, the Morley rank of a definable set depends on the language L. Notice that for an infinite definable set, the Morley rank is always at least 1. An infinite definable set X is called *strongly minimal*, if every definable subset is either finite or co-finite, so its complement is finite, i.e. X does not contain disjoint definable subsets of Morley rank 1. So a strongly minimal set X has Morley rank and Morley degree 1. If you think of the Morley rank as measuring the complexity of the definable subsets of a given set, then the strongly minimal sets are the simplest infinite sets to be considered.

In the context of algebraic geometry, Morley rank and algebraic dimension coincide, and any infinite field definable in a structure of finite Morley rank is necessarily algebraically closed. This analogy has lead to the so-called Cherlin-Zil'ber Conjecture, viz. that the infinite simple groups of finite Morley rank are exactly the simple linear algebraic groups over algebraically closed fields, see also the survey by Altınel [A] about work on the Cherlin-Zil'ber Conjecture following the outline of (the revision of) the classification of the finite simple groups.

3   BN-pairs

If the Cherlin-Zil'ber Conjecture is true, then any infinite simple group of finite Morley rank must have a definable BN-pair, namely the standard BN-pair of an algebraic group G where B is a Borel subgroup of G, i.e. a maximal connected solvable subgroup of G, and N is the normalizer of a maximal torus T of G inside B. The *Weyl group* $W = B/T$ is a finite group generated by involutions.

As in the classification of the finite simple groups it hence is of particular importance to classify groups of finite Morley rank having a definable BN-pair. For this, we use the close connection between *buildings of spherical type* and BN-pairs, which was established by Tits [Ti].

3.1 Definition The subgroups B and N of a group G form a BN-pair if the following holds:

(i) $\langle B, N \rangle = G$

(ii) $T = B \cap N \lhd N$ and the *Weyl group* $W = N/T$ is generated by a set of distinguished involutions $S = \{s_1, \ldots, s_k\}$, where k is defined to be the *rank* of the BN-pair.

(iii) For all $v_i \in N$ with $v_i H \in S$ and $v \in N$, we have $vBv_i \subseteq BvB \cup Bvv_iB$.

(iv) $B^v \neq B$ for all $v \in N$ with $vH \in S$.

The BN-pair is called *spherical* if W is finite, and in this case obviously, $|W| = 2n$ for some n.

It is easy to see that the property of two definable subgroups of a given group to form a BN-pair is a first-order property (so the fact that they form a BN-pair is expressible by a first order sentence) if and only if W is finite, and it was shown in [TVM1] that for groups of finite Morley rank (and more generally, for *stable* groups) the Weyl group is necessarily finite. This allows us to talk about *definable BN-pairs* in groups of finite Morley rank.

Tits classified the buildings corresponding to spherical BN-pairs of Tits rank at least 3. He proved that these buildings are uniquely determined by their rank 2 residues, i.e. by polygons, and showed that only a very restricted class of polygons – all satisfying the Moufang condition – can come up as residues in buildings of higher rank. For polygons such a general classification is not possible. However, the Moufang condition is a heavy restriction, which allows Tits and Weiss to give a complete list, see [TW].

Since the Tits rank 2 case is crucial in the classification, we give a brief outline of the connection between spherical BN-pairs of Tits rank 2 and their corresponding buildings, which are exactly the *generalized polygons* see also [VM].

3.2 Polygons Let $n \geq 2$ be an integer. A bipartite graph $\mathfrak{P} = (\mathcal{P}, \mathcal{L}, \mathcal{I})$ with incidence relation $\mathcal{I}$ and corresponding distance function d is called a *generalized n-gon*, if it satisfies the following three axioms:

(*i*) For all elements $x, y \in \mathcal{P} \cup \mathcal{L}$ we have $d(x,y) \leq n$.

(*ii*) If $d(x,y) = k < n$, then there is a unique k-chain $(x_0 = x, x_1, \ldots, x_k = y)$ joining x and y (and we denote $x_1 = \mathrm{proj}_x y$).

(*iii*) $\mathfrak{P}$ is *thick*, i.e. every element $x \in \mathcal{P} \cup \mathcal{L}$ is incident with at least three other elements.

A pair (p, ℓ) with $p \mathcal{I} \ell$ is called a *flag*. It is easy to see that for $n = 2$, these axioms just define a complete bipartite graph, and for $n = 3$ these are precisely the axioms of a projective plane.

An *ordered ordinary n-gon* is a $2n$-chain $(x_0, \ldots, x_{2n} = x_0)$ of distinct elements, i.e. instead of (*iii*) above we require that every element is incident with exactly two elements. The set of elements at distance k from x is denoted by $D_k(x)$.

3.3 Connection between groups and polygons Let G be a group with an irreducible BN-pair of rank 2, and suppose that the associated Weyl group $N/B \cap N$ has order $2n$, for $n \geq 3$. Let G_a and G_ℓ be the proper parabolic subgroups of G containing B. We define an incidence structure on the coset spaces $\mathcal{P} = G/G_a$ and $\mathcal{L} = G/G_\ell$ by defining a point gG_a to be incident with a line $g'G_\ell$ if and only if $gG_a \cap g'G_\ell \neq \emptyset$. The axioms of a BN-pair yield that the incidence structure defined in this way is a generalized n-gon.

Clearly, G acts as a group of automorphisms on this generalized n-gon, and one can show that it acts transitively on the set of ordered ordinary n-gons contained in it.

For the converse direction, i.e. for getting from a generalized polygon to the BN-pair of Tits rank 2, let $\mathfrak{P} = (\mathcal{P}, \mathcal{L}, \mathcal{I})$ be a generalized n-gon and let (a, ℓ) be a flag. Suppose that a group $G \leq Aut(\mathfrak{P})$ acts transitively on the set of ordered ordinary n-gons. The corresponding BN-pair of G can be seen as follows: Let (a, ℓ) be a flag, and Γ an ordinary n-gon containing (a, ℓ). Then B is the stabilizer of (a, ℓ), N is the setwise stabilizer of Γ, and $\mathfrak{P}$ is isomorphic to the coset geometry $(G/G_a, G/G_\ell, \{(gG_a, gG_\ell); g \in G\})$, where G_a and G_ℓ denote the stabilizer in G of the elements a and ℓ, respectively. As before G_a and G_ℓ then form the parabolic subgroups of G containing B, and $T = B \cap N$ is the pointwise stabilizer of Γ. Note that since G is transitive on the set of ordered ordinary n-gons, B is transitive on the set of flags opposite (a, ℓ) (i.e., flags (x, y) such that x and y have distance n from either a or ℓ). The BN-pair *splits* if there is a normal nilpotent subgroup U of B with $B = U.T$.

3.4 Lemma *With the previous notation, the BN-pair splits if and only if there is a normal nilpotent subgroup U of B acting transitively on the set of flags opposite (a, ℓ).*

Proof. First assume the BN-pair splits as $B = U.T$ where T is the stabilizer of an ordinary n-gon containing (a, ℓ). Then U must be transitive on the flags opposite (a, ℓ) since B is. Conversely, suppose there is a nilpotent subgroup U of B acting transitively on the set of flags opposite (a, ℓ) and let T be the stabilizer of an ordinary n-gon Γ containing (a, ℓ). There is a unique flag (x, y) contained in Γ opposite (a, ℓ). By assumption. for every element $g \in B$, there is some $u \in U$ with $g(x, y) = u(x, y)$. Hence gu^{-1} fixes (x, y) and thus all of Γ. So $gu^{-1} \in T$, as required.

$\square$

We have thus seen that a BN-pair of Tits rank 2 with $|W| = 2n$ is equivalent to a generalized n-gon with an automorphism group transitive on ordered ordinary n-gons.

The *Moufang condition* is an even stronger condition on the automorphism group of a generalized n-gon.

3.5 Definition A generalized n-gon $\mathfrak{P}$ satisfies the Moufang condition if for every path $\alpha = (x_0, \ldots, x_n)$, the group of automorphisms U_α of $(\mathcal{P}, \mathcal{L}, \mathcal{I})$ fixing all elements incident with $x_1, \ldots, x_{n-1}$ acts transitively on the elements incident with x_0 different from x_1. In this case $\mathfrak{P}$ is called a *Moufang polygon*.

Similar to the case of projective planes, one can use coordinatization to see that U_α has to act freely on the elements incident with x_0 different from x_1 and that the definition is symmetric in the sense that U_α acts transitively on the elements incident with x_0 different from x_1 if and only if it acts transitively on the elements incident with x_n different from x_{n-1}.

Eventually, one would like to show that a *split* BN-pair of Tits rank 2 with $|W| = 2n$ is equivalent to a *Moufang* generalized n-gon. For the finite case, this was proved by Fong and Seitz, who also obtained a classification of the *finite* generalized Moufang polygons.

These geometric concepts are used in showing the following

3.6 Theorem [KTVM] *If G is an infinite simple group of finite Morley rank with an irreducible BN-pair of Tits rank ≥ 3, then G is (interpretably) isomorphic to a simple algebraic group over an algebraically closed field. If the BN-pair of G has Tits rank 2 and the associated polygon is a Moufang polygon, then the same is true. More precisely, in this case we have $G \cong PSL_3(K), PSp_4(K)$ or $G_2(K)$ for some algebraically closed field K.*

Since spherical irreducible buildings of Tits rank ≥ 3 are uniquely determined by their rank 2 residues, the crucial step in the proof of Theorem 3.6 was in fact the following theorem which uses the general classification of Moufang polygons by Tits and Weiss:

3.7 Theorem [KTVM] *If $\mathfrak{P}$ is an infinite Moufang polygon of finite Morley rank, then $\mathfrak{P}$ is either the projective plane, the symplectic quadrangle, or the split Cayley hexagon over some algebraically closed field.*

3.8 Remark: In particular, $\mathfrak{P}$ is an algebraic polygon as classified in [KT].

4   The Tits rank 2 case

Theorem 3.6 shows that it is left to consider the Cherlin-Zil'ber Conjecture for BN-pairs of Tits rank at most 2, without assuming the Moufang condition.

The conjecture implies in particular that any simple group must have a definable 'split' BN-pair. In the standard BN-pair, the solvable linear group B is triangulizable and splits as $B = U \rtimes T$, where U is a nilpotent normal subgroup of B consisting of strictly upper triangular matrices in B and the diagonalizable part T. More generally, a BN-pair is called *split* if there is a normal nilpotent subgroup U of B such that $B = U.(B \cap N)$.

Fong and Seitz showed in [FS] that a finite group with a split BN-pair of Tits rank 2 is essentially a group of Lie type. Their proof relies heavily on the result due to Feit and Higman that finite generalized polygons exist only for $n = 3, 4, 6$ and 8 and hence for the corresponding Weyl group W, we must have $|W| = 2n$ with $n \in \{3, 4, 6, 8\}$. Furthermore, it uses the classification of finite split BN-pairs of Tits rank 1 (i.e. the finite split 2-transitive groups) due to Kantor, Hering and Seitz, an important result in the theory of finite groups which has no known analog in the finite Morley rank situation and even less so in the general infinite case.

In fact, as far as the Feit-Higman result on finite polygons is concerned one can rather show that there does not exist an analogue for the infinite case - not even if one adds the condition of finite Morley rank:

4.1 Theorem [Te1] *For all $n \geq 3$, there are $2^{\aleph_0}$ many generalized n-gons of finite Morley rank not interpreting any infinite group. Furthermore, there are generalized n-gons of finite Morley rank for all $n \geq 3$ whose automorphism groups act transitively on the set of ordered $n + 1$-gons.*

In these examples the point rows and line pencils are strongly minimal, and the examples are in fact almost strongly minimal; i.e. in the definable closure of a strongly minimal set. The Hrushovski-style construction proceeds via a combination of free extensions and amalgamations, see also [Ba] and [Po] using the fact that generalized n-gons for fixed n are first-order structures in a natural way. As explained in 3.3, the automorphism group of such a strongly homogeneous polygon has a BN-pair. However, since these examples do not interpret any infinite group, the automorphism groups are not definable and do not have finite Morley rank. In particular all root groups must be finite, showing these examples to be as far as possible from the Moufang polygons.

This is a new class of examples for polygons and BN-pairs even without the condition on finite Morley rank, extending examples obtained by Tits [Ti] who constructed generalized polygons having BN-pairs for all n. However, the attempt eg. in [FKS] to refine that construction to obtain $n + 1$-gon transitive generalized n-gons had failed, see [Was]. These new examples thus show once again that without further restrictions one cannot expect to classify infinite groups with BN-pairs of rank 2.

In view of the close connection between generalized polygons and their automorphism groups it is natural to consider the question whether the construction can be modified in such a way to obtain new infinite groups of finite Morley rank - albeit no new simple ones: it follows from a result of Pillay that groups obtained from Hrushovski's amalgamation construction will be nilpotent-by-finite (see [Bau]). The following result shows that, except in the algebraic cases one cannot obtain definable automorphism groups for generalized polygons as long as the point rows and line pencils have Morley rank 1.

4.2 Theorem [TVM1] *If $\mathfrak{P}$ is a generalized n-gon with strongly minimal point rows and line pencils, $n \geq 3$, and $G \leq Aut(\mathfrak{P})$ is a group of finite Morley rank which acts transitively and definably on the set of ordered ordinary n-gons contained in $\mathfrak{P}$, then one of the following holds:*

   (i) $(n = 3)$, G is definably isomorphic to $PSL_3(K)$ for some algebraically closed field K, and $\mathfrak{P}$ is the projective plane over K;

   (ii) $(n = 4)$, G is definably isomorphic to $PSp_4(K)$ and $\mathfrak{P}$ is the symplectic quadrangle over K;

   (iii) $(n = 6)$, G is definably isomorphic to $G_2(K)$ and $\mathfrak{P}$ is the split Cayley hexagon over K.

For the proof, we make essential use of a result of Hrushovski's which characterizes transitive effective permutation groups of finite Morley rank on strongly minimal sets: by his result, the *Levi factors* of G, i.e. the permutation groups induced by G on the point rows and line pencils of $\mathfrak{P}$, must be permutation equivalent to either the affine group $K^+ \rtimes K^*$ or to the simple algebraic group $PSL_2(K)$ for some algebraically closed field K. We use this information to derive the Moufang property for $\mathfrak{P}$. While in light of Theorem 3.6 this would already be sufficient to obtain the result, we here use geometric information to show directly that $\mathfrak{P}$ is isomorphic to one of the three cases mentioned in the theorem. We would like to stress the fact that our approach does not need the full classification of Moufang polygons (in particular the yet unpublished parts about quadrangles can be avoided).

In the language of groups we can state the theorem slightly more generally as follows:

4.3 Theorem [TVM1] *If G is a group with an irreducible BN-pair of (Tits-) rank 2, such that the Levi factors are permutation equivalent to either PSL_2 or AGL_1 for a field with at least 4 elements, then the corresponding generalized polygon is Moufang and G contains its little projective group.*

In particular, if G has finite Morley rank and the BN-pair is definable, then G is definably isomorphic to $PSL_3(K)$, $PSp_4(K)$, or $G_2(K)$ for some algebraically closed field K.

5 'Split' BN-pairs

In order to make progress on BN-pairs of Tits rank 2 and of finite Morley rank, we use the connections between BN-pairs and generalized polygons.

As the examples in Theorem 4.1 show, for infinite groups with BN-pairs of Tits rank 2 and $|W| = 2n$, all n are possible for the order of the corresponding Weyl group, even in the finite Morley rank situation. This contrasts sharply with the situation where the BN-pair is split or even just 'weakly split' as in the following theorem:

5.1 Theorem [TVM2] *Let G be a simple group with a split BN-pair of rank 2 where $B = U.T$ for $T = B \cap N$ and a normal subgroup U of B with $Z(U) \neq 1$. Then the Weyl group $W = N/B \cap N$ has cardinality $2n$ with $n = 3, 4, 6, 8$ or 12.*

Furthermore, if G acts transitively on the ordinary $(n+1)$-gons of the corresponding generalized n-gon $\mathfrak{P}$, then $\mathfrak{P}$ is a Moufang n-gon for $n = 3, 4$ or 6.

Note that by the second part of the theorem, the examples in 4.1 can never have split BN-pairs. The proof consists of two steps. First one shows that if B is the stabilizer of the flag (p, ℓ) in G, then elements of $Z(U)$ must fix $D_k(p) \cup D_k(\ell)$ for $k < \frac{n}{2}$. The existence of such elements is then sufficient to imply that $n = 3, 4, 6, 8$ or 12. The second part is proved using the transitivity of the root elations coming from the transitivity assumption.

Theorem 5.1 can be applied to obtain the following results for the weakly split case:

5.2 Theorem (cf. [Te2]) *Let G be a simple group of finite Morley rank with a definable BN-pair of rank 2 where either B is solvable or $B = U.T$ for*

$T = B \cap N$ and a normal subgroup U of B with $Z(U) \neq 1$. Let $n = |W|/2$ and let Z denote the center of the commutator subgroup of the connected component of B or $Z(U)$, respectively.

(i) If $n = 3$, then G is definably isomorphic to $PSL_3(K)$ for some algebraically closed field K.

(ii) If Z contains a B-minimal subgroup A with $RM(A) \geq RM(P_i/B)$ for both parabolic subgroups P_1 and P_2, then $n = 3, 4$ or 6 and G is definably isomorphic to $PSL_3(K), PSp_4(K)$ or $G_2(K)$ for some algebraically closed field K.

Note that all of these conditions are necessarily satisfied if the Cherlin-Zil'ber Conjecture is true.

Finally, we obtain the following classification of split BN-pairs:

5.3 Theorem [Te2] *Let G be a simple group of finite Morley rank with a definable split BN-pair of Tits rank 2. Then $|W| = 2n$ with $n \in \{3, 4, 6, 8\}$ and if $|W| \neq 16$, then G is definably isomorphic to $PSL_3(K), PSp_4(K)$ or $G_2(K)$ for some algebraically closed field K.*

Most of the arguments also work in the general context and so we can almost completely delete the finiteness assumption in the theorem of Fong and Seitz:

5.4 Theorem [TVM3] *Let G be a group with an irreducible spherical split BN-pair of rank 2. Let $\mathfrak{P}$ be the generalized n-gon associated to this (B,N)-pair and let W be the associated Weyl group. If $|W| \neq 16$, $\mathfrak{P}$ is a Moufang polygon and G/R contains its little projective group, where R denotes the kernel of the action of G on $\mathfrak{P}$.*

With the Cherlin-Zil'ber Conjecture in mind, the next step should be to show that a BN-pair of finite Morley rank where B is solvable splits necessarily.

A first result in that direction is the following.

5.5 Proposition [Te3] *Let G be a simple infinite group of finite Morley rank with a definable BN-pair of Tits rank 2 where B is solvable. If G acts transitively on the ordinary $(n+1)$-gons in the corresponding generalized n-gon $\mathfrak{P}$, then the BN-pair splits and G is definably isomorphic to $PSL_3(K), PSp_4(K)$ or $G_2(K)$ for some algebraically closed field K.*

Proof. Let G and $\mathfrak{P}$ be as in the statement of the proposition, and let (p, ℓ) be the flag in $\mathfrak{P}$ stabilized by B. Because the set of flags opposite (p, ℓ)

has Morley degree 1 by [KTVM] 2.11, the connected component B^0 of B acts also transitively on this set. As B^0 is solvable, the commutator subgroup $(B^0)'$ is nilpotent. We will show the BN-pair splits by showing that $(B^0)'$ acts transitively on the flags opposite (p, ℓ), so we can apply the previous theorem and Theorem 3.6 to obtain the desired conclusion.

Let (x, y) be a flag opposite (p, ℓ), and let T be the stabilizer of (x, y) in B, so T stabilizes the ordinary n-gon $\Gamma = (p = x_0, \ell = x_1, \ldots x_n = x, x_{n+1} = y, \ldots x_{2n} = x_0)$ defined by the flags (p, ℓ) and (x, y). Note that the transitivity of G on ordinary $n + 1$-gons implies that T acts transitively on $D_1(x_i) \setminus \{x_{i-1}, x_{i+1}\}$ for $i = 1, \ldots 2n$, where $x_{2n+1} = x_1$. (As with the following arguments it might be helpful to draw a sketch.) Again, because the Morley degree of $D_1(x)$ is 1 by [KTVM] 2.11, we conclude that also T^0 acts transitively on $D_1(x_i) \setminus \{x_{i-1}, x_{i+1}\}$ for $i = 1, \ldots 2n$.

Let $t \in T^0$ and flags $(x_1, y_1), (x_2, y_2)$ opposite (p, ℓ) with $t(x_1, y_1) = (x_2, y_2)$ be given. Since B^0 is transitive on flags opposite (p, ℓ), there is some $g \in B^0$ with $g(x, y) = (x_1, y_1)$, and so $g^{-1}t^{-1}gt \in (B^0)'$ with $t(g(t^{-1}(g^{-1}(x_1, y_1))))) = (x_2, y_2)$.

Now let $(p_1, q_1), (p_2, q_2)$ be arbitrary flags opposite (p, ℓ) determining ordinary n-gons Γ_1 and Γ_2, respectively. Let ℓ_1 and ℓ_2, respectively, be the elements of Γ_1 and Γ_2, respectively, incident with p and different from ℓ, and let p^1, p^2 be the elements of Γ_1 and Γ_2, respectively, incident with ℓ and different from p. By conjugating T if necessary, we may assume that p and ℓ are the only elements that Γ has in common with either Γ_1 and Γ_2. As T^0 acts transitively on $D_1(p) \setminus \{\ell, x_{2n-1}\}$ and on $D_1(\ell) \setminus \{x, x_2\}$, there is some element $t \in T^0$ such that $t(\ell_1) = \ell_2$ and $t(p^1) = p^2$. If $t(p_1, q_1) \neq (p_2, q_2)$, we now take a conjugate T' of T fixing an ordinary n-gon Γ_3 having exactly the path (ℓ_2, p, ℓ, p^2) in common with Γ_2. Then by the same transitivity argument, T'^0 contains some element t_0 with $t_0(t(p_1, q_1)) = (p_2, q_2)$.

By the argument in the previous paragraph, there are elements $g, g' \in (B^0)'$ with $g(p_1, q_1) = t(p_1, q_1)$ and $g'(t(p_1, q_1)) = t_0(t(p_1, q_1))$. Thus we have $gg' \in (B^0)'$ with $gg'((p_1, q_1)) = (p_2, q_2)$, showing the transitivity of $(B^0)'$ on flags opposite (p, ℓ).

$\square$

References

[A] T. Altınel, Simple groups of finite Morley rank of even type, this volume.

[Ba] J. Baldwin, Rank and homogeneous structures, this volume.

[Bau] A. Baudisch, CM-trivial stable groups, this voloume.

[FKS] M. Funk, O.H.Kegel, K.Strambach, Gruppenuniversalität und Homogenisierbarkeit, *Ann. Mat. Pura Appl* 141 (1985)1–126.

[FS] P. Fong, G. Seitz, Groups with a (B, N)-pair of rank 2. I, II. *Invent. Math.* **21** (1973), 1 – 57; ibid. **24** (1974), 191 – 239.

[KT] L. Kramer, K. Tent, Algebraic polygons, *J. Algebra* (1996) 182, 435–447.

[KTVM] L. Kramer, K. Tent, H. Van Maldeghem, Simple groups of finite Morley Rank and Tits buildings, *Israel Journal of Mathematics* 109 (1999) 189–224.

[Po] B. Poizat, Amalgames de Hrushovski, this volume.

[Te1] K. Tent, Very homogeneous generalized n-gons of finite Morley rank, *J. London Math. Soc.* (2) **62** (2000), no. 1, 1–15.

[Te2] K. Tent, Split BN-pairs of finite Morley Rank, to appear in *Ann. Pure Appl. Logic.*

[Te3] K. Tent, BN-pairs of finite Morley Rank where B is solvable, in preparation.

[TVM1] K. Tent, H. Van Maldeghem, BN-pairs with affine or projective lines, to appear in *J. Reine Angew. Math.*

[TVM2] K. Tent, H. Van Maldeghem, Irreducible BN-pairs of rank 2, to appear in *Forum Math.*

[TVM3] K. Tent, H. Van Maldeghem, Moufang Polygons and Irreducible Spherical BN-pairs of rank 2, submitted.

[Ti] J. Tits, *Buildings of spherical type and finite BN-pairs*, Lecture Notes in Math. **386**, Springer (1974).

[Ti] J. Tits, Endliche Spiegelungsgruppen, die als Weylgruppen auftreten, Inventiones Mathematicae 43 (1977), 283–295.

[TW] J. Tits, R. Weiss, *The classification of Moufang polygons*, in preparation.

[VM] H. Van Maldeghem, Introduction to polygons, this volume.

[Was] J. Wassermann, Freie Konstruktionen von verallgemeinerten n-Ecken, Diplomarbeit, Universität Tübingen 1993.

CM-trivial stable groups

Andreas Baudisch

1 Introduction

Let T be a complete countable first order theory. We work in a monster model $\mathcal{C}$ of T (see [5]). All the models of T that we shall consider will be small elementary submodels of $\mathcal{C}$; similarly, we shall consider only subsets of $\mathcal{C}$ which are small in comparison to its power. We assume that T is stable. This provides us an independence relation $B \underset{A}{\downarrow} C$ between subsets A, B, C of $\mathcal{C}$: that we read B and C are independent over A. CM-triviality is a geometrical property of this relation $\downarrow$. In this short survey we will consider the role of CM-triviality with respect to Zil'ber's Conjecture, the influence of CM-triviality on the algebraic structure of stable groups and the preservation of CM-triviality by Mekler's construction.

2 Preliminaries and examples

We denote $\mathrm{Aut}_{(A)}(\mathcal{C})$ the set of all automorphisms of $\mathcal{C}$ that fix A pointwise. Then T is stable if and only if there exists an independence relation $B \underset{A}{\downarrow} C$ between subsets B, A, and C of $\mathcal{C}$ characterized by the following properties

0. $\downarrow$ is invariant by automorphism.

1. <u>Symmetry</u> $B \underset{A}{\downarrow} C$ iff $C \underset{A}{\downarrow} B$.

2. <u>Transitivity</u> $B \underset{A}{\downarrow} C_1 \cup C_2$ iff $B \underset{A \cup C_1}{\downarrow} C_2$ and $B \underset{A}{\downarrow} C_1$.

3. <u>Extension</u> For A, B, C there is some $f \in \mathrm{Aut}_{(A)}(\mathcal{C})$ such that $f(B) \underset{A}{\downarrow} C$.

4. <u>Local Character</u> Assume a finite sequence $\bar{b}$ and C be given. Then there is some $A \subseteq C$ such that $|A| \leq |T| \ (= \aleph_0)$ and $\bar{b} \underset{A}{\downarrow} C$.

5. <u>Finite Character</u> $B \underset{A}{\downarrow} C$ iff for all finite sequences $\bar{b}$ in B and $\bar{c}$ in C we have $\bar{b} \underset{A}{\downarrow} \bar{c}$.

6. <u>Boundedness</u> Let $\bar{b}$ be finite. $\Omega = \{f(\bar{b}) : f \in \mathrm{Aut}_{(A)}(\mathcal{C}) \text{ and } f(\bar{b}) \underset{A}{\downarrow} C\}$ splits at most into $2^{|T|}$ many orbits under the action of $\mathrm{Aut}_{(C)}(\mathcal{C})$ on the same orbit. If $A = M$ is a model of T, then all elements of Ω are on one orbit.

If $f \in \mathrm{Aut}(\mathcal{C})$, then we define $f(\mathrm{tp}(B/C))$ to be

$$\{\varphi(\bar{x}_{\bar{b}}, f(\bar{c})) : \mathcal{C} \vDash \varphi(\bar{b}, \bar{c}) \text{ for } \bar{b} \text{ in } B \text{ and } \bar{c} \text{ in } C\}$$

where $\bar{b} \rightsquigarrow \bar{x}_{\bar{b}}$ is a bijection between B and a set of variables. If T is stable, then $B \underset{A}{\downarrow} C$ can be defined by the following property:

For every sequence of automorphisms $f_0, f_1, \dots$ in $\mathrm{Aut}_{(A)}(\mathcal{C})$ ($i < \omega$) such that $\{f_i(C) : i < \omega\}$ is indiscernible over A, then $\bigcup_{i<\omega} f_i(\mathrm{tp}(B/A \cup C))$ is consistent.

The definition above rests on the fact that, in the stable case forking is equivalent to dividing. Furthermore note that $\mathrm{acl}(A)$ is the union of all finite sets that are definable over A. Next we will give some examples to illustrate the notions above.

Example 1. To formulate the elementary theory T of an infinite dimensional vector space over a field F we use a language with $+$, $-$, 0 and a unary function $f_a(x)$ for every element a of the field F. In the models of T $f_a(v)$ is the scalar product of the vector v with the field element a.

Let A, B, C subsets of the monster model $\mathcal{C}$ of T. Then $\mathrm{acl}(A)$ is the linear hull of A. The theory T is ω-stable and $B \underset{A}{\downarrow} C$ is the linear independence. It is a well-known property of vector spaces that

$$B \underset{\mathrm{acl}(B)\cap\mathrm{acl}(C)}{\downarrow} C$$

for all B and C. Therefore T is 1-based. For the general case we have to modify this definition.

Example 2. We choose a language of a binary relation $E(x, y)$. Let T be the theory of an equivalence relation with infinitely many classes, each class infinite. Then T is ω-stable and $\mathrm{acl}(A) = A$ for all A.

Assume we have two different elements $a \neq b$ in $\mathcal{C}$ with $\mathcal{C} \vDash E(a, b)$. Then $\mathrm{acl}(a) = \{a\}$, $\mathrm{acl}(b) = \{b\}$, $\mathrm{acl}(a) \cap \mathrm{acl}(b) = \emptyset$, and $a \underset{\emptyset}{\not\downarrow} b$. But T is also 1-based since this property is defined in T^{eq}. (In the first example it was sufficient to consider T because of the weak elimination of imaginaries for T.)

T^{eq} was introduced by S. Shelah [14]. For every equivalence relation $\Delta(\bar{x}, \bar{y})$ on $\mathcal{C}^n$ ($n < \omega$), that is definable without parameters, a new sort $\mathcal{C}^n/\Delta$ of the Δ-classes and a unary function f_Δ with $f_\Delta(\overline{a}) = \overline{a}/\Delta$ for $\overline{a}$ in $\mathcal{C}$ are introduced. In this way we obtain a many-sorted structure $\mathcal{C}^{\mathrm{eq}}$ from $\mathcal{C}$. T^{eq} is the theory of C^{eq} in a suitable many-sorted language L^{eq}. In our Example 2 we have a new sort $\mathcal{C}/E$. For the chosen $a \neq b$ with $\mathcal{C} \vDash E(a,b)$ we have $\{a/E\} = \{b/E\}$ as the intersection of $\mathrm{acl}^{\mathrm{eq}}(\{a\})$ and $\mathrm{acl}^{\mathrm{eq}}(\{b\})$ and $a \underset{a/E}{\overset{\smile}{\mid}} b$.

T is 1-based in the sense of the following definition:

Definition. T is 1-based, if for all sets B and C in $\mathcal{C}$ we have

$$B \underset{\mathrm{acl}^{\mathrm{eq}}(B) \cap \mathrm{acl}^{\mathrm{eq}}(C)}{\overset{\smile}{\mid}} C \,.$$

Note that we can show that we get the same notion if B and C are from $\mathcal{C}^{\mathrm{eq}}$.

Furthermore it is well-known that a stable theory T is 1-based if and only if for all models $M \preceq \mathcal{C}$ and $\overline{c}$ in $\mathcal{C}$ we have $\mathrm{Cb}(\overline{c}/M) \subseteq \mathrm{acl}^{\mathrm{eq}}(\overline{c})$.

Example 3. We choose the language of field theory with $+$, $-$, $\cdot$, 0, 1. Let T be the elementary theory of an algebraically closed field. T is again ω-stable. Then $\mathrm{acl}(A)$ is the algebraic closure of A in the sense of algebraic geometry. The relation $B \underset{A}{\overset{\smile}{\mid}} C$ is algebraic independence. As in Example 1 T^{eq} is not necessary. (We have elimination of imaginaries.) T is the standard example of a theory that is not 1-based. To show this we choose a, b, c in $\mathcal{C}$ algebraically independent. Now we consider $\mathcal{C}^2$. We can consider $\{a, b\}$ as a code of the line $y = ax + b$. If $d = ac + b$, then (c, d) is a generic point of this line. We can show $\mathrm{acl}(\{a, b\}) \cap \mathrm{acl}(\{c, d\}) = \emptyset$. Furthermore we have $\{a, b\} \underset{\emptyset}{\overset{\smile}{\nmid}} \{c, d\}$.

3 Zil'ber's Conjecture

Definition. T is λ-categorical, if all models of T of power λ are isomorphic.

λ-categorical theories with $\lambda > \aleph_0$ are ω-stable of finite Morley rank. This was shown in M. Morley's seminal paper [9].

Theorem 3.1 (M. Morley) *T is λ-categorical for some $\lambda > \aleph_0$ if and only if T is λ-categorical for all $\lambda > \aleph_0$.*

The theories in Examples 1 and 3 are uncountably categorical, but not in Example 2.

For a long time the theories of algebraically closed fields and theories that are closely associated to them were the only known examples of uncountably categorical theories that are not 1-based. Therefore B. Zil'ber conjectured [16]:

Zil'ber's Conjecture. Let T be an uncountably categorical theory that is not 1-based. Does T allow the interpretation of an infinite field?

E. Hrushovski refuted this conjecture:

Theorem 3.2 (E. Hrushovski [6]) *There are $2^{\aleph_0}$ many uncountably categorical theories of structures that are not 1-based and do not even allow the interpretation of an infinite group.*

To prove Theorem 3.2 E. Hrushovski considers structures with a ternary relation. He equips them with a predimension that is essential for the forking-geometry of the final structure. He obtains this desired structure by a modified version of Fraïssé's amalgamation procedure. His ideas have been used by many authors to produce structures with unusual model-theoretic behaviour (cf. [13] in this volume). The only counterexample to Zil'ber's conjecture in group theory which is known so far is also constructed by the use of a complicated amalgamation of bilinear maps:

Theorem 3.3 ([2]) *There are $2^{\aleph_0}$ many uncountably categorical theories of groups that are not 1-based and do not allow the interpretation of an infinite field.*

4   Definition of CM-triviality

All known counterexamples to Zil'ber's Conjecture are CM-trivial. CM-triviality is a geometric property of $\underset{}{\cancel{\smile}}$ introduced by E. Hrushovski in [6]. If T allows the interpretation of a field, then T is not CM-trivial. Hence CM-triviality can be used to show the absense of a field.

Definition (in T^{eq}). T is CM-trivial, if after naming any set of parameters for all a, b, c in $\mathfrak{C}^{\mathrm{eq}}$
if $\operatorname{acl}^{\mathrm{eq}}(a) \cap \operatorname{acl}^{\mathrm{eq}}(b) = \operatorname{acl}^{\mathrm{eq}}(\emptyset)$,
     $\operatorname{acl}^{\mathrm{eq}}(ab) \cap \operatorname{acl}^{\mathrm{eq}}(ac) = \operatorname{acl}^{\mathrm{eq}}(a)$, and $c \underset{b}{\cancel{\smile}} a$,

then $c \underset{\emptyset}{\cancel{\smile}} a$.

Lemma 4.1 *If T is 1-based, then T is CM-trivial.*

Proof. Assume the situation above. Since T is 1-based $\mathrm{acl}^{\mathrm{eq}}(a) \cap \mathrm{acl}^{\mathrm{eq}}(b) = \mathrm{acl}^{\mathrm{eq}}(\emptyset)$ implies $a \underset{\emptyset}{\downarrow} b$. By transitivity $a \underset{b}{\downarrow} c$ and $a \underset{\emptyset}{\downarrow} b$ implies $a \underset{\emptyset}{\downarrow} cb$ and therefore $a \underset{\emptyset}{\downarrow} c$.

It is easily seen that a stable structure with a definable field F is not CM-trivial. We get a counterexample if we choose

a as a code of a generic plane in F^3,

b as a code of a generic line in the plane that belongs to a, and

c as a code of a generic point of that line.

For the details look at A. Pillay's paper [10]. There is also shown:

Lemma 4.2 *Let T be stable. T is CM-trivial if and only if for all $M \preceq N \preceq \mathfrak{C}$ and all $\bar{c}$ in $\mathfrak{C}$ with $\mathrm{acl}^{\mathrm{eq}}(\bar{c}M) \cap N^{\mathrm{eq}} = M^{\mathrm{eq}}$ we have $\mathrm{Cb}(\mathrm{tp}(\bar{c}/M)) \subseteq \mathrm{acl}^{\mathrm{eq}}(\mathrm{Cb}(\mathrm{tp}(\bar{c}/N)))$.*

Since all counterexamples to Zil'ber's conjecture are CM-trivial it is an open question whether uncountably categorical theories (or theories of finite Morley rank) that are not CM-trivial allow the interpretation of an infinite field. We only know:

Theorem 4.3 ([4] with A. Pillay) *There is a ω-stable structure that is not CM-trivial and does not allow the interpretation of a field. But it has infinite Morley rank.*

5 CM-trivial stable groups

Now we will concentrate on stable groups. They are stable structures with a group operation. But they can have much more structure. The geometrical properties defined above have a strong influence on the algebraic properties of these groups:

Theorem 5.1 (E. Hrushovski, A. Pillay [7]) *Stable 1-based groups are abelian by finite. In such a group definable sets are boolean combinations of cosets of $\emptyset$-definable subgroups.*

Theorem 5.2 (A. Pillay [10]) *A ω-stable CM-trivial group of finite Morley rank is nilpotent by finite and of bounded exponent up to an abelian direct summand.*

Theorem 5.3 (F. Wagner [15]) *Let G be a group with a stable CM-trivial theory. If G is soluble, then G is nilpotent by finite. If G is torsionfree and $\mathrm{Th}(G)$ is small, then G is abelian and connected.*

The counterexamples in Theorem 3.3 are the only known examples of ω-stable groups of finite Morley rank that are CM-trivial but not abelian by finite. By Theorem 5.2 all such groups are nilpotent by finite.

We give a short outline of A. Pillay's proof of Theorem 5.2.
Let G be a ω-stable CM-trivial group of finite Morley rank that is not nilpotent by finite. W.l.o.g. G is connected. If G were solvable, then an infinite field would be definable in it by B. Zil'ber [17], contradicting CM-triviality. Hence G is non-solvable. By the same argument every solvable connected subgroup of G is nilpotent. Therefore G is a bad group. Let us choose G with these properties of minimal finite Morley rank. By the known results G is simple (see [12]). By the analysis of simple bad groups by Corredor, Borovik and Poizat, and Cherlin and Nesin we obtain the existence of a proper definable, connected and selfnormalizing subgroup H of G such that distinct conjugates H^g of H are pairwise disjoint over $\{1\}$. Then A. Pillay uses these properties of H to construct a counterexample to CM-triviality.

6   Mekler's Construction

For every structure M of finite signature A.H. Mekler [8] has constructed a group $G(M)$ with the same stability spectrum. Since the theory of every structure of finite signature is biinterpretable (see [5]) with a theory of a nice graph he has restricted himself to these graphs.

A nice graph is a structure with only one binary symmetric and irreflexive relation $R(x, y)$ such that

   i) $\exists x_0 x_1 (x_0 \neq x_1)$,

   ii) $\forall x_0 x_1 \exists y (x_0 \neq x_1 \rightarrow y \neq x_0 \wedge y \neq x_1 \wedge R(x_0, y) \wedge \neg R(x_1, y))$,

   iii) no triangles,

   iv) no squares.

Now let p be prime greater 2. Let S^* be the theory of nice graphs.

Theorem 6.1 (A.H. Mekler [8]) *There are a uniform construction of groups $G(M)$ for every nice graph M, a group theory T^*, and an interpretation Γ of S^* in T^* such that:*

i) T^* *is a theory of nilpotent groups of class* 2 *and of exponent* p.

ii) $G \vDash T^*$ *if and only if there is some* $M \vDash S^*$ *such that* $G(M) \equiv G$.

iii) *For* S^*-*models* M *and* N *we have* $M \equiv N$ *if and only if* $G(M) \equiv G(N)$.

iv) $\Gamma(G(M)) \cong M$.

v) $\operatorname{Th}(M)$ *is* λ-*stable if and only if* $\operatorname{Th}(G(M))$ *is* λ-*stable*.

Note that the theories of Mekler groups are never $\aleph_0$-categorical. Furthermore they are all of infinite rank. On the other side we can show [3]:

Theorem 6.2

vi) $\operatorname{Th}(M)$ *is simple if and only if* $\operatorname{Th}(G(M))$ *is simple*.

vii) $\operatorname{Th}(M)$ *is stable and* CM-*trivial if and only if* $\operatorname{Th}(G(M))$ *is stable and* CM-*trivial*.

Note that by v) $\operatorname{Th}(M)$ is stable if and only if $\operatorname{Th}(G(M))$ is stable. In the proof of vii) it is easily seen that CM-triviality of $\operatorname{Th}(G(M))$ implies the CM-triviality of $\operatorname{Th}(M)$. This follows since for every saturated model G of T^* any automorphism of the nice graph $\Gamma(G)$ can be lifted to an automorphism of G. The proof of the other direction is rather complicated. We use the definition of CM-triviality given by Lemma 4.2.

We assume that $G \preceq H \preceq \mathcal{C}$ are saturated models of $\operatorname{Th}(G(M))$ and A is a finite subgroup of the monster model $\mathcal{C}$ such that $\operatorname{acl}^{\operatorname{eq}}(A \cup G) \cap H^{\operatorname{eq}} = G^{\operatorname{eq}}$. For all possible positions of A with respect to G we describe $\operatorname{Cb}(\operatorname{tp}(A/G))$ and show that it is contained in $\operatorname{acl}^{\operatorname{eq}}(\operatorname{Cb}(\operatorname{tp}(A/H)))$. An essential part of $\operatorname{Cb}(\operatorname{tp}(A)/G)$ is the canonical base of the graph-type $\operatorname{tp}(\operatorname{acl}^{\operatorname{eq}}(A) \cap \Gamma(\mathcal{C})/\Gamma(G))$ in the nice graph $\Gamma(\mathcal{C})$. Here $\operatorname{acl}^{\operatorname{eq}}(A) \cap \Gamma(\mathcal{C})$ is used to denote a nice graph that is naturally assigned to A by the interpretation Γ. Further difficulties arise from the fact that we have to work with imaginary elements in the groups.

Theorem 6.2 shows that the class of stable CM-trivial groups is as rich as the class of all stable CM-trivial structures. But if we insist on the finiteness of the Morley rank, then only few examples are known: Abelian groups of finite Morley rank and the groups from Theorem 3.3.

We can use the ideas of the proof of Theorem 6.2vii) to show the following.

Let $F_2(p, \omega)$ be the free group in the variety of nilpotent groups of class 2 and of exponent p (prime > 2) with ω free generators.

Corollary 6.3 $\operatorname{Th}(F_2(p, \omega))$ *is* ω-*stable and* CM-*trivial*.

In [1] ω-stability of $\operatorname{Th}(F_c(p, \omega))$ is proved.

References

[1] A. Baudisch, *Decidability and stability of free nilpotent Lie algebras and free nilpotent p-groups of finite exponent*, Annals of Math. Logic 23 (1982), 1-25.

[2] A. Baudisch, *A new uncountably categorical group*, Transactions of the AMS 348 (1996), 3889-3940.

[3] A. Baudisch, *Mekler's construction preserves CM-triviality*, preprint 2000.

[4] A. Baudisch and A. Pillay, *A free pseudospace*, Journal of Symbolic Logic 65(2000), 443-460.

[5] W. Hodges, *Model theory*, Cambridge 1993.

[6] E. Hrushovski, *A new strongly minimal set*, Ann. Pure Appl. Logic 62 (1993), 147-166.

[7] E. Hrushovski and A. Pillay, *Weakly normal groups*, Logic Colloquium'85 in Paris, edited by the Paris Logic Group, North-Holland, Amsterdam 1987, 233-244.

[8] A.H. Mekler, *Stability of nilpotent groups of class 2 and prime exponent*, Journal of Symbolic Logic 46 (1981), 781-788.

[9] M.D. Morley, *Categoricity in power*, Trans. Amer. Math. Soc. 114, 514-538.

[10] A. Pillay, *The geometry of forking and groups of finite Morley rank*, Journal of Symbolic Logic 60 (1995), 1251-1259.

[11] A. Pillay, *CM-triviality and the geometry of forking*, to appear in Journal of Symbolic Logic.

[12] B. Poizat, *Groupes stables*, Villeurbanne (Nur al-Mantig wal-Ma'vifah), 1987.

[13] B. Poizat, *Amalgames de Hrushovski* (Une tentative de classification), this volume.

[14] S. Shelah, *Simple unstable theories*, Ann. of Math. Logic 19(1980), 177-203.

[15] F.O. Wagner, *CM-triviality and stable groups*, Preprint 1996.

[16] B. Zil'ber, *Structural properties of models of aleph-one-categorical theories*, Intern. Cong. Math.'83 Warsaw, 359-368.

[17] B. Zil'ber, *Groups and rings whose theory is categorical* (in Russian), Fundamenta Math. 95, 173-188, 1977, (or A.M.S. Translations 149/1991, 1-16).

AMALGAMES DE HRUSHOVSKI

Une tentative de classification

Bruno POIZAT

Institut Girard Desargues, Mathématiques, bâtiment 101,
Université Claude Bernard (Lyon-1), 43,
boulevard du 11 novembre 1918,
69622 Villeurbanne cedex, France
poizat@desargues.univ-lyon1.fr

Contrairement aux apparences, cet exposé a été rédigé à l'intention d'analphabètes en Logique. C'est pour cela qu'il contient un court paragraphe d'introduction aux amalgames de FRAISSE, et qu'il insiste avec lourdeur sur les plans projectifs, qui font partie du patrimoine commun aux participants de cette rencontre.

Cependant, je m'attends à ce qu'il ait des lecteurs logiciens, et comme je ne peux ici réécrire [POIZAT 1985], il y aura ça et là des appartés confidentiels de Théorie des modèles pure et dure. J'espère que ces développements n'agaceront pas les lecteurs étrangers à cette discipline, qui ne lâcheront pas le fil conducteur de l'argumentation et qui verront en quoi consistent ces nouveaux objets fabriqués par HRUSHOVSKI, même s'il est difficile de leur expliquer pourquoi les théoriciens des modèles tenaient tant à les construire.

Introduction

En 1988, au cours d'un congrès qui se tenait à Durham, Ehud HRUSHOVSKI annonça qu'il disposait d'une méthode très simple, disait-il, pour fabriquer de nouvelles structures fortement minimales (i.e. "de dimension un"), dont l'existence contredisait une conjecture de Boris ZIL'BER hantant depuis plusieurs années l'imaginaire collectif des théoriciens des modèles. Cette méthode s'est révélée être un véritable procédé de fabrication industrielle d'exemples et de contre-exemples de toutes sortes, si bien qu'il me semble que le temps est venu de jeter un regard global sur toutes les structures qu'elle a produites, de les grouper par familles et de les classer suivant divers critères que je vais discuter immédiatement; il serait également tentant d'évaluer les possibilités et les limites de la méthode, d'essayer de préciser ce qu'elle peut faire et ne peut pas faire: n'étant pas très doué de prophétie, je m'en abstiendrai par peur d'être contredit par les événements du futur immédiat.

Chacune de ces constructions repose sur le choix préliminaire d'une notion de "dimension", ou plus exactement de prédimension. Pour n'être pas inutilement formel, je me contente d'illustrer la notion de dimension par trois exemples :

(i) une simple cardinalité: si M est un ensemble, et A une partie finie de M, $d(A)$ est le nombre d'éléments de A;

(ii) une dimension projective: si M est un espace vectoriel sur un corps k, et A une partie finie de M, $d(A)$ est la dimension k-linéaire de l'espace vectoriel qu'il engendre;

(iii) quelque chose de bien plus complexe: si M est un corps, k un sous-corps de M, et A une partie finie de M, $d(A)$ est le degré de transcendance sur k du corps $k(A)$.

La dimension apparaissant en (i) est un exemple de dimension triviale, tandis que celle qui apparaît en (ii) est un exemple de dimension modulaire; leur propriété caractéristique s'exprime ainsi: "des parties closes sont toujours indépendantes au-dessus de leur intersection"; pour des ensembles A et B, cela signifie que la dimension relative de A sur B est la même que la dimension relative de A sur $A \cap B$, c'est-à-dire que le nombre des éléments de A qui ne sont pas dans B est le même que le nombre des éléments de A qui ne sont pas dans $A \cap B$ (ou encore que $d(A) + d(B) = d(A \cup B) + d(A \cap B)$); pour des espaces vectoriel U et V, cela signifie également que la dimension relative de U sur V est la même que la dimension relative de U sur $U \cap V$, c'est-à-dire que des éléments de U sont k-linéairement indépendants modulo V si et seulement s'ils le sont modulo $U \cap V$ (ou encore que $d(U) + d(V) = d(U+V) + d(U \cap V)$). Cette propriété n'est pas du tout vérifiée dans le troisième cas: le degré de transcendance de K sur L est en général strictement inférieur au degré de trancendance de K sur $K \cap L$, même si ces corps sont algébriquement clos; autrement dit, des éléments de K peuvent être algébriquement indépendants au-dessus de $K \cap L$, et ne pas l'être sur L .

La conjecture de ZIL'BER que la première construction de HRUSHOVSKI a démolie, c'était que seules ces trois sortes de dimension, triviale, modulaire, ou mettant en jeu un corps, pouvaient apparaître dans un contexte abstrait de Théorie des modèles (la notion modèle-théorique générale de dimension étant qualifiée de "rang de Morley"). Les nouvelles dimensions de HRUSHOVSKI se présentent le plus souvent sous la forme $\delta(A) = \alpha \cdot d_1(A) - \beta \cdot d_2(A)$, où d_1 et d_2 sont des dimensions du genre de celles ci-dessus, et α et β des coefficients, entiers sauf exception. Il faut bien sûr une certaine intuition de ce qu'on cherche à construire pour trouver une formule adaptée: nous en discuterons lors de l'examen des cas particuliers. Par ailleurs, si nous espérons pour δ le comportement d'une dimension, une exigence minimale est de se restreindre à des classes de structures où $\delta(A)$ est toujours positif. Et même là, ce n'est pas si simple: ce serait bien naïf de croire que cette formule donne bille-en-tête la dimension; le schéma général, c'est qu'on peut définir une notion de *clôture autosuffisante*, et que la dimension $d(A)$ de A, c'est la prédimension $\delta(\underline{A}) = d(\underline{A})$ de sa clôture $\underline{A}$. C'est déjà assez surprenant que ça marche si bien comme ça!

Pour établir les propriétés les plus élémentaires de l'autosuffisance, il semble indispensable que la partie négative d_2 de la formule de prédimension soit de type (i) ou (ii); notre première dichotomie sera donc de distinguer le cas où elle est **triviale** du cas où elle est **modulaire**. La deuxième concerne les objets sur lesquels elle porte: si d_2 dénombre des relations qui sont ajoutées à la structure A, nous dirons qu'elle est **du deuxième ordre**; sinon, si elle porte, comme c'est toujours le cas pour d_1, sur les éléments de A, nous dirons qu'elle est **du premier ordre**; dans ce dernier cas, il est nécessaire que des éléments indépendants au sens de d_1 le soient au sens de d_2 (autrement dit, $d_1(A) \le d_2(A)$).

Comme l'indique le titre de cet article, cette dimension permet de construire des amalgames, suivant un procédé que je décrirai en détail; l'amalgame le plus général est dit **non-collapsé**. Souvent, il n'a pas toutes les propriétés qu'on attend de lui, car il est engraissé de morceaux parasites, étrangers à la prédimension δ; pour le faire maigrir, il convient de conduire l'amalgamation dans une classe plus restreinte de structures, et l'amalgame (ou plutôt les amalgames) ainsi obtenu, qui est une sous-structure rétrécie de l'amalgame général, est dit **collapsé**. Il convient donc de distinguer le collapsé du non-collapsé, ou peut-être plus essentiellement le **collapsable** du **non-collapsable**!

Les critères typologiques que je viens de décrire conditionnent la construction même des amalgames; les suivants influent sur leurs propriétés logiques. Si les constituants A de l'almagame sont des objets finis, nous dirons qu'il est **localement fini**; si c'est seulement la dimension d_2 qui s'exprime finiment (par exemple si c'est une cardinalité, ou une dimension linéaire sur un corps fini), nous dirons qu'il est **semi localement fini**; dans le cas **non localement fini**, la construction de l'amalgame ne peut se décrire dans la logique usuelle (finitiste du premier ordre), ou bien, si elle fait, c'est au prix d'arguments mathématiques non-triviaux.

J'admets bien volontiers que cette introduction a un côté brutal pour un lecteur ou une lectrice qui ne sait pas encore de quoi je parle; tout s'éclairera, je l'espère, dans les exemples qui suivent.

0   La préhistoire: Fraïssé et Jonsson.

Ce n'est pas HRUSHOVSKI qui a eu le premier l'idée de construire des structures par amalgamation: c'est un terrain qui a été systématiquement exploré, au début des années 50, par Roland FRAISSE, puis par Bjarni JONSSON (voir [FRAISSE 1986]). Il nous suffit ici de décrire quelques exemples.

Pour nous, un graphe est une relation binaire $g(x,y)$, entre éléments d'un ensemble appelé son support; nous lui demandons d'être anti-réflexive et symétrique, soit encore de satisfaire la condition suivante:

$$(\forall x)(\forall y)g(x,y) \to (x \neq y \land g(y,x)).$$

On convient de dire que les points a et b sont liés par le graphe $g(x,y)$ quand $g(a,b)$ est satisfaite.

On peut amalgamer les graphes: si B et C sont deux graphes ayant une restriction commune A, on peut en particulier les amalgamer librement au-dessus de A; le support de l'amalgame libre $B \oplus_A C$, c'est la réunion disjointe de ceux B et de C au-dessus de celui A; ce graphe est extension de B comme de C, c'est-à-dire que deux points de B, ou deux points de C, y sont liés si et seulement s'il le sont au sens de B ou au sens de C; quant à la liberté, elle signifie qu'on n'introduit pas de liens entre $B - A$ et $C - A$.

En itérant systématiquent, sans jamais oublier personne, cette amalgamation sur des graphes finis, on obtient à la limite un graphe M dénombrable ayant la propriété suivante: si B est un graphe fini, ayant une restriction A, alors tout plongement de A dans M s'étend en un plongement de B dans M. Ce graphe est qualifié d'homogène-universel: homogène parce que tout isomorphisme entre deux sous-graphes finis A et A' de M s'étend en un automorphisme de M; universel parce que tout graphe fini (et même dénombrable) a une copie isomorphe parmi les restrictions de M. A l'isomorphie près, il n'existe qu'un seul graphe homogène-universel *dénombrable*. On l'appelle aussi graphe aléatoire, parce que, si on joue les liens à pile-ou-face, on obtient presque sûrement un graphe qui lui est isomorphe; on encore graphe générique, parce que c'est aussi le graphe majoritaire pour la catégorie de Baire. Sa théorie se trouve également être celle de la majorité des graphes finis.

Comme les morceaux amalgamés sont finis, il est clair que cette condition d'homogénéité et d'universalité s'exprime dans la logique finitiste usuelle. Il suffit même d'une liste d'axiomes très simples (un par entier n):

$$(\forall x_1,\ldots,x_n)(\forall y_1,\ldots,yn)(\exists z)\bigwedge x_i \neq y_j \to \left(\bigwedge g(x_i,z) \land \ \bigwedge \neg g(y_j,z)\right)$$

exprimant des cas particuliers de cette condition. Pour ce qui est de ses propriétés modèle-théoriques, c'est une structure sans grand mystère: elle est instable simple, de rang un, avec forking trivial.

Sur cet aria on peut jouer beaucoup de variations, par exemple en renonçant à la symétrie et en produisant la relation binaire homogène-universelle; ou bien en exigeant la transitivité pour obtenir la chaîne homogène-universelle, isomorphe à l'ordre des nombres rationnels. Plus proche de nos préoccupations sera le graphe biparti homogène-universel, un graphe (explicitement) biparti étant par définition la donnée d'une relation unaire (= un prédicat) $p(x)$ et d'un graphe $g(x,y)$ ne pouvant lier que des éléments de p à des éléments de $\neg p$. On considérera de même des relations ternaires, ou d'arité supérieure.

1 La première construction.

Je traite ici en détail l'exemple initial de HRUSHOVSKI, découvert à la fin des années 80, et publié bien plus tard dans [HRUSHOVSKI 1993]. Je m'inspire d'un exposé de séminaire précoce, [GOODE 1989]; je profite avec empressement de l'occasion qui m'est offerte d'exprimer profusément ma reconnaissance au Professeur John B. GOODE, qui m'a admis aux rangs des disciples d'un esprit profond, d'un acteur éminent et d'un observateur subtil des progrès de la Logique mathématique contemporaine, et qui a bien voulu me faire partager ses visions inspirées et tonifiantes.

Pour éviter de n'obtenir à la fin qu'un amalgame trivial, nous devons introduire des liens (au moins) ternaires. Nous considérons donc des relations à trois arguments $r(x,y,z)$ antiréflexives symétriques, c'est-à-dire satisfaisant à la condition suivante :

$$(\forall x)(\forall y)(\forall z) \; r(x,y,z) \to (x \neq z \wedge r(y,x,z) \wedge r(x,z,y)).$$

Nous appelons lien un ensemble $\{a,b,c\}$ tel que (a,b,c) satisfasse r et, si A est une telle relation de support fini, nous posons:

$$\delta(A) = \textit{nombre de points de } A - \textit{nombre de liens de } A.$$

La philosophie de cette formule est transparente: n points libres sont sans liens, leur δ vaut n; donc le point générique de A est de poids un: ce que nous voulons construire, c'est un objet de dimension un! S'ils ne sont pas libres, chaque lien fait décroître la dimension. Naturellement, nous avons fait des choix arbitraires dans notre façon de compter les liens: il n'y a par exemple aucune obligation de se limiter à des liens symétriques; je l'ai fait pour fixer la présentation.

Relativement à la classification proposée dans l'introduction, le d_2 de cette prédimension est trivial et du second ordre. Elle est localement finie, si bien qu'il n'y a aucune difficulté à axiomatiser, par une liste d'axiomes universels excluant les restrictions finies interdites, la classe C_0 des relations ternaires (symétriques, antiréflexives) dont toute restriction finie A a un $\delta(A)$ positif ou nul. A partir de maintenant, toutes les relations que nous considérerons seront dans cette classe.

Il est inévitable, sauf à ne considérer que des cas triviaux, de rencontrer des exemples d'une A ayant une extension B telle que $\delta(B) < \delta(A)$, ce qui est contraire à l'idée qu'on se fait d'une dimension. Pour corriger ce défaut, nous introduisons la notion d'autosuffisance: A est autosuffisant dans M si, quel que soit B intermédiaire entre A et M, alors $\delta(B) \geq \delta(A)$; autrement dit, le nombre de points de $B - A$ est supérieur au nombre de liens ajoutés, qui sont les liens entre éléments de B faisant intervenir au moins un point de $B - A$.

En dénombrant points et liens, on établit facilement les propriétés essentielles de l'autosuffisance:

Existence. Si A est une restriction finie de M, elle *a* toujours une extension B finie qui soit autosuffisante dans M (prendre B de δ minimal parmi les extensions de A).

Transitivité. Si A est autosuffisant dans B et B est autosuffisant dans M, alors A est autosuffisant dans M.

Intersection. Si A et B sont autosuffisants dans M, il en est de même de leur intersection.

Clôture. Il existe une plus petite restriction de M contenant A et autosuffisante dans M, appellée clôture autosuffisante de A dans M.

Limite. Si $A_0, \ldots, A_i, A_{i+1}, \ldots$ est une suite croissante de relations, chacune étant autosuffisante dans la suivante, alors elles sont toutes autosuffisantes dans leur limite $M = \cup A_i$.

Nous définissons ensuite l'amalgame libre de B et de C au-dessus de leur restriction commune A comme dans le cas binaire, et montrons sans douleur le:

Lemme d'amalgamation. *Si A est autosuffisant dans B et dans C, alors l'amalgame libre $B \oplus_A C$ est bien dans C_0, et B et C en sont des restrictions autosuffisantes.*

et même le:

Lemme d'amalgamation asymétrique. *Si A est autosuffisant dans B, alors l'amalgame libre $B \oplus_A C$ est bien dans C_0, et C est autosuffisant dans $B \oplus_A C$.*

Il ne reste plus qu'a reproduire la construction de FRAISSE, mais en la limitant aux plongements autosuffisants; on obtient alors une homogène-universelle dénombrable, unique à l'isomorphie près, caractérisée par la propriété suivante:

Si A est une restriction autosuffisante finie de M, et si par ailleurs A est autosuffisante dans B également finie, alors il existe un A-plongement de B dans M, dont l'image est autosuffisante dans M.

Nous voyons donc que toute relation finie A (de C_0!) se plonge autosuffisamment dans M, que tout isomorphisme entre deux restrictions finies et autosuffisantes de M se prolonge en un automorphisme de M. Le type de A dans M, c'est la description de sa clôture autosuffisante à l'isomorphie près.

Puisque je me suis permis une confidence modèle-théorique, je poursuis un court instant cette voie. Comme la classe C_0 est élémentaire, la clôture autosuffisante de A est préservée quand on remplace M par une extension élémentaire, si bien qu'elle est incluse dans la clôture algébrique de A au sens modèle-théorique (et en fait, dans le cas de cet amalgame non collapsé, elle lui est égale). Si donc nous voulons analyser les rangs des types dans M, il est légitime de le faire sur des ensembles autosuffisants.

Si A est autosuffisant dans B, lui-même autosuffisant dans M, nous dirons que cette extension est minimale s'il n'y a pas de B' autosuffisant dans B strictement compris entre A et B. Il est clair que toute extension autosuffisante finie se décompose en une tour non raffinable de telles extensions minimales. Elles sont de deux sortes: ou bien $B - A$ est composé d'un seul point b, sans aucun lien avec A, et donc $\delta(B) = \delta(A) + 1$: on dit alors que le point b est générique sur A; ou sinon $\delta(B) = \delta(A)$, et il faudra que le lecteur, en faisant des petits dessins, se convainque qu'il y a beaucoup d'extensions minimales de cette sorte, avec des cardinalités de $B - A$ aussi grandes qu'on veut.

Soit donc A une restriction autosuffisante de M, A' une restriction auto-suffisante de M contenant A, et B une extension autosuffisante minimale de A, $\delta(B) = \delta(A)$; combien y a-t-il de façons de placer B par rapport à A'? Eh bien, si B n'est pas inclus dans A', la seule possibilité est de constituer l'amalgame libre $A' \oplus_A B$ et de le plonger autosuffisamment dans M: toute autre façon est incompatible avec l'autosuffisance de A! Le type de B sur A n'a qu'une seule extension non algébrique: c'est ce qu'on appelle un type de rang un; il est également très facile de voir qu'il est trivial, c'est-à-dire que l'indépendance par paires implique l'indépendance globale (pour que des copies $B_1, \ldots B_n$ de B soient en position d'amalgame libre au-dessus de A, il suffit qu'elles le soient deux-à-deux).

Pour calculer le rang de b au-dessus de A, $b \notin A$, on distingue deux cas; nous notons B la clôture autosuffisante de $A \cup \{b\}$; si $\delta(A) = \delta(B)$, par additivité le rang de b est la hauteur de la tour d'extensions minimales en lesquelles se décompose B/A; sinon, b est générique sur A (il n'a pas de liens avec A et $A \cup \{b\}$ est autosuffisant dans M), et comme il faut qu'il y ait un type de rang ω, c'est lui.

Un lecteur novice en Théorie des modèles, qui aura médité avec une admiration mèlée d'effroi l'obscurité des paragraphes précédents, et qui n'a jamais vécu dans la géométrie d'un type (régulier) de rang ω, pourra faire l'exercice suivant: on dit que b est ciscendant sur A si la clôture autosuffisante de A et celle de $A \cup \{b\}$ ont même δ; montrer que cette relation de ciscendance satisfait les conditions de Van der Waerden (autrement dit, c'est un matroïde, comme la dépendance linéaire, ou la dépendance algébrique).

Il semble donc que ce qui a été construit, c'est quelque chose de rang ω, mais pour en être sûrs il reste un détail à vérifier: il faut voir que nous avons bien analysé tout les types d'une théorie complète, soit encore que l'homogène-universel est ω-saturé. Pour cela, il convient d'en axiomatiser la théorie de la manière suivante: on considère une extension autosuffisante minimale non générique B de A; on note $\phi(\underline{x})$ le diagramme de A, et $\psi(\underline{x}, \underline{y})$ celui de B, et on considère l'axiome suivant:

$$(\forall \underline{x})(\forall \underline{y}) \ \phi(\underline{x}) \rightarrow \psi(\underline{x}, \underline{y}).$$

La liste T de tous ces axiomes, ajoutés à ceux de C_0, décrit la théorie complète de M. Ce qui est curieux, c'est que ni la protase, ni l'apodose ne contiennent de traces d'autosuffisance: l'axiome exprime simplement que, si A est plongé dans M, ce plongement s'étend à B. C'est une situation qui se rencontre dans tous les amalgames non collapsés, qui s'explique ainsi: tout d'abord, l'axiome est vérifié par M, car si $\underline{A}$ est la clôture autosuffisante de A dans M, le lemme d'amalgamation asymétrique assure que $\underline{A} \oplus_A B$ est une extension autosuffisante de $\underline{A}$, qu'on peut donc plonger dans M; par ailleurs, si A est autosuffisant dans M, l'image de B devra l'être nécessairement, puisque $\delta(B) = \delta(A)$. Ces axiomes assurent donc les propriétés d'homogénéité et d'universalité autosuffisantes quand $\delta(B) = \delta(A)$; quant au générique, c'est la compacité qui s'en charge, et un théoricien des modèles expérimenté n'aura pas de peine à voir que M est le modèle ω-saturé dénombrable de T.

Ce qui a été construit, c'est une structure de rang de Morley oméga, dont la géométrie combinatoire est originale. Elle n'est pas triviale, en effet, puisque, si on prend trois points liés qu'on plonge autosufisamment dans M, on obtient ainsi trois points génériques non indépendants, mais deux-à-deux indépendants. Par ailleurs, il est peu vraisemblable qu'en amalgamant des relations finies on construise un groupe ou un corps! Pour confirmer cette impression, on montre qu'aucun groupe n'est interprétable dans M: une façon de faire est de prendre trois points a, b, c génériques et indépendants dans un groupe, et de montrer, en comptant points et liens, qu'on ne peut obtenir dans

M la configuration correspondant aux six points $a, b, c, a + b, a + c, a + b + c$. Mais ce qu'on voudrait, c'est que M fût un objet de dimension un, c'est-à-dire que son matroïde correspondît à une notion d'algébraïcité. Autrement dit, il faudrait que les types triviaux parasites, au lieu d'être de rang un, devinssent algébriques, c'est-à-dire ne se réalisassent qu'un nombre fini de fois. Pour obtenir cela, il faut travailler dans une classe plus restreinte que la classe C_0, qui est une classe C_μ associée à une fonction μ attribuant à chacune des extensions autosuffisantes minimales de δ relatif nul un entier qui en limite le nombre de réalisations. La classe C_μ est définie par des axiomes universels. On a beaucoup de liberté pour choisir μ: la seule chose qui importe, c'est que ses valeurs restent assez grandes pour permettre l'amalgamation.

Précisons: appelons amalgame économique un amalgame libre $B \oplus_A C$ pour lequel il n'existe pas de B_1 strictement compris entre A et B, ni de C_1 strictement compris entre A et C, autosuffisants et A-isomorphes. En commençant par identifier ce qui peut l'être, on peut systématiquement ne faire que des amalgames économiques. On montre alors que la classe C_μ est stable par amalgame économique (nous verrons en détail un exemple de collapsage à propos des plans projectifs).

L'homogène-universel, pour les plongements autosuffisants, de la classe C_μ est une structure de rang de Morley (= dimension) un non classique. L'axiomatisation de la théorie de cet amalgame collapsé est un peu plus subtile, car les axiomes inductifs (en $\forall \exists$) doivent incorporer la part d'autosuffisance que nécessite la version asymétrique du lemme d'amalgamation économique.

1' Variantes de la première construction.

La première variante est de HRUSHOVSKI lui-même: le problème est de construire des structures de rang de Morley un non triviales, dont le groupe d'automorphismes est n fois transitif pour un entier n arbitraire (dans le cas classique, le maximum de transitivité est atteint par la droite projective, qui est 3-transitive). Pour cela, on introduit des liens $(n+1)$-aires, et on amalgame dans la classe définie par $\delta(A) = n$ dès que A a plus de n éléments; dans les homogènes-universels, n points sont toujours sans liens et autosuffisants.

Une autre variante est due à Viktor VERBOVSKII, qui remarque qu'un petit collapsage fait de la relation r le graphe d'une fonction: reprenant l'exemple de notre lien ternaire $r(x, y, z)$, il suffit d'amalgamer dans la classe des relations pour lesquelles, a et b étant donnés, il existe au plus un c satisfaisant $r(a, b, c)$. Elle est définie par l'axiome:

$$(\forall x)(\forall y)(\forall z)(\forall t) \ r(x, y, z) \wedge \ r(x, y, t) \rightarrow z = t,$$

et, comme on le voit facilement, elle est préservée par amalgame économique.

En introduisant simultanément une infinité $r_n(x_1, \ldots, x_n)$ de graphes de fonctions de cette sorte, [VERBOVSKII xxxx] fabrique des structures de rang de Morley un éliminant (fortement) les imaginaires; malgré ce langage infini, la construction reste localement finie (et triviale, du second ordre, collapsée), puisqu'un lien de type r_n ne peut intervenir que sur un ensemble d'au moins n éléments.

La troisième, due à [IKEDA 2001], résoud un vieux problème de ZIL'BER: il s'agit de construire des structures faiblement, mais non fortement, minimales qui ne sont pas le modèle premier de leur théorie. L'exemple s'obtient par un collapsage partiel de la construction de HRUSHOVSKI; on fabrique ainsi une structure de rang de Morley ω, dont les n premiers modèles premiers, sur la suite de Morley du générique, sont minimaux, où n est arbitraire.

2　Les plans projectifs de Baldwin.

Dans un plan projectif, il y a des points et il y a des droites, et une relation d'incidence: un point peut être sur une droite, ou à côté. On demande que par deux points il passe une droite et une seule, et que deux droites se coupent en un seul point.

Nous considérons un plan projectif comme un graphe explicitement biparti: le prédicat $p(x)$ représente l'ensemble des points, son complément celui des droites, et le graphe $g(x, y)$ la relation d'incidence. Nous noterons $\iota(x, y)$ la fonction qui associe à deux points distincts l'unique droite qui passe par ces deux points, qui associe leur unique point commun à deux droites distinctes, et qui, pour fixer les choses, vaut x dans les autres cas. Dans ce contexte, un lien (c'est-à-dire la figure formée d'un point incident à une droite) s'appelle un drapeau. L'ensemble D des points qui sont sur une droite δ, qu'il vaut mieux éviter de confondre avec la droite δ elle-même, ça s'appelle une rangée de points, tandis que l'ensemble P des droites qui passent par un point α, c'est un pinceau de droites.

On dit que le plan est non dégénéré (ou épais) s'il comprend au moins quatre points trois-à-trois non alignés. On observe alors que le plan est coordonné par une quelconque de ses rangées de points D, avec l'aide de deux points a et b qui ne sont pas dans D: un couple (x, y) de points de D distincts représente le point α du plan qui est à l'intersection des droites $\iota(a, x)$ et $\iota(b, y)$.

A un corps, commutatif ou non, K est associé le plan projectif $\Pi(K)$ dont les droites sont les plans vectoriels (i.e. passant par l'origine) et les points les droites vectorielles de $K^3 = K \times K \times K$, l'incidence étant l'inclusion. On peut aussi l'obtenir en ajoutant une droite à l'infini au plan affine. Les plans projectifs de cette sorte sont dits arguésiens: ce sont exactement ceux qui satisfont la loi de perspective des triangles, illustrée sur l'en tête du courrier de l'Institut Girard Desargues; ils forment donc une classe élémentaire. Une copie du corps K est définissable dans le plan $\Pi(K)$, une fois qu'on a fixé une droite et trois points sur cette droite comme paramètres.

Les plans arguésiens satisfont la condition de Moufang, qui s'exprime ainsi: considérons le groupe G des automorphismes du plan qui fixent les points de la rangée D associée à la droite δ, ainsi que les droites d'un pinceau P dont le sommet π est situé sur δ. Ce groupe G est définissable en termes de la relation d'incidence du plan; en effet, il ne peut exister qu'une seule application τ_{ab} de G envoyant le point a sur le point b, où a et b sont alignés avec π (dans le cas d'un plan arguésien, quand on envoie la droite δ à l'infini, on constate que τ_{ab} est une translation); en effet, si nous prenons c hors de la droite $\iota(b,y)$, et notons ω le point d'intersection de D et de la droite $\iota(c,a)$, $\tau_{ab}(c)$ ne peut être que le point commun à $\iota(\omega, b)$ et $\iota(\pi, c)$; nous pouvons donc paramétrer par des paires de points les applications ainsi définies qui sont des automorphismes du plan! Il s'agit d'un cas particulier d'une construction modèle théorique générale: le plan projectif est $D \cup P$-interne, et G est son groupe de liaison. La condition de Moufang, c'est que ce groupe définissable G agisse transitivement sur chacune des rangées de points associée à une droite du pinceau P; elle est élémentaire, et en fait correspond à une configuration que Van MALDEGHEM, qui m'a expliqué tout ça, appelle le "Petit Desargues".

Fabriquer des plans projectifs finis, c'est un casse-tête épouvantable pour combinatoristes. Par contre, il est facile d'en construire qui sont infinis (et qui ne satisfont pas la condition de Moufang!). Appelons sous-plan projectif un graphe biparti comme ci-dessus, tel que par deux points distincts il passe au plus une droite, et que deux droites distinctes aient au plus un point commun. Tout sous-plan s'étend en un plan; la façon la plus simple, c'est de faire ça librement: si deux droites n'ont pas d'intersection, on ajoute un nouveau point incident à chacune, *et qui n'est sur aucune des autres droites*; on fait dualement si deux points n'ont pas leur droite, et on répète.

Une très belle observation de Madame Katrin TENT, c'est qu'on peut construire des plans projectifs homogènes, éliminant les quanteurs si on ajoute au langage la fonction ι, qui sont les analogues des graphes homogènes-universels. Dans un tel plan, les rangées de points ne portent pas d'autre structure que celle de l'égalité: elles forment des ensembles totalement indiscernables, sur lesquels le groupe d'automorphismes agit n-transitivement pour tout n (on ne confondra pas homogénéité - qui ne produit aucun groupe définissable - et liaison). En effet, la théorie des plans projectifs admet une modèle complétion, dont les axiomes sont les suivants: on considère un sous-plan fini B, et une partie A de B qui soit relativement close dans B pour la fonction ι; soit $\phi(\underline{x})$ et $\psi(\underline{x}, \underline{y})$ les diagrammes respectifs de A et de B, dans le langage des graphes bipartis, d'où est exclue la fonction ι; il faut mettre tous les axiomes de la forme $(\forall \underline{x})(\forall \underline{y})\; \phi(\underline{x}) \to \psi(\underline{x}, \underline{y})$.

Ça marche parce que si A est contenu dans le plan projectif Π, on obtient un sous-plan en amalgamant librement B et Π au-dessus de A, qu'il ne reste plus qu'à compléter, par exemple librement, en un plan projectif (on peut amalgamer les plans, mais pas les sous-plans!). Madame TENT a également observé qu'on obtenait un modèle homogène dénombrable de cette théorie en amalgamant des plans qui sont librement engendrés par un sous-plan fini.

A un sous-plan fini, ou plus généralement à un graphe biparti fini, [BALD-WIN 1994] associe la prédimension suivante:

$$
\begin{aligned}
\delta(A) \;=\; & 2 \textit{ fois le nombre de points} \\
& + 2 \textit{ fois le nombre de droites} \\
& - \textit{ le nombre de drapeaux.}
\end{aligned}
$$

L'idée, c'est qu'un plan est un objet de dimension deux, donc qu'il faut attribuer un rang égal à deux à son point générique; cet objet de dimension deux est coordinatisé par un objet de dimension un, à savoir une rangée de points: il convient donc qu'un point incident à une droite soit de rang au plus un sur cette droite.

Toute la théorie de la section 1 se transpose immédiatement, et il ne reste plus qu'à construire l'homogène-universel pour les amalgames autosuffisants dans la classe C_0. Ce premier amalgame n'est pas un plan projectif: deux droites s'y coupent en un ensemble infini, certes, mais qui est petit; il est de rang un, ce qui est tout-à-fait négligeable quand on songe qu'on a construit un objet de rang $\omega \cdot 2$, coordinatisé par un objet de rang ω, dont la géométrie est originale. Pour avoir un plan, il faut faire un léger collapsage: il suffit d'observer qu'un amalgame autosuffisant économique de sous-plans dans C_0 est un sous-plan de C_0. On obtient alors un vrai plan projectif de rang de Morley $\omega \cdot 2$, dont les rangées de points ont pour rang ω. Si on le collapse totalement, on obtient des plans de rang de Morley deux, dont les rangées (et les pinceaux!) sont de rang de Morley un, et dans lesquels aucun groupe infini n'est définissable (ce sont des "anti-Moufang"!).

Pourquoi on a voulu faire ça? Parce que construire un nouvel objet de rang de Morley deux revient à construire le nouvel objet de rang un qui le coordinatise. Reprenons notre rangée de points D, avec deux points a et b qui lui sont extérieurs; le carré cartésien D^2 de D permet de représenter les points du plan, à l'exception de ceux qui sont alignés avec a et b. On peut donc définir dans le langage du plan une relation 6-aire $\rho(u, v, w, x, y, z, a, b)$ sur la droite D exprimant l'alignement des trois points de coordonnées (u, v), (w, x) et (y, z). Comme le type de (a, b) sur D est définissable, on trouve un n-uple $\underline{p}$ de points de D permettant d'exprimer cet alignement par une relation $r(u, v, w, x, y, z, \underline{p})$ de la structure induite par le plan sur D. D muni de cette relation est un nouvel ensemble de rang un; en effet, il permet de redéfinir le plan, qui est le prototype de structure non monobasée (dans un drapeau générique (π, δ), δ est canonique pour $tp(\pi/\delta)$ sans être algébrique sur π), ce qui interdit la modularité de la géométrie de D; par ailleurs, il n'interprète pas de corps (ni même de groupe). Vu la signification que prennent les plans projectifs quand le rang de Morley est fini, il est bien naturel de procéder ainsi: si le plan est l'objet de rang deux par excellence, le tour de force est ici de faire coïncider la dimension combinatoire avec la dimension modèle-théorique, c'est-à-dire le rang de Morley.

Comme nous l'avons remarqué, la stabilité interdit l'homogénéité totale des rangées de points (le groupe d'automorphismes ne peut être $n+6$ fois transitif sur les rangées; je ne sais si n est borné). Pour ce qui est de la construction de Baldwin, on peut, en restreignant la classe où l'on amalgame, obtenir la transitivité sur des petites configurations, mais cette possibilité est limitée. Par exemple, prenons A formé de quatre points génériques et indépendants dans un plan de Baldwin; $\delta(A) = 8$, le plan projectif $\Pi(A)$ engendré par A est libre (sinon la prédimension décroîtrait!), et autosuffisant dans l'univers; à l'intérieur de ce plan, on trouve aisément cinq points ($\delta = 10$), qui engendrent un plan libre non autosuffisant: le groupe d'automorphisme a donc nécessairement plusieurs orbites sur les 5-uplets de points engendrant des plans libres. De même, dans $\Pi(A)$ on trouve 7 points alignés, qui avec leur droite forment un motif de prédimension 9: c'est un obstacle à la transitivité sur les 7-uplets de points alignés.

J'ai calculé, j'espère sans me tromper, les valeurs possibles de la prédimension pour les sous-plans projectifs de petite cardinalité. Si $\#(X) = 1$, $\delta(X) = 2$; si $\#(X) = 2$, $3 \leq \delta(X) \leq 4$; si $\#(X) = 3$, $4 \leq \delta(X) \leq 6$; si $\#(X) = 4$, $5 \leq \delta(X) \leq 8$; si $\#(X) = 5$ ou 6, le minimum de $\delta(X)$ est 6; si $\#(X) = 7$ ou 8, le minimum de $\delta(X)$ est 7; si $9 \leq \#(X) \leq 13$, le minimum de $\delta(X)$ est 8. Pour $\#(X) = 14$, le minimum de $\delta(X)$ vaut 7: il est atteint par le plan projectif sur le corps à deux éléments. Ensuite, le minimum de δ redescend; par exemple le plan projectif du corps à trois éléments, composé de treize points et de treize droites, a un δ nul. Les plans projectifs finis plus gros ont un δ négatif.

La lecture de ce tableau montre qu'on peut amalgamer économiquement dans la classe des sous-plans définie par la condition "$\delta(X) \geq 8$ dès que $\#(X) \geq 14$", qui équivaut à "$\delta(X) \geq 8$ dès que $\#(X) \geq 9$". Dans les plans de Baldwin de cette classe, toutes les configurations d'au plus trois points-droites sont autosuffisantes, à deux exceptions près: celle de trois points ($\delta = 6$) qui sont alignés (quand on ajoute la droite, $\delta = 5$; les trois points plus leur droite sont autosuffisants), ainsi que la configuration duale de trois droites concourantes. On a de même transitivité sur les sextuplets de points alignés.

2' Variantes du plan projectif.

Les plans projectifs (ou trigones) sont les plus simples des polygones généralisés de TITS, qui eux-mêmes sont les moins compliqués de ses immeubles. [De BONIS & NESIN 1998] ont étendu la construction de BALDWIN aux $(2n+1)$-gones; [TENT 2000] répare une de leurs maladresses, et fabrique pour tout n des n-gones de rang de Morley fini, qui ont toutes les propriétés de transitivité qu'il est raisonnable d'espérer. Ils n'ont pas la propriété de Moufang, et même ne permettent de définir aucun groupe infini: leurs groupes d'automorphismes ont des BN-paires (j'espère que vous que savez ce que c'est), qui n'ont aucune raison d'être de rang de Morley fini puisqu'il ne sont pas définissables (contrairement au "petit groupe projectif" d'un polygone de Moufang, engendrés par toutes les translations).

On peut mettre ici les "trigonométries" de [SUDOPLATOV xxxx]; le but ultime de ces trigonométries est de produire des types puissants, ce qui demandera de sortir du contexte ω-stable; du moins, la construction de SUDO-PLATOV a le mérite de montrer que ces choses-là existent et ne sont pas incompatibles avec la stabilité.

3   Un cas à part:
structures stables ω-catégoriques.

Je signale ici un cas tout-à-fait à part d'amalgame de Hrushovski, pour lequel il n'y a pas de meilleure (pré)publication de l'auteur que [HRUSHOVSKI 1989]; à son propos, consulter [WAGNER 1994]. Pour cet amalgame, les points et les liens (binaires) sont comptés avec des coefficients qui donnent une prédimension à valeurs non entières! Il faut alors choisir soigneusement la classe C_0 où l'on amalgame de manière à obtenir la propriété suivante: la taille de la clôture autosuffisante de A est bornée en fonction de celle de A (ce n'était pas du tout le cas pour les objets construits dans les sections précédentes). Le but de cette dimension à valeurs réelles est de produire une notion d'indépendance, si bien que l'amalgame homogène-universel finalement obtenu est stable (non superstable), et ω-catégorique à cause du controle de l'autosuffisance.

Dans notre classification, il serait trivial, du second ordre, localement fini et non collapsé.

On a longtemps espéré qu'un amalgame de Hrushovski permettrait de construire un groupe stable, ω-catégorique et non abélien par fini, mais toutes les tentatives se sont heurtées à des obstacles insurmontables; on trouvera une explication de ces échecs dans [BAUDISCH 2000].

Il faut raccrocher à cet exemple la construction par [HERWIG 1995] d'un type de poids oméga dans un contexte menu; cet exemple est obtenu comme une limite de structures du type précédent, et, bien que son langage comporte une infinité de relations binaires, il reste localement fini; il est trivial, du second ordre, non collapsé.

4 L'amalgame de deux structures.

L'amalgame le plus fantastique a été obtenu dans [HRUSHOVSKI 1992]: HRUSHOVSKI considère deux structures M_1 et M_2 de rang un, dans des langages disjoints L_1 et L_2, et il en fait une seule dans le langage $L_1 \cup L_2$! Il faut que ces structures satisfassent une petite hypothèse technique (définissabilité du degré de Morley), et aussi qu'elles soient (dénombrables et) saturées (dans le cas des corps algébriquement clos, il faut les prendre de degré de transcendance dénombrable).

C'est ainsi qu'on peut obtient des structures de rang de Morley un interprétant d'une part un corps algébriquement clos de caractéristique 2 et d'autre part un deuxième de caractéristique 3 !

HRUSHOVSKI amalgame des $L_1 \cup L_2$-structures M dont la réduite à L_1 est isomorphe à M_1, et la réduite à L_2 est isomorphe à M_2. Il associe à une partie finie A de M la prédimension:

$$\delta(A) = d_1(A) + d_2(A) - \textit{nombre de points de } A,$$

où d_1 désigne la dimension au sens de M_1 et d_2 celle au sens de M_2.

On demande comme d'habitude que ce δ soit positif. L'idée est de compenser la somme des deux dimensions (par exemple, le degré de transcendance des deux corps) par le nombre d'éléments: un n-uple de points indépendants, qui sont donc indépendants au sens de M_1 comme au sens de M_2, et autosuffisamment plongé dans l'univers, est de prédimension n.

Cette fois, la dimension est triviale, mais du premier ordre; son caractère semi-localement fini permet une axiomatisation facile de la classe C_0. L'amalgame général des deux structures est de rang de Morley ω.

En fait, HRUSHOVSKI n'en parle pas, et construit directement des collapsés. La définition de la classe $C\mu$ est bien plus délicate que dans les exemples précédents; en outre, tous les choix de fonction μ ne donnent pas des amalgames collapsés saturés, ni même stables: de fait, ces collapsés peuvent interpréter n'importe quoi, et n'ont une théorie des modèles décente que dans $L_{\infty,\omega}$. Pour obtenir des amalgames de rang de Morley un, il faut collapser uniformément, en respectant ce que [BALDWIN-HOLLAND 2000] ont appelé la condition "finite to one".

La complexité de la construction des amalgames, et de la description des motifs intervenant dans la définition de la fonction μ, fait de la lecture de cet article un exercice très éprouvant.

5 Les corps colorés.

Bien plus facile à appréhender sont les amalgames de corps algébriquement clos colorés, c'est-à-dire munis de prédicats unaires.

Quand la couleur est blanche/noire, on compte ainsi la prédimension d'un corps coloré k de degré de transcendance fini: $\delta(k)$ = deux fois le degré de transcendance de k moins son nombre de points noirs. Il s'agit d'une dimension triviale, du premier ordre, semi-localement finie, et la positivité de δ s'axiomatise sans problèmes.

Pour amalgamer librement K et L au-dessus de k, on les mets en position d'indépendance, ou linéaire disjonction, au-dessus de k; $K \oplus_k L$ est la clôture algébrique du corps engendré par $K \cup L$, et ses points noirs sont ceux de $K \cup L$. Le premier amalgame donne un corps de rang de Morley $\omega \cdot 2$, dont l'ensemble de points noirs est de rang ω; si on le collapse en respectant la condition "finite to one", on obtient des structures de rang 2 dont l'ensemble des points noirs est de rang un (et coordonne le corps: par exemple on peut imposer que tout point s'écrive de manière unique comme somme de deux points noirs).

Il est donc possible d'ajouter de la structure à un corps algébriquement clos tout en lui conservant un rang de Morley fini. Plus généralement, cette méthode permet d'obtenir des corps de rang $\omega^\alpha \cdot n$ avec α et n arbitraires, ce qui répond à une question de [BERLINE-LASCAR 1986].

Une construction qui ressemble beaucoup à celle-ci, c'est le "corps avec courbe" de [CHKP xxxx]. On y considère des corps algébriquement clos avec une relation binaire $C(x,y)$, qu'on appelle "la courbe". La prédimension est $\delta(k)$ = degré de transcendance de k moins le nombre de points de la courbe de k. L'amalgame non collapsé est de rang de Morley ω; sa théorie est la limite, lorsque le degré D tend vers l'infini, des courbes algébriques planes génériques, définies par une équation polynomiale $P(x,y) = 0$ de degré D dont les coefficients sont algébriquement indépendants. C'est le seul amalgame de Hrushovski défini jusqu'à présent qui ait une interprétation naturelle en termes d'objets mathématiques classiques. Si on le collapse, finite-to-onement, on obtient un corps de rang de Morley un portant une "courbe" exotique non-algébrique.

Maintenant, nous entrons dans le domaine des dimensions modulaires, du premier ordre. Si la couleur est rouge, les points rouges forment un sous-groupe additif des corps considérés, et plus exactement un sous-κ-espace vectoriel où κ est un corps fixé. Si la couleur est verte, les points verts forment un sous-groupe multiplicatif divisible des corps considérés, par exemple sans torsion, et constituent donc un Q-espace vectoriel (noté multiplicativement). La formule de prédimension est $\delta(k)$ = deux fois le degré de transcendance de k moins la dimension linéaire (sur κ ou Q) de ses points colorés. Dans la définition .des amalgames libres, il faut prendre comme points colorés le groupe, additif ou multiplicatif, engendré par les points colorés de K et ceux de L .

La construction même de l'homogène-universel non collapsé ne cause aucune difficulté. Par contre, son axiomatisation pose divers problèmes.

Dans le cas additif, si le corps κ est fini, tout se passe comme pour les points noirs, et on, obtient en caractéristique p, un corps de rang $\omega \cdot 2$ avec un sous-groupe additif rouge de rang ω. En caractéristique nulle, le corps κ est infini; s'il est algébriquement clos, on peut se tirer d'affaire en amalgamant des paires de corps, ce qui pose plusieurs problèmes techniques dus au côté flottant de la clôture autosuffisante.

Dans le cas multiplicatif, c'est encore plus compliqué, ne serait-ce pour voir que la classe C_0 est élémentaire. Il faut faire appel à des résultats de ZIL'BER, dérivés des travaux de James AX sur la conjecture de Shanuel pour les séries, qui déclarent que certaines familles de tores sont finies. Et ça ne marche qu'en caractéristique nulle.

Mais la vraie difficulté, c'est qu'on ne sait pas les collapser, par suite de la difficulté de la localisation des points colorés dans les amalgames. Nous devons nous contenter (pour l'instant?), de ces corps de rang $\omega \cdot 2$ avec un sous-groupe additif de rang ω en caractéristique p, de rang $\omega^2 \cdot 2$ avec un sous-groupe additif de rang ω^2 en caractéristique nulle, de rang $\omega \cdot 2$ avec un sous-groupe additif de rang ω en caractéristique nulle. En particulier, on n'est pas assuré de l'existence des mauvais corps qui pourraient obstruer la classification des groupes simples aleph-un-catégoriques, qui sont des corps de rang de Morley fini avec un sous-groupe multiplicatif infini propre et définissable: pour en fabriquer, il faudrait collapser l'homogène-universel vert.

6 L'exponentielle de Zil'ber.

La dernière en date des conjectures de Boris ZIL'BER, c'est une tentative extrêmement audacieuse de caractériser en termes abstraits l'exponentielle complexe. Il amalgame hrushovskiennement des corps K de caractéristique nulle, augmentés d'un homomorphisme $e(x)$ de K^+ dans K^*, $e(x + y) = e(x) \cdot e(y)$, dont le noyau est cyclique. La formule de prédimension est $\delta(A) =$ degré de transcendance de $A \cup e(A)$ moins la dimension Q-linéaire de A. C'est donc modulaire, du premier ordre, non localement fini et non collapsé.

Ces modèles interprètent l'Arithmétique, mais cela ne gêne pas ZIL'BER puisqu'il ne travaille pas dans une classe élémentaire; il se demande si, dans sa classe, le modèle premier sur une suite de Morley générique continupotente est isomorphe à l'exponentielle complexe! La première étape, ça serait de résoudre la conjecture de Shanuel, qui affirme précisément que l'exponentielle complexe satisfait à la positivité de la prédimension.

7 Le groupe de Baudisch.

A ma connaissance, Andreas BAUDISCH est le seul qui ait réussi le collapsage d'un amalgame modulaire. Il est localement fini, et, à ce qu'il me semble, du second ordre. En effet, BAUDISCH, dont le but est d'obtenir des groupes nilpotents de classe deux et de rang de Morley deux qui n'interprètent pas de corps, commence par amalgamer des formes bilinéaires qui définiront l'application commutateur du groupe.

Il considère le corps $k = F_p$, $p \neq 2$, des k-espaces vectoriels V et W et une application bilinéaire alternée $[x, y]$ de $V \times V$ sur W; elle se factorise par une application linéaire λ de $\wedge^2 V$ dans W, lequel est isomorphe à $\wedge^2 V / ker(\lambda)$. La prédimension est $\delta(V) =$ dimension k-linéaire de V moins celle du noyau de λ; on voit que c'est une façon modulaire de compter les liens de la relation $[x, y] = 0$.

BAUDISCH a aussi décrit l'analogue de l'amalgame de FRAISSE dans ce contexte, à la Cave des groupes, Villeurbanne, le 5 novembre 2000.

Conclusion

Si nous dressons le tableau de classification de tous ces amalgames nous obtenons le panorama suivant (on excusera les oublis et omissions), qui, je l'espère, convaincra le lecteur de la pertinence de la typologie.

Amalgame originel de 1988, publié dans [HRUSHOVSKI 1993], et variante de [VERBOVSKII xxxx]: trivial, second ordre, localement fini, collapsé. Collapsage partiel chez [IKEDA 2001]. De même nature sont les amalgames de graphes systématiquement étudiés par [ABM 1999].

Structure ω-catégorique de [HRUSHOVSKI 1989], et sa variante menue de [HERWIG 1995]: trivial, second ordre, localement fini, non collapsé.

Plans projectifs de [BALDWIN 1994] et polygones de [De BONIS-NESIN 1998], [TENT 2000]: trivial, second ordre, localement fini, collapsé. Les trigonométries de [SUDOPLATOV xxxx] ne sont pas collapsées.

Mélange de deux structures, [HRUSHOVSKI 1992]: trivial, premier ordre, semi localement fini, collapsé.

Corps avec points noirs: trivial, premier ordre, semi localement fini, collapsé. Corps avec courbe: non collapsé, mais collapsable.

Corps avec points rouges, caractéristique p: modulaire, premier ordre, semi localement fini, non collapsé. Corps avec points verts, caractéristique nulle: modulaire, premier ordre, non localement fini, non collapsé.

Exponentielle de ZIL'BER: modulaire, premier ordre, non localement fini, non collapsé.

Groupes de [BAUDISCH 1996]: modulaire, second ordre, localement fini, collapsé.

Références

[ABM 1999] Roman AREF'EV, John T. BALDWIN, Marco MAZUCCO, Classification of δ-invariant amalgamation classes, The Journal of Symbolic Logic, 64, 1743-1750

[AX 1971] James AX, On Schanuel's Conjectures, Annals of Mathematics, 93, 252-258

[BALDWIN 1994] John T. BALDWIN, An almost strongly minimal non-desarguesian (sic) projective plane, Transactions American Mathematical Society, 342, 695-711

[BALDWIN-HOLLAND 2000] John T. BALDWIN & Kitty HOLLAND, Constructing ω-stable structures: fields of rank 2, The Journal of Symbolic Logic, 65, 371-391

[BAUDISCH 1996] Andreas BAUDISCH, A new uncountably categorical group, Transactions American Mathematical Society, 48, 3889-3940

[BAUDISCH 2000] Andreas BAUDISCH, Closure in aleph-0-categorical bilinear maps, The Journal of Symbolic Logic, 65, 914-922

[BERLINE-LASCAR 1986] Chantal BERLINE & Daniel LASCAR, Superstable groups; a partial answer to conjectures of Cherlin and Zil'ber, Annals of Pure and Applied Logic, 30, 1-43

[CHKP xxxx] Olivier CHAPUIS, Ehud HRUSHOVSKI, Pascal KOIRAN & Bruno POIZAT, La limite des théories de courbes génériques, à paraître au Journal of Symbolic Logic

[De BONIS-NESIN 1998] M.J. de BONIS & Ali NESIN, There are 2-to-aleph-0 almost strongly minimal generalized n-gons that do not interpret an infinite group, The Journal of Symbolic Logic, 65, 485-508

[FRAISSE 1986] Roland FRAISSE, Theory of Relations, North Holland

[GOODE 1989] John B. GOODE, Hrushovski's geometries, Seminarberichte, Humboldt-Universitt zu Berlin, 104, 106-117

[HERWIG 1995] Bernhard HERWIG, Weight ω in stable theories with few types, The Journal of Symbolic Logic, 60, 353-373

[HRUSHOVSKI 1989] Ehud HRUSHOVSKI, A stable ω-categorical pseudo-plane, preprint

[HRUSHOVSKI 1992] Ehud HRUSHOVSKI, Strongly minimal expansions of algebraically closed fields, Israel Journal of Mathematics, 79, 129-151

[HRUSHOVSKI 1993] Ehud HRUSHOVSKI, A new strongly minimal set, Annals of Pure and Applied Logic, 62, nb. 2, 147-166

[IKEDA 2001] Koichiro IKEDA, Minimal, but not strongly minimal structures with arbitrary finite dimensions, The Journal of Symbolic Logic, 66, 117-126

[POIZAT 1985] Bruno POIZAT, Cours de Théorie des Modèles, Nur al-Mantiq wal-Ma'arifah (Traduction par Moses KLEIN, A Course in Model Theory, Springer, 2000)

[POIZAT 1997] Bruno POIZAT, Corps de rang deux, The 3rd Barcelona Logic Meeting, Quaderns num. 9, juliol 1997, Centre de Recerca Matematica, Barcelona, 31-44

[POIZAT 1999] Bruno POIZAT, Le carré de l'égalité, The Journal of Symbolic Logic, 64, 1339-1355

[POIZAT xxxx] Bruno POIZAT, L'égalité au cube, à paraître au Journal of Symbolic Logic

[SUDOPLATOV xxxx] Sergei Vladimirovich SUDOPLATOV, Communication au 5° Colloque franco-qazaq de Théorie des Modèles, Qaragandy, juin 2000

[TENT 2000] Katrin TENT, Very homogeneous n-gons of finite Morley rank, Journal of the London Mathematical Society, 62, 1-15

[VERBOVSKII xxxx] Viktor V. VERBOVSKII, On elimination of imaginaries for the strongly minimal sets of E. Hrushovski, preprint

[WAGNER 1994] Frank Olaf WAGNER, Relational structures and dimensions, dans Kaye/Macpherson (eds), Automorphisms of first-order structures, Oxford Science Publications, 153-180.

[ZIL'BER xxxx] Boris Iosifovich ZIL'BER, Communication à "la Cave des Groupes", Villeurbanne, 04-11-2000

Rank and Homogeneous Structures

John T. Baldwin

Department of Mathematics, Statistics and Computer Science

University of Illinois at Chicago*

The notion of constructing a homogenous structure was introduced in the 1950's by Fraïssé. It was extended by Jonsson from countable to uncountable relational structures. In 1969, Grzegorczyk asked how many theories are categorical in $\aleph_0$. The amalgamation construction was quickly used [24, 27, 21] to show there are continuum many such theories. These constructions clearly gave rise to unstable structures. As stability theory developed, the problem of classifying stable $\aleph_0$-categorical structures arose[1]. Lachlan [40] and Shelah[51] showed that every superstable $\aleph_0$-categorical theory was ω-stable. Work of Cherlin, Harrington, Lachlan, Hrushovski and Zilber showed there were only countably many ω-categorical, ω-stable structures (and much more). But, Lachlan's conjecture that there was no strictly stable $\aleph_0$-categorical structure remained open until the late 80's.

Mainly in infinitary contexts Shelah (e.g. [52]) had studied variants of the construction which strengthen amalgamation to 'free amalgamation'. The freeness of the amalgamation corresponds to a stability condition. But in [36] Hrushovski provided a concrete way of constructing such amalgams [36]. With this method he refuted both Lachlan's conjecture and Zilber's conjecture that every strongly minimal set was 'bi-interpretable' with a discrete set, a vector space, or a field.

As we outline below these are two of a large family of variants of the amalgamation construction, which are determined by what we call here a 'rank' function on a class of models. Unexpectedly, the counterexample to the Lachlan conjecture is intimately related with the almost sure theories of random graphs. For irrational $\alpha, 0 < \alpha < 1$, let μ_n^α be the probability measure on

*The author thanks the organizers of the Wurzburg conference for a valuable meeting. This paper was prepared at the Mittag-Leffler institute during the Fall of 2000 and the support of the institute is greatfully acknowledged. Partially supported by NSF grant DMS-9510377.

[1]Stability theory provides a hierarchy of theories which provides a framework for dividing questions in model theory. This is described in detail in such texts as [4, 51, 42]; for an overview see the table at the end.

graphs of size n given by edge probability $n^{-\alpha}$. For every first order sentence ϕ, $\lim_{n\to\infty} \mu_n^\alpha(\phi)$ is 0 or 1. The theory T^α of the almost sure sentences is stable and has the finite model property. It was constructed by variants on the Hrushovski construction in [12], improving the argument of [54]. More expectedly, there have been a large number of variants on the construction to construct groups, expansions of fields and various kinds of geometries with specific model theoretic properties. Our purpose here is to partially systematize this family of constructions and to provide a comprehensive list of the applications during the first 10 years of its history.

Each application of this method depends on the choice of a class $\boldsymbol{K}$ of models and a rank function δ on the members of $\boldsymbol{K}$. This rank induces a notion $\leq$ of strong substructure on some (usually all) substructures of members of $\boldsymbol{K}$. Under appropriate hypotheses a homogeneous model for the pair $(\boldsymbol{K}, \leq)$ is constructed. The rank reenters the argument in several ways to reduce the quantifer complexity of the theory of the homogeneous model and to prove the stability conditions.

As in some papers ([11]) too many adjectives were being piled on the word homogeneous, the term 'generic' (first applied in this context in [39]) is often used for the model constructed. It has long been a curiosity to me that the structure constructed in random graph case [12] is a countable first order homogeneous model which is neither saturated nor prime. In his first papers on this subject Hrushovski referred to the difficulty of determining that the model constructed was a 'first order structure'. One rendering of this metaphor is that it asks whether the structure is ω-saturated. We describe in Section 5 another explanation, which was at least implicit in [50] and clarified in [33, 48, 37, 44, 19]. The question becomes, 'Is the class of existentially closed models in $\boldsymbol{K}$ (in a suitably expanded language) an elementary class?'.

In the first section of the paper we lay out a general context for these examples. In the second, we rehearse a large number of examples showing their dependence on the choice of the class of structures and the rank. Many papers [14, 22, 59, 48, 25] have laid out the basics of the method of construction so we don't expound all of those details here. In Section 3 we discuss the definablility requirements on the rank functions and connections with algebraic geometry which arise in constructing expansions of fields. Section 4 describes how the rank produces the stability requirement. These arguments work best when the homogeneous structure is saturated; in Section 5, we examine the situation when saturation fails.

1 Setting

Let $\langle K(N), \wedge, \vee \rangle$ be a lattice of substructures of a model N. For purposes of this paper a rank is a function δ from $K(N)$ to a discrete subgroup of the reals $(\mathcal{R})$, which is defined on each N in a class $\mathbf{K}$. We write $\delta(A/B) = \delta(A \vee B) - \delta(B)$ to indicate the relativization of the rank. We demand *only* that δ is monotonic: if $B \subseteq A, C \subseteq N$ and $A \wedge C = B$,

$$\delta(B/A) \geq \delta(B/C).$$

This requirement can be rephrased as asserting that δ is lower semimodular:

$$\delta(A \vee B) - \delta(B) \leq \delta(A) - \delta(A \wedge B).$$

We say δ is modular if the inequality is an equality. Examples of δ include cardinality, relation size (number of instances of a relation), vector space dimension, and transcendence degree. All of these but the last are modular. The simplest example of 'relation size' is just the number of edges in a (symmetric) graph.

This notion of 'rank' is much weaker than any other of the rank notions used in stability theory. Several authors [14, 61] have used the term predimension; this seems inappropriate in the general context since the resulting combinatorial structure may not satisfy exchange and thus have nothing like a dimension.

Let T_{-1} be a theory such that any subset X of a model N of T_{-1} is contained in a minimal submodel of N; this implies there is a natural notion of a finitely generated model. We denote this submodel $\langle X \rangle_N$, dropping the subscript N when the choice of N is evident. This condition is clearly satisfied if T_{-1} is universally axiomatized or strongly minimal and almost all of our examples fall into one of these two classes. This condition is closely connected (see [29] 6.4 and 6.5) with the requirement that T_{-1} is axiomatized by universal-existential sentences and thus that the class of its models is closed under unions of chains. Let $\overline{\mathbf{K}}_{-1} = \mathrm{mod}(T_{-1})$; $\mathbf{K}_{-1}$ is the *finitely generated* members of $\overline{\mathbf{K}}_{-1}$. Some of the choices for T_{-1} include: T_{-1} is a universal theory in a finite relational language; $\mathbf{K}_{-1}$ is the finite models of T_{-1}; T_{-1} is Acf_p; $\mathbf{K}_{-1}$ contains those algebraically closed fields of finite transcendence degree; more generally, T_{-1} is a strongly minimal, inductive theory with elimination of quantifiers and imaginaries and the definable multiplicity property; $\mathbf{K}_{-1}$ contains the models generated by finitely many independent elements.

The construction of the homogeneous model is made with respect to a notion of strong substructure. We define the most used notion now. In Application 2.2, we discuss a variant which produces simple theories.

Definition 1.1 *For $N \models T_{-1}$, $K(N)$ is the substructures of N which are in*

$\boldsymbol{K}_{-1}$. *For $A, B \in \overline{\boldsymbol{K}}_{-1}$, we say A is a* strong substructure *of B and write $A \leq B$ if for every $B' \in \boldsymbol{K}_{-1}$ with $B' \subseteq B$, $\delta(B'/B' \cap A) \geq 0$.*

The definition easily implies that the operation of assigning to a set the smallest superset containing it that is strong in the universe satisfies all the axioms of a combinatorial pregeometry except exchange. This is a crucial exception. In most cases that we discuss there is no way to assign a single 'dimension' to a model.

Definition 1.2 *We denote by $\overline{\boldsymbol{K}}_0$ the members of $\overline{\boldsymbol{K}}_{-1}$ which have hereditarily positive rank and by $\boldsymbol{K}_0$ those which are finitely generated and have hereditarily positive rank. T_0 denotes the theory of $\overline{\boldsymbol{K}}_0$,*

(In some cases, showing $\overline{\boldsymbol{K}}_0$ is an elementary class requires some effort.) Usually, it makes little difference whether one constructs homogeneous models wtih respect to $\boldsymbol{K}_0$ or $\overline{\boldsymbol{K}}_0$. We explore the sitation where it does matter in Section 5.

The following result follows easily from the monotonicity property.

Theorem 1.3 *The notion of strong substructure has the following properties.*

- **A1**. *If $M \in \overline{\boldsymbol{K}}_{-1}$ then $M \leq M$.*

- **A2**. *If $M \leq N$ then $M \subseteq N$.*

- **A3**. *If $A, B, C \in \overline{\boldsymbol{K}}_{-1}$, $A \leq B$, and $B \leq C$ then $A \leq C$.*

- **A4**. *If $A, B, C \in \overline{\boldsymbol{K}}_{-1}$, $A \leq C$, $B \leq C$ and $A \subseteq B$ then $A \leq B$.*

Since $\leq$ is imposed by δ, the following property holds, which is more special than the general case, e.g. [39].

- **A5**. *If $A, B, C \in K(N)$, $A \leq C$, $B \subseteq C$, then $A \cap B \leq B$.*

We restrict to $\overline{\boldsymbol{K}}_0$ precisely to obtain:

- **A6**. *$\emptyset \in \overline{\boldsymbol{K}}_0$ and $\emptyset \leq A$ for all $A \in \overline{\boldsymbol{K}}_0$.*

Definition 1.4 *The pair $(\boldsymbol{K}, \leq)$ has the* amalgamation property *if for $N, M \in \boldsymbol{K}$ with $A \leq M, N$, there exists $N_1 \in \boldsymbol{K}$ and $M, N \leq N_1$.*

We have not required $A \in \boldsymbol{K}$. When applied to theories T_{-1} which are universally axiomatized, clearly $A \in \overline{\boldsymbol{K}}_{-1}$. for other T_{-1}, allowing $A \notin \overline{\boldsymbol{K}}_{-1}$ strengthens the amalgamation hypothesis. In the presence of **A6**, the amalgamation property for $(\overline{\boldsymbol{K}}_{-1}, \leq)$, implies $\overline{\boldsymbol{K}}_{-1}$ has both the amalgamation and the joint embedding (any two members of $\overline{\boldsymbol{K}}_{-1}$ have a common strong extension) properties.

Definition 1.5 *The model M is* $(K, \leq)$*-homogeneous (or generic) if $A \leq M, A \leq B \in K$ implies there exists $B' \leq M$ such that $B \cong_A B'$.*

Theorem 1.6 *If a class* $(K, \leq)$ *has the amalgamation property and the joint embedding property then there is a* $(K, \leq)$*-homogeneous structure.*

If $(K, \leq)$ has only countably many countable members then there is a countable $(K, \leq)$-homogeneous structure. If $(K, \leq)$ is closed under unions of chains (as in all the examples considered here) there is a $(K, \leq)$-homogeneous structure in K.

If $(K_{-1}, \leq)$ has the amalgamation property, both $(K_0, \leq)$ and $(\overline{K}_0, \leq)$ have the amalgamation and joint embedding properties and thus a homogeneous model. We let M denote the K_0-homogenous model and M_1 denote the $\overline{K}_0$-homogeneous model. We distinguish several notions that might be termed 'locally finite' in this discussion.

Definition 1.7

1. *The class* $\overline{K}_{-1}$ *is* locally finite *if for every $A \in \overline{K}_{-1}$ and every finite $A_0 \subset A$ there is a finite $A_1 \in K$ with $A_0 \subseteq A_1 \subset A$. A_1 is finitely generated over $A_0 \in \overline{K}_{-1}$, if $A_1 = A_0 \vee B$ where B is in K_0.*

2. *The class* $(\overline{K}_0, \leq)$ *is* locally closed *if for every $A \in \overline{K}_0$ and every $A_0 \subset A$ there is an $A_1 \in \overline{K}_{-1}$ with $A_0 \subseteq A_1 \leq A$ and A_1 is finitely generated over A_0.*

With this vocabulary we now describe some of the main applications of the method. Later we will discuss in more detail the techniques to establish quantifier reduction and stability conditions.

2 Combining rank functions

This section lists many applications of the method. Each application depends on the choice of class K and a rank function on members of K. Many of the ranks are obtained by standard combinations of ones that are already known. If δ_1, δ_2 are ranks defined on a class K, so are

$$\delta = \alpha \delta_1 + \beta \delta_2$$

for any *positive* reals α, β and

$$\delta = \alpha \delta_1 - \beta \delta_2$$

for any *positive* reals α, β if δ_2 is modular!

With this observation, most of the examples of this construction can be seen as built up from the examples given at the beginning. We describe a collection of examples which all use the relation:

$$\delta = \alpha\delta_1 - \beta\delta_2$$

In many cases it is more convenient to define the rank function of finite sequequences from models of T_{-1} rather than on substructures; we expand on this point in Section 3.

Applications 2.1 *Many of the examples are obtained with δ_1 the cardinality of a finite relational structure and δ_2 the number of tuples which satisfy a fixed list of symmetric relations on the structure. We refer to these as* ab initio *examples.*

1. *ab initio* finite relational language: δ_1 is cardinality, δ_2 is 'relation size'.

  (a) $\alpha = \beta = 1$. This is the dimension function for the first application of the method: Hrushovski's new strongly minimal set [36].

    i. The class K_0^μ depends on a function bounding multiplicity as described in Section 4 and yields [36] a strongly minimal set. If the μ-function is relaxed to allow even one infinite value, the rank is infinite [10]. There are continuum many different theories of this sort depending on the choice of μ.

    ii. Working with the class of all structures K_0 with hereditarily non-negative rank yields a theory of rank ω [25]. There are countably many classes which satisfy a certain 'δ-invariance' condition; they are classified in [1, 2].

    iii. It is straightforward that Hrushovski's example does not admit elimination of imaginaries but Verbovskiy [58] provides a variant which does.

    iv. Hedman [26] has shown that none of the almost strongly minimal theories which can be constructed in this way can be axiomatized with finitely many variables.

  (b) $\alpha = 2$, $\beta = 1$. Baldwin [5] varied the method to construct almost strongly minimal projective planes which have no infinite definable groups of automorphisms. In [6] he showed these planes had the least possible structure in the sense of the Lenz-Barlotti classification.

  (c) $\alpha = n - 1$, $\beta = n - 2$. Debonis and Nesin (for odd n) [41] and Tent [57] (uniformly for all n) constructed almost strongly minimal generalized n-gons. The automorphism groups of Tent's structures were highly transitive even though they were not Moufang. Thus she showed that the analog of the Feit-Higman theorem [23] did not hold for finite Morley rank n-gons.

(d) $\alpha = 1$, β irrational;

    i. K_0^1: Hrushovski [34] constructed a strictly stable $\aleph_0$-categorical theory thereby refuting Lachlan's conjecture.

    ii. Herwig [28] varied the construction by allowing an infinite language to find a theory with infinite p-weight. This paper also contains the best published exposition of Hrushovski's $\aleph_0$-categorical stable theory.

    iii. K_0^2. Baldwin and Shi [14] modified the second Hrushovski construction only by changing the class of finite structures. The resulting theory is uneventful model theoretically: another non-small strictly stable theory. But, it turned out [12] that this is the almost sure theory of random graphs with edge probability $n^{-\beta}$ which had originally been constructed by [54]. Baldwin and Shelah [13] show this theory has the dimensional order property but not the finite cover property.

    iv. Baldwin [3] (see also Shelah [53]) has generalized this argument to show a 0-1-law for expansions of successor by graphs with edge probablility $n^{-\beta}$.

(e) $\alpha = 1$, β rational; This gives rise [14] to a class of ω-stable theories. There is no longer a connection with random graphs.

The other cases involve more 'algebraic' dimensions as ingredients for δ and we treat these as separate cases.

2. *Baudisch groups*:

  (a) δ_1, δ_2 are the vector space dimension of a vector space E and an associated subspace of $\bigwedge^2 E$. $\delta = \delta_1 - \delta_2$. Baudisch [15] constructs a nilpotent $\aleph_1$-categorical group which does not interpret a field.

  (b) In [16], Baudisch analyzes some obstructions to extending Hrushovski's construction of a strictly stable structure to find a strictly stable $\aleph_0$-categorical group.

3. *fusions*: δ_1, δ_2 are Morley rank on two strongly minimal sets which share the same universe. Let,

$$\delta(\mathbf{x}) = \delta_1(\mathbf{x}) + \delta_2(\mathbf{x}) - \lg(\mathbf{x}).$$

The resulting amalgam is a strongly minimal expansion of both of the original structures [36]. Holland [31, 30] clarifies this construction and in [32] proves that these theories (as well as the Hrushovski strongly minimal set) are model complete.

4. *enriched fields* $\alpha = 2$, $\beta = 1$, Work in $L = (+, 0, -, *, 1, B)$, the language of fields. N is an algebraically closed field and $K(N)$ is its algebraically closed subfields. Let $\delta_1 = d_f$ be transcendence degree and let δ_2 vary as indicated in the subcases. In each case, $(\boldsymbol{K}_0, \leq)$ has the amalgamation property. (Take linearly disjoint copies of the fields and paint all possible new points in the join white.) These constructions were motivated by the search for first any (expansion of a) field with finite Morley rank and then for so-called bad fields: fields with 'new' definable subgroups of the multiplicative group.

   (a) *bicolored fields*: $\delta_2(\boldsymbol{a}) = |B \cap \boldsymbol{a}|$. Poizat [46] constructs a bicolored field of rank $\omega \times 2$ working with the class $\boldsymbol{K}_0$; Baldwin and Holland [9] find a rank 2 field by showing (for appropriate choice of μ) the $(\boldsymbol{K}_0^{\mu}, \leq)$-generic structure is ω-saturated. A corrected version of their proof is available on line and will appear in their paper [7], which considers the slightly tricky technicalities which arise for $\alpha > 2$).

   (b) *additive bad fields*: δ_2 additive linear dimension. Poizat [47] has completed the infinite rank case.

   (c) *multiplicative bad fields*: δ_2 multiplicative linear dimension. Again, Poizat [47] has completed the infinite rank case.

5. *A nondefinability result*: [20]: Using a dimension on algebraically closed fields which assigns to a sequence of ordered pairs twice the transcendence degree of its union minus its length, the authors show there is no first order formula $\phi(R)$ which holds of the relation R on an algebraically closed field just if R is Zariski closed.

6. Ikeda [38] has used variants of this method to settle a question raised in [18]. He shows that there are structures of every finite dimension which are minimal (every definable subset is finite or cofinite) but not strongly minimal (i.e. in some elementary extension there is an infinite/coinfinite definable set.)

7. Sudoplatov [56, 55] has used variants of this method to show there exists an ω-stable *group trigonometry* on a projective plane. Group trigonometries and some generalizations were devised by Sudoplatov to study combinatorial geometries arising in model theory.

All the *ab initio* examples are locally finite in the sense of Definition 1.7. When the parameter α is a rational number or in the $\aleph_0$-categorical example, the class is locally closed. But the strictly stable example of Baldwin-Shi is not locally closed. The Baudisch group is also locally finite.

Applications 2.2 *Simple theories:* Hrushovski introduced a variant on the notion of strong substructure which allows the construction of strictly simple theories. The key is to make the inequality in the definition of strong substructure strict. For $A, B \in \overline{\boldsymbol{K}}_{-1}$, we say A is a *∗-strong substructure* of B and write $A \leq^* B$ if for every $B' \in \boldsymbol{K}_{-1}$ with $B' \subseteq B$, $\delta(B'/B' \cap A) > 0$. With this notion Hrushovski constructed an $\aleph_0$-categorical strictly simple theory where forking is not locally modular. This argument was sketched in [34], treated more expansively in [33] and has a full exposition in [48]. Pourmahdian [48], following Pillay [44] also develops the appropriate notions of simplicity for the class of existentially closed models even if it is not an elementary class.

Applications 2.3 *Pseudoexponentiation:* Zilber [61] defines a rank on two sorted structures (D, ex, R) where D is a field of characteristic 0, R is a field of characteristic p and ex is a homomorphism of the additive group of D onto the multiplicative group of R. He defines a 'rank' or 'predimension'

$$\delta(X) = d_f(X) + d_f(ex(X)) - d_{vs}(X)$$

and investigates, for example, connections with the Schanuel conjecture.

Pillay and Baudisch ([17, 45, 43] have investigated more intensively the geometric conditions (such as CM-triviality) which arise in these constructions and also studied higher dimensional analogues.

3 Definability Conditions on δ

The Hrushovski construction produces a countable model. We want to draw conclusions about all models of the theory of this model. For this, we need to impose several conditions on δ. The first of the following just establishes notation, the second reaffirms our commitment to monotonicity; the third and fourth specify definability requirements on δ. We earlier described δ as a function on models. In order to develop the definability constraints we must consider a variant where δ is defined on finite sequences. In the *ab initio* situation these definability requirements are so obvious they go unnoticed. They become increasingly more difficult to fulfill for various kinds of enriched fields.

Definition 3.1 *A class $\boldsymbol{K}$ has δ-formulas over parameters if the following conditions hold. ($\boldsymbol{K}$ has δ-formulas over the empty set if they hold when $\mathbf{b} \in \mathrm{acl}(\emptyset)$.)*

1. *If* $\text{Diag}(\boldsymbol{a}) = \text{Diag}(\mathbf{b})$ *then* $\delta(\boldsymbol{a}) = \delta(\mathbf{b})$.

2. *(monotonicity)* $\delta(\boldsymbol{a}/\mathbf{b}) \geq \delta(\boldsymbol{a}/\mathbf{bc})$ *whenever* $\mathbf{c} \cap \boldsymbol{a} = \mathbf{c} \cap \mathbf{b}$.

3. *(definability) For any integer* k *and* $\boldsymbol{a}, \mathbf{b} \subseteq C \in \boldsymbol{K}$ *with* $\delta(\boldsymbol{a}/\mathbf{b}) = k$, *there is a* δ-**formula for** $\boldsymbol{a}$ **over** $\mathbf{b}$, $\varphi(\overline{x}; \overline{y}) \in \text{Diag}(\boldsymbol{a}; \mathbf{b})$, *such that the following hold for any* $\boldsymbol{a}', \mathbf{b}' \subseteq B \in \boldsymbol{K}$: *if* $B \models \varphi(\boldsymbol{a}'; \mathbf{b}')$, *then* $\delta(\boldsymbol{a}'/\mathbf{b}') \leq k$.

4. *(q.f. type determined) For any* $N \in \boldsymbol{K}$ *and* $\boldsymbol{a}, \boldsymbol{a}', \mathbf{b} \subseteq N$, *if* $N \models \phi(\boldsymbol{a}; \mathbf{b}) \wedge \phi(\boldsymbol{a}'; \mathbf{b})$ *and*

$$\delta(\boldsymbol{a}/\mathbf{b}) = \delta(\boldsymbol{a}'/\mathbf{b}) = k$$

and

$$\mathbf{b}\boldsymbol{a} \leq \langle \mathbf{b}\boldsymbol{a} \rangle,$$
$$\mathbf{b}\boldsymbol{a}' \leq \langle \mathbf{b}\boldsymbol{a}' \rangle,$$

then

$$diag(\mathbf{b}\boldsymbol{a}) = diag(\mathbf{b}\boldsymbol{a}').$$

Definition 3.2 *Let* $\boldsymbol{a}, \mathbf{b} \in N \in \boldsymbol{K}$. $\boldsymbol{a}$ *is a* **minimal intrinsic extension** *of* $\mathbf{b}$ *if* $\delta(\boldsymbol{a}/\mathbf{b}) < 0$ *but for every* $\boldsymbol{a}'$ *properly contained in* $\boldsymbol{a}$, $\delta(\boldsymbol{a}'/\mathbf{b}) \geq 0$.

Now suppose the $\boldsymbol{K}$ in Definition 3.1 is one of the $\overline{\boldsymbol{K}}_{-1}$ described above. Note that if δ-formulas can be found over the empty set, then there is a first order theory T_0 axiomatizing $\overline{\boldsymbol{K}}_0$; just assert that the intrinsic closure of the empty set is empty. Such δ-formulas are found easily in the *ab initio* case; Hrushovski's construction for fusions in [35] extends painlessly to bicolored fields enriched only by a predicate. When the predicate is required to define a group very delicate questions arise. A lemma of Zilber [62] (with ideas from Hrushovski) provides a partial solution to this difficulty for the case of multiplicative bad fields.

Definition 3.3 *Let* T_W *be the minimal torus containing a variety* W. *Let* U *be an irreducible subvariety of an irreducible variety* W. V *is an* atypical component *of* $W \cap T_V$ *if*

$$d_f(W) - d_f(V) < d_f(T_W) - d_f(T_V).$$

Lemma 3.4 (Zilber) *Fix a variety* W_1 *defined over* $\overline{Q}$ *by the equations* $\psi(\mathbf{x}, \mathbf{y})$. *There is a finite set* $T_1, \dots T_n$ *of proper subtori of* T_W *such that: for any* $\mathbf{b}$, *and* $W = W(\mathbf{b}) = \{\boldsymbol{a} :\models \psi(\boldsymbol{a}, \mathbf{b})\}$ *if* V *is an irreducible subvariety of* W *and* V *is an atypical component of* $W \cap T_V$ *with then for some* $i < n$ *and some* $\mathbf{f}$, $V \subseteq T\mathbf{f}$.

The following corollary (and a proof of the lemma incorporating ideas of Marker) can be found both in [8] and [47].

Corollary 3.5 *Let T_{-1} be as in Application 2.1 4c). For any sequence $\boldsymbol{a}$ with $\delta(\boldsymbol{a}/\emptyset) > -\infty$, there is a formula $\phi(\mathbf{x})$ which is a δ-formula for $\boldsymbol{a}$ over $\emptyset$.*

In order to find a finite rank expansion of the complex numbers, with a predicate for a subgroup of the multiplicative group either an improvement or a clever application of Lemma 3.4 is required. This project only seems really possible in characteristic 0 both because the proof of Lemma 3.4 uses that hypothesis heavily and because Wagner [60] shows a bad field of finite characteristic is extremely unlikely.

4 Stability and Finite Rank

The main purpose of marrying ranks to the Fraïssé-construction is to control the stability class and more particularly the Morley Rank of the resulting structure. How is this control exercised? In one approach, the map $\delta(a/X)$ is made monotone in the second argument by replacing δ with $d_M(a/X) = \inf\{\delta(a/Y) : X \subseteq Y \subseteq_{fin} M\}$ and it is shown that a notion of independence built on 'a is independent from A over B if $d_M(a/AB) = d_M(a/B)$' (with another condition) satisfies most of the axioms of forking. This works for forking with respect to formulas of low quantifier complexity.

There are several strategies to prove stability or (ω-stability). First get the result for formulas with low quantifier complexity by combinatorial means: counting types [46], or following the unpublished Hrushovski paper using d-independence [14] or by checking the order property [37]. If T^* (the theory of the generic model) has been shown model complete, the 'stability' result follows immediately. If not, specific technical arguments can be given [14].

Hrushovski [33] suggested the following device to simplify the description of the situation.

Definition 4.1 *Form the language L^+ by adding a relation symbol $R_{AB}(\mathbf{x})$ for each pair (A, B) where B is a minimal intrinsic extension of A. For any of our theories, T^0, T^0_{nat} is the L^+-theory extending T^0 which asserts:*

$$[\exists \mathbf{y} \Delta_{AB}(\mathbf{x}, \mathbf{y})] \leftrightarrow R_{AB}(\mathbf{x}).$$

We denote the natural expansion of an L-structure N to L^+ by N^+ and the collection of expansions of models in a class $\boldsymbol{K}$ by $\boldsymbol{K}^+$.

Baldwin and Holland [9] provide a necessary and sufficient condition for the homogeneous structure to be ω-saturated. We need a little vocabulary.

Definition 4.2 *For any* $\mathbf{b} \in A \in \mathbf{K}$, *let* $I^*(\mathbf{y})$ *be a collection of formulas so that if* $B \models I^*(\mathbf{b}')$, *then* $\mathbf{b}' \leq B$ *and* $\langle \mathbf{b} \rangle \cong \langle \mathbf{b}' \rangle$ *(I.e. the* L^+*-diagram of* $\mathbf{b}$*.)*

Definition 4.3 $(\mathbf{K}, \delta)$ admits strong separation of quantifiers *if for any* $\mathbf{b} \leq \mathbf{ab} \leq \langle \mathbf{ab} \rangle \in \mathbf{K}$ *with* $\mathbf{a}$ *minimal strong over* $\langle \mathbf{b} \rangle$, *the following holds: For any formula* $\tau(\mathbf{x}; \mathbf{y})$ *in* $I^*(\mathbf{a}, \mathbf{b})$, *there is* $\sigma(\mathbf{y}) \in I^*(\mathbf{b})$ *such that whenever* $\mathbf{b}' \subseteq C \in \mathbf{K}$ *and* $C \models \sigma(\mathbf{b}')$, *there is* $D \in \mathbf{K}$ *with* $C \leq D$ *and* $\mathbf{a}' \in D$ *such that*

$$D \models \tau(\mathbf{a}'; \mathbf{b}').$$

Theorem 4.4 ([9]) $(\overline{\mathbf{K}}_0, \delta)$ *admits strong separation of quantifiers iff the generic is* ω*-saturated.*

With each of the dimension functions for the ω-stable examples above two classes of theories can be constructed. If one amalgamates over all structures with hereditarily nonnegative dimension, the theory has infinite Morley rank. To investigate the finite rank case requires a different choice of the amalgamation class. In order to describe it we need other notions.

Definition 4.5 $\mathbf{a}$ *is* primitive *over* $\mathbf{b}$ *if*

1. $\delta(\mathbf{a}/\mathbf{b}) = 0$

2. $\delta(\mathbf{a}/\mathbf{b}') < 0$ *if* $\mathbf{b} \subset \mathbf{b}' \subset \mathbf{ab}$.

Suppose $A/B \in \mathbf{K}_0$ is primitive, let M be $(\mathbf{K}_0, \leq)$-homogeneous and let $\chi_M(A/B)$ denote the number of copies of A over B in M. There are three possibilities.

- $\delta(A/B) < 0$ implies $\chi_M(A/B)$ is finite.

- $\delta(A/B) > 0$ implies $\chi_M(A/B)$ is infinite.

- $\delta(A/B) = 0$ implies $\chi_M(A/B)$ is undetermined.

A key point for the completeness of the theories T^α (the random graph case) is that when α is irrational, the third case cannot occur. For rational α (e.g. $\alpha = \beta = 1$ in the original strongly minimal construction), in the full class $\mathbf{K}_0$, the primitive structures also occur infinitely often and the theory has infinite rank. To guarantee that the theory has finite rank we make the following restriction; we remove the ambivalence by changing 'undetermined' in the third case to finite. But to axiomatize this restriction requires great care.

A *primitive code* is a sequence of parameters $\mathbf{c}$ which completely describes a primitive pair. It includes a formula specifying the quantifier free type of

a/b. For any function μ from codes to natural numbers we define $\mathbf{K}^\mu$ to be the members of $\overline{\mathbf{K}}_0$ which have less than $\mu(\mathbf{c})$ independent realizations over the same canonical parameter of each code $\mathbf{c}$.

In all the examples so far considered it is fairly straightforward to transfer amalgamation from $\mathbf{K}_0$ to $\mathbf{K}_0^\mu$ provided only that $\mu(A/B)$ is sufficiently large ([36, 5, 46]). In order that $\mathbf{K}_0^\mu$ admit strong separation of variables, a stronger condition on μ is required: it suffices that μ be finite-to-one [30, 35, 9]. In fact, this condition is essential:

Theorem 4.6 *[9] For $\mathbf{K}_0$, the class of bicolored fields, $(\mathbf{K}_0, \delta)$ and if μ is finite-to-one the class of bicolored fields, $(\mathbf{K}_0^\mu, \delta)$ admits strong separation of quantifiers.*

Theorem 4.7 *[9] For appropriate μ which is not finite to one,*

1. *The generic is not saturated.*

2. *In fact, T^μ is not ω-stable, nor even small.*

5 Infinitary Logic and Finite Diagrams

There are two examples which are particularly intriguing. For irrational α, the Baldwin-Shi homogenous model is not saturated; indeed the theory has no countable saturated model. Again, Theorem 4.7 shows that if μ is not finite-to-one the theory T^μ of a bicolored field may not have a countable saturated model. Infinitary logic and Shelah's notion [49] of finite diagram are useful for explaining this situation. In fact, the reason is different for the two cases.

The *finite diagram* of a model M, $D = D(M)$ is the collection of finite types over the empty set realized in M. A D-model is a structure N with $D(N) \subseteq D$. The class of D-models is $L_{\omega_1,\omega}$ definable but may not be an elementary class. Note that 'N is existentially closed' (for $\mathbf{K}$) is a property of $D(N)$.

Consider any class $\overline{\mathbf{K}}_0$ with associated theories T_0 in L and T_{nat}^0 in L^+ (see Definition 4.1). Let M be generic for $\mathbf{K}_0$; we denote its theory by T^*. In all our examples, $(T_{nat}^0)_\forall$ is a universal theory with amalgamation and joint embedding. In particular, in the language of [33], it is a Robinson theory. Let $Ex(T_{nat}^0)$ denote the class of existentially closed (for $\overline{\mathbf{K}}_0$) models of $(T_{nat}^0)_\forall$. Since T_{nat}^0 is a Robinson theory, on $Ex(T_{nat}^0)$, every formula is equivalent to a (possibly infinite) Boolean combination of quantifier-free formulas. So if $Ex(T_{nat}^0)$ is first order, it admits quantifier elimination in L^+. Since the added relation symbols represented existential formulas this means that in L, the theory of the generic is *nearly model complete*: every formula is a Boolean combination of existentials.

Consider three class of models of T^*: all models of T^*, $Ex(T_{nat}^0)$ and $Mod(D)$ where D is the finite diagram of M^+:

$$Mod(D) \subseteq Ex(T_{nat}^0) \subseteq Mod(T^*).$$

The first containment holds as a model N is existentially complete if it omits (incomplete) types of the form $\{\neg\theta(\mathbf{y}) : (\forall\mathbf{xy})[\phi(\mathbf{x},\mathbf{y}) \to \theta(\mathbf{y})\} \cup \{\neg(\exists\mathbf{x})\phi(\mathbf{x},\mathbf{y})\}$, where θ and ϕ are quantifier-free in L^+. Thus, any model of a complete theory whose diagram is contained in the diagram of an existentially closed model is existentially closed.

In general, $D(M^+)$ and $Ex(T_{nat}^0)$ may not be first order. Consider T^μ as in Lemma 4.7 and T^α, the almost sure theory of Application 2.1 1.d.iii.

The first order theory T^μ of Lemma 4.7 is not ω-stable nor even small. But it is easy to see from condition 4) on the definition of δ-formulas (Definition 3.1) that $Ex(T_{nat}^\mu)$ is ω-stable as a finite diagram and so ω-stable as a first order order theory if it is axiomatiable. This implies $Ex(T_{nat}^\mu) \neq Mod(T^\mu)$.

On the other hand taking T^* as T^α, $Mod(D) \neq Ex(T_{nat}^0)$ while $Ex(T_{nat}^0) = Mod(T^*)$. The equality follows from the proof in [12] that T^α is nearly model complete; alternatively, Hyttinen proved it directly [37]. The inequality then follows since D is countable while $Mod(T^*)$ has continuum many types over the empty set.

As M^+ is existentially closed, if M^+ is ω-saturated then T^* is model complete (in L^+!). Thus the three classes coincide if the generic is ω-saturated. This happens for bicolored fields if μ is finite-to-one. If not, we can either investigate the first order theory or the infinitary classes, which are well behaved, using the technology of [19]. Several interesting questions arise from such considerations. For which μ (beyond the finite-to-one), if any, is T^μ stable or even simple? Let M_1 be the homogeneous model for $\overline{K}_0$. Where does $D(M_1)$ fit into the picture above? Can $Ex(T_{nat}^\mu)$ always (in the contexts discussed in this paper) be realized as the class of models of $D(N)$ for some N?

References

[1] R.D. Aref'ev. $K_{(H,|A|-e(A),\leq)}$-homogeneous-universal graphs. In B.S. Baizhanov M.B. Aidarkhanov, editor, *Proceedings of Informatics and Control Problems Institute, Almaty*, pages 27–40. 1995. In Russian.

[2] R.D. Aref'ev, J.T. Baldwin, and M. Mazzucco. δ-invariant amalgamation classes. *Journal of Symbolic Logic*, 64:1743–1750, 1999.

[3] J.T. Baldwin. Random expansions of geometries. submitted.

[4] J.T. Baldwin. *Fundamentals of Stability Theory.* Springer-Verlag, 1988.

[5] J.T. Baldwin. An almost strongly minimal non-desarguesian projective plane. *Transactions of the American Mathematical Society*, 342:695–711, 1994.

[6] J.T. Baldwin. Some projective planes of Lenz Barlotti class I. *Proceedings of the A.M.S.*, 123:251–256, 1995.

[7] J.T. Baldwin and K. Holland. Constructing ω-stable structures: Infinite rank. preprint, 199x.

[8] J.T. Baldwin and K. Holland. Constructing ω-stable structures: Bad fields. in progress, 200?

[9] J.T. Baldwin and K. Holland. Constructing ω-stable structures: Rank 2 fields. *Journal of Symbolic Logic*, 65:371–391, 2000.

[10] J.T. Baldwin and M. Itai. K-generic projective planes have Morley rank two or infinity. *Mathematical Logic Quarterly*, 40:143–152, 1994.

[11] J.T. Baldwin and S. Shelah. Abstract classes with few models have 'homogeneous-universal' models. *Journal of Symbolic Logic*, 60:246–266, 1995.

[12] J.T. Baldwin and S. Shelah. Randomness and semigenericity. *Transactions of the American Mathematical Society*, 349:1359–1376, 1997.

[13] J.T. Baldwin and S. Shelah. DOP and FCP in generic structures. *Journal of Symbolic Logic*, 63:427–439, 1998.

[14] J.T. Baldwin and Niandong Shi. Stable generic structures. *Annals of Pure and Applied Logic*, 79:1–35, 1996.

[15] A. Baudisch. A new uncountably categorical group. *Transactions of the American Mathematical Society*, 348:889–940, 1995.

[16] A. Baudisch. Closures in $\aleph_0$-categorical bilinear maps. *Journal of Symbolic Logic*, 65:914–922, 2000.

[17] A. Baudisch and Pillay Anand. A free pseudospace. *Journal of Symbolic Logic*, 65:443–460, 2000.

[18] O. Belegradek. On minimal structures. *J. Symbolic Logic*, 63:421–425, 1998.

[19] S. Buechler and O. Lessmann. Simple homogeneous models. preprint, 200?

[20] O. Chapuis, E. Hrushovski, P. Koiran, and B. Poizat. La limite des théories de courbes génériques. preprint, 2000.

[21] A. Ehrenfeucht. There are continuum ω_0-categorical theories. *Bulletin de l'Acadamie Polonaise des sciences math., astr., et phys.*, XX:425–427, 1972.

[22] D. Evans. $\aleph_0$-categorical structures with a predimension. preprint.

[23] W. Feit and G. Higman. The nonexistence of certain generalized polygons. *Journal of Algebra*, 1:114–131, 1964.

[24] W. Glassmire. There are $2^{\aleph_0}$ countably categorical theories. *Bulletin de l'Acadamie Polonaise des sciences math., astr., et phys.*, XIX:185–190, 1971.

[25] J. Goode. Hrushovski's Geometries. In Helmut Wolter Bernd Dahn, editor, *Proceedings of 7th Easter Conference on Model Theory*, pages 106–118, 1989.

[26] S. Hedman. Finitary axiomatizations and local modularity of strongly minimal theories. preprint, 2000.

[27] W. Henson. Countable homogeneous relational structures and $\aleph_0$-categorical theories. *The Journal of Symbolic Logic*, 37:494–500, 1972.

[28] B. Herwig. Weight ω in stable theories with few types. *J. Symbolic Logic*, 60:353–373, 1995.

[29] W. Hodges. *Model Theory*. Cambridge University Press, 1993.

[30] K. Holland. An introduction to the fusion of strongly minimal sets: The geometry of fusions. *Archive for Mathematical Logic*, 6:395–413, 1995.

[31] K. Holland. Strongly minimal fusions of vector spaces. *Annals of Pure and Applied Logic*, 83:1–22, 1997.

[32] K. Holland. Model completeness of the new strongly minimal sets. *J. Symbolic Logic*, 64:946–962, 1999.

[33] E. Hrushovski. Simplicity and the Lascar group. preprint.

[34] E. Hrushovski. A stable $\aleph_0$-categorical pseudoplane. preprint, 1988.

[35] E. Hrushovski. Strongly minimal expansions of algebraically closed fields. *Israel Journal of Mathematics*, 79:129–151, 1992.

[36] E. Hrushovski. A new strongly minimal set. *Annals of Pure and Applied Logic*, 62:147–166, 1993.

[37] T. Hyttinen. Canonical finite diagrams and quantifier elimination. preprint.

[38] K. Ikeda. On minimal structures. preprint.

[39] D.W. Kueker and C. Laskowski. On generic structures. *Notre Dame Journal of Formal Logic*, 33:175–183, 1992.

[40] A.H. Lachlan. Two conjectures regarding the stability of ω-categorical theories. *Fundamenta Mathematicae*, 81:133–145, 1974.

[41] A. Nesin M. J. De Bonis. There are $2^{\aleph_0}$ many almost strongly minimal generalized n-gons that do not interpret an infinite group. *J. Symbolic Logic*, 63:485–508, 1998.

[42] A. Pillay. *An introduction to stability theory*. Clarendon Press, Oxford, 1983.

[43] A. Pillay. The geometry of forking and groups of finite Morley rank. *Journal of Symbolic Logic*, 60:1251–1259, 1995.

[44] A. Pillay. Forking in the category of existentially closed structures. preprint, 1999.

[45] A. Pillay. the geometry of forking and cm-triviality. *Journal of Symbolic Logic*, 65:474–480, 2000.

[46] Bruno Poizat. Le carré de l'egalité. *The Journal of Symbolic Logic*, 64:1339–1356, 1999.

[47] Bruno Poizat. L'egalité au cube. preprint, 1999.

[48] M. Pourmahdian. *Simple Generic Theories*. PhD thesis, Oxford University, 2000.

[49] S. Shelah. Finite diagrams stable in power. *Annals of Mathematical Logic*, 2:69–118, 1970.

[50] S. Shelah. The lazy model-theoretician's guide to stability. *Logique et Analyse*, 18:241–308, 1975.

[51] S. Shelah. *Classification Theory and the Number of Nonisomorphic Models*. North-Holland, 1978.

[52] S. Shelah. Universal classes: Part 1. In J. Baldwin, editor, *Classification Theory, Chicago 1985*, pages 264–419. Springer-Verlag, 1987. Springer Lecture Notes 1292.

[53] S. Shelah. 0-1 laws. preprint 550, 199?

[54] S. Shelah and J. Spencer. Zero-one laws for sparse random graphs. *Journal of A.M.S.*, 1:97–115, 1988.

[55] Sudoplatov S.V. On hypergraphs of minimal prime models. In *International Conference devoted to the 90 years of birthday of A.I.Maltsev. Novosibirsk: IDMI,*, pages 112–113, 1999.

[56] Sudoplatov S.V. On type identifications in trigonometrical theories. In *International Conference devoted to the 90 years of birthday of A.I.Maltsev. Novosibirsk: IDMI*, pages 111–112, 1999.

[57] Katrin Tent. Very homogeneous generalized n-gons of finite morley rank. *J.London Math. Soc. (2)*, 62:1–15, 2000.

[58] V. Verbovskiy. On the elimination of imaginaries for the strongly minimal sets of Hrushovski. preprint.

[59] F. Wagner. Relational structures and dimensions. In *Automorphisms of first order structures*, pages 153–181. Clarendon Press, Oxford, 1994.

[60] F. Wagner. Fields of finite morley rank. preprint, 200x.

[61] B.I. Zilber. Fields with pseudoexponentiation. preprint, 2000.

[62] B.I. Zilber. Intersecting varieties with tori. preprint, 2000.

THE GEOGRAPHY OF STABLE THEORIES

The Number of countable models

T	$2^{\aleph_0}$	$\aleph_1$	$\aleph_0$	$2 < n < \aleph_0$	1
not simple	discrete order	?	contrive	Ehrenfeucht	dense order
strictly simple	contrive	?	contrive	?	contrive
supersimple	contrive	?	contrive	NO	random graph
strictly stable	A: Th (ω^ω, E_i)	?	vary A	?	Hrushovski
str. superstable	B: Th $(2^\omega, E_i)$	1/2 done	vary B	NO	NO
ω-stable	contrive	NO	contrive	NO	vector sp.
$\aleph_1$-cat	NO	NO	ACF$_0$	NO	vector sp.

'contrive' means the example is easily contrived. E_i is the equivalence relation $\sigma E_i \tau$ if $\tau|i = \sigma|i$.

CONSTRUCTIONS OF SEMILINEAR TOWERS OF STEINER SYSTEMS

KEITH JOHNSON,

116 Kirkstall Hill, Leeds, England

1 Introduction

In this paper, model-theoretic methods based on Fraïssé amalgamation are used to build structures whose automorphism groups are infinite Jordan permutation groups. The structures themselves are Steiner systems, or 'towers' of Steiner systems in the sense of [2] or [3]. These are defined in Section 2.

Throughout the paper, (G, Ω) will denote a permutation group G on a set Ω. Recall that if (G, Ω) is transitive and $\Gamma \subset \Omega$, then Γ is a *Jordan set* if $|\Gamma| > 1$ and the pointwise stabiliser in G of $\Omega \setminus \Gamma$ is transitive on Γ. The Jordan set is *improper* if for some $n \in \mathbf{N}$, G is $(n + 1)$-transitive on Ω and $|\Omega \setminus \Gamma| = n$, and is *proper* otherwise. A *Jordan group* is a transitive permutation group with a proper Jordan set.

By a result of Jordan (see [12] Theorem J1) any finite primitive Jordan group is 2-transitive, and using the classification of finite simple groups the finite primitive Jordan groups have been classified - see for example [12] or Kantor [10]. A structure theorem for infinite primitive Jordan groups is given in [3], based on earlier work in [4]. If such a group is not *highly transitive* (that is, k-transitive for all positive integers k), then it must be a group of automorphisms of some relational structure on Ω which is non-trivial in the sense that not every permutation is an automorphism. In [3] it is shown that every primitive infinite Jordan group which is not highly transitive is a group of automorphisms of a structure of one of the following types: a Steiner system (possibly with infinite blocks); a linear order (or circular order, or linear betweenness relation, or separation relation); one of four kinds of 'treelike' relational structures, examined in detail in [5] (semilinear orders, B-relations, C-relations, or D-relations); or a *limit* of B-relations, D-relations,

1991 *Mathematics Subject Classification:* 20B27

A final draft of this paper was prepared by Dugald Macpherson (Department of Pure Mathematics, University of Leeds, Leeds LS2 9JT, UK).

or Steiner systems.

Of the above classes of structures, all except the Steiner systems and the limit structures are pretty well understood (though we do not claim a *classification* of the structures admitting primitive Jordan automorphism groups), and the automorphism groups are at most 3-transitive. Jordan groups preserving limits of B-relations and D-relations are at most 2-transitive, and examples are given by Adeleke in [1]. There are further examples of the last two kinds in recent work of Bhattacharjee and Macpherson, with Fraïssé amalgamation used to build relational structures.

The situation for Jordan groups acting on Steiner systems and on limits of Steiner systems is less clear. By the above remarks, any 4-transitive but not highly transitive Jordan group preserves a structure of one of these kinds. For Steiner systems, the classical examples are projective and affine spaces over arbitrary fields, and the Jordan sets are the complements of subspaces. But model theory gives many other constructions: if M is any saturated strongly minimal set, then its automorphism group is a 2-transitive Jordan group preserving a Steiner system, acting on the quotient modulo $\sim$ of $M \setminus \mathrm{cl}(\emptyset)$, where $x \sim y$ if $\mathrm{acl}(x) = \mathrm{acl}(y)$. This means that the construction technique of Hrushovski [8] gives many new Jordan groups acting on Steiner systems. It was probably already well-known, but was shown in [11] (see also Ch. 15 of [6]) that a very simple version of the Hrushovski construction gives, for any positive integer k, a k-transitive but not $(k+1)$-transitive Jordan group acting on a Steiner k-system. Concerning *limits* of Steiner systems, there is a construction in [2] of a 3-transitive Jordan group acting on a limits of Steiner 2-systems, but some details seem a little unclear.

The present paper gives constructions of both these kinds, closely related to each other. We prove the following theorems. As noted above, the first is a consequence of the Hrushovski construction, but here a different construction is given, also based on amalgamation but using 'free' constructions of Steiner systems. The main new points in the second theorem (compared to [2]) are that the degree of transitivity is arbitrary, there is an explicit invariant relational structure, and the construction is by amalgamation so it should be easier to read off other transitivity properties.

Theorem 1.1 *For every positive integer k there is a Steiner k-system whose automorphism group is a k-transitive but not $(k+1)$-transitive Jordan group.*

Theorem 1.2 *For every positive integer k, there is a $(k+1)$-transitive but not $(k+2)$-transitive Jordan group which preserves a limit of Steiner k-systems.*

Theorem 1.2 is proved by constructing an 'upper semilinear tower' of Steiner k-systems, where each individual Steiner k-system is in fact one of

the ones constructed in Theorem 1.1. The automorphism group of this structure, in its induced action on the sort whose elements are the points lying
in the various Steiner systems, is then a Jordan group preserving a limit of
Steiner k-systems. Section 2 of the paper consists of background definitions
and notation. Theorem 1.1 is proved in Section 3. In Section 4, a semilinear
order in built and then in Section 5 this construction is combined with that in
Section 3 to produce an upper semilinear tower of Steiner k-systems. Finally
in Section 6, we verify certain extra properties of the automorphism group of
this tower.

Throughout this paper we assume that we are dealing with countable
structures, and hence with permutation groups of countable degree. However by Löwenheim–Skolem arguments, for all infinite cardinals κ, there exist
permutation groups (G, Ω) with essentially the same properties and $|\Omega| = \kappa$.

Question 1. Is the theory of the Steiner system constructed in Theorem 1.1
stable? (The analogous construction based on the Hrushovski technique gives
a superstable example).

Question 2. Is there a permutation group satisfying the conclusion of
Theorem 1.2 which is oligomorphic, that is, has finitely many orbits on t-
sets for all positive integers t? And what can be said about ω-categorical
Steiner k-systems with $k > 3$ (perhaps with a Jordan group condition on the
automorphism group).

The results in this paper can also be found in the author's PhD thesis [9].
The author wishes to thank Bernhard Herwig for many helpful conversations,
and in particular for some key insights involved in the free constructions in
Section 3. This work was funded by an EPSRC Research Studentship.

2 Background information

This section consists mainly of definitions of (partial) Steiner systems and
free constructions of Steiner systems. We state the version of Fraïssé amalgamation used, and also define the notion of a *tower* of Steiner systems, and
say what it means for a group to preserve a limit of Steiner systems.

First, a small piece of notation. If $\phi(x_1, \ldots, x_n)$ is a first order formula,
then we write $\exists^{\neq} x_1 \ldots x_n \phi(x_1, \ldots, x_n)$ as an abbreviation for

$$\exists x_1 \ldots \exists x_n (\bigwedge_{1 \leq i < j \leq n} x_i \neq x_j \wedge \phi(x_1 \ldots, x_n)).$$

We write $\forall^{\neq} x_1 \ldots x_n$ analogously.

Definition 2.1 A *Steiner system* is an incidence structure consisting of a
set Ω of *points*, a set $\mathcal{B}$ of *blocks*, and an incidence relation between them (a

subset of $\Omega \times \mathcal{B}$). It has parameters k and λ, where k is an integer greater than 1, and λ is a cardinal (possibly infinite) with $k \leq \lambda$, such that the following conditions hold.

(i) If $x_1, \ldots, x_k$ are distinct elements of Ω then there is a unique element of $\mathcal{B}$ incident with each of $x_1, \ldots, x_k$.

(ii) Any element of $\mathcal{B}$ is incident with exactly λ elements of Ω.

We also call such a structure a *Steiner k-system*. It is *non-trivial* if $k < \lambda$, and $\mathcal{B}$ contains more than one element.

In the constructions in this paper, the parameter λ takes value ω, that is each block is incident with countably infinite many points.

A *partial Steiner k-system*, S, is a set of points and blocks along with an incidence relation such that every block is incident with at least k and at most ω many points, and any two blocks are incident with at most $k - 1$ common points. It is *non-trivial* if it consists of more than one block. If $k + 1$ points of a (partial) Steiner k-system are not all incident with a single block then we say they are *non-incident*. We can consider a (partial) Steiner k-system, S, as an ordered triple $(\Omega_S, \mathcal{B}_S, I)$ where Ω_S is the set of points of S and $\mathcal{B}_S$ the set of blocks of S, and $I \subseteq \Omega_S \times \mathcal{B}_S$ is an incidence relation. If a point x is incident with a block Σ we shall denote this by $xI\Sigma$ (rather than $I(x, \Sigma)$). The notation $M \leq N$ for partial Steiner systems M, N then means that M is (model-theoretically) a substructure of N.

Definition 2.2 Given (partial) Steiner k-systems M, N with $M \leq N$, we define $\mathrm{cl}_N(M)$, the *closure* of M in N, to be the smallest substructure, S of N, containing M that is closed under

(1) $\forall \Sigma \in \mathcal{B}_S \ \forall x \in \Omega_N \ (xI\Sigma \rightarrow x \in \Omega_S)$,

(2) $\forall^{\neq} x_1, \ldots, x_k \in \Omega_S \ \forall \Sigma \in \mathcal{B}_N \ ((x_1 I\Sigma \wedge \ldots \wedge x_k I\Sigma) \rightarrow \Sigma \in \mathcal{B}_S)$

If $M = \mathrm{cl}_N(M)$ we say that M is *closed* in N.

We shall now give two versions of a notion of *free construction* for (partial) Steiner systems. In each case, we consider (partial) Steiner k-systems, M and N, with $M \leq N$ and look at $\mathrm{cl}_N(M)$. By the theorem which follows, the definitions can be used interchangeably.

Definition 2.3 We say $\mathrm{cl}_N(M)$ *is free around* M if $\mathrm{cl}_N(M)$ can be obtained from M by a step by step construction. At odd steps we adjoin arbitrarily many points of N that are not already adjoined, and are incident with just a single block that was adjoined at some previous step, so that points adjoined corresponding to distinct blocks are disjoint. At even steps, for each k-set of points that are not incident with a block, we adjoin a single block that is incident with just these k points, so that distinct such sets of k points give rise to distinct blocks.

Definition 2.4 We say $\text{cl}_N(M)$ *is free around* M if there is a partial function Parent : $\mathcal{B}_S \to \Omega_S^{\{k\}}$ (the set of all unordered k-sets of Ω_S) that satisfies the following:

(i) $\forall \Sigma \in \mathcal{B}_M \; \text{Parent}(\Sigma)$ is undefined;

(ii) $\forall \Sigma \in \mathcal{B}_S \backslash \mathcal{B}_M \; \text{Parent}(\Sigma)$ is defined;

(iii) $\forall \Sigma \in \mathcal{B}_S \backslash \mathcal{B}_M \; \forall x \in \Omega_M \; (xI\Sigma \to x \in \text{Parent}(\Sigma))$;

(iv) $\forall x \in \Omega_S \backslash \Omega_M \; \forall \Sigma_1, \Sigma_2 \in \mathcal{B}_S \; ((\Sigma_1 \neq \Sigma_2 \wedge xI\Sigma_1 \wedge xI\Sigma_2) \to (x \in \text{Parent}(\Sigma_1) \vee x \in \text{Parent}(\Sigma_2)))$;

(v) $\forall x \in \Omega_S \backslash \Omega_M \; \exists \Sigma \in \mathcal{B}_S \; (xI\Sigma \wedge x \notin \text{Parent}(\Sigma))$.

If for some block Σ, $\text{Parent}(\Sigma)$ is undefined then we say that for all $x \in \Omega_S$ with $xI\Sigma$ we have $x \notin \text{Parent}(\Sigma)$.

By (iv) the block Σ in (v) is unique. We will denote it by $\text{Birth}_S(x)$, the *birth block* of x in S. For each block $\Sigma \in \mathcal{B}_S \backslash \mathcal{B}_M$, we call the points $x \in \text{Parent}(\Sigma)$, the *parent points* of Σ, or simply the *parents* of Σ. We shall call such a function a *parent* function.

Theorem 2.5 *Suppose* M, N *are partial Steiner* k-*systems with* $M \leq N$. *Then* $\text{cl}_N(M)$ *is free around* M *in the sense of Definition 2.3 if and only if it is in the sense of Definition 2.4.*

Proof. This (and the equivalence with several other conditions) is due to B. Herwig and Johnson and can be found in Johnson [9].

We now give an account of a generalised version of Fraïssé's Theorem for constructing countable structures with rich automorphism groups. It is taken from Evans [7], and used throughout this paper.

Definition 2.6 Let $\mathcal{C}$ be a class of $\mathcal{L}$-structures for some first order relational language $\mathcal{L}$. A *class of* $\mathcal{C}$-*embeddings* is a collection $\mathcal{E}$ of embeddings $f : M \to N$ (where $M, N \in \mathcal{C}$) such that:

(i) any isomorphism is in $\mathcal{E}$;

(ii) $\mathcal{E}$ is closed under composition;

(iii) if $f : M \to P$ is in $\mathcal{E}$ and $N \leq P$ is a substructure in $\mathcal{C}$ such that $f(M) \subseteq N$, then the map obtained by restricting the range of f to N is also in $\mathcal{E}$.

Suppose $\mathcal{E}$ is a class of $\mathcal{C}$-embeddings. If M is a $\mathcal{L}$-structure and N a finite substructure which is in $\mathcal{C}$, then we say that N is $\mathcal{E}$-*embedded* in M if whenever P is a finite substructure of M which is in $\mathcal{C}$ and contains N, the inclusion map $N \to P$ is in $\mathcal{E}$.

Definition 2.7 Suppose $\mathcal{E}$ is a class of $\mathcal{C}$-embeddings. We define the *joint embedding property* (JEP) and the *amalgamation property* (AP).

(JEP) If $M, N \in \mathcal{C}$ there exists $P \in \mathcal{C}$ and embeddings $f : M \to P$ and $g : N \to P$ such that $f, g \in \mathcal{E}$.

(AP) Suppose $M, N_1, N_2 \in \mathcal{C}$ and $f_i : M \to N_i$ are embeddings in $\mathcal{E}$. Then there exists $P \in \mathcal{C}$ and embeddings $g_i : N_i \to P$ in $\mathcal{E}$ such that $g_1 f_1 = g_2 f_2$.

Theorem 2.8 *Suppose $\mathcal{C}$ is a collection of finite $\mathcal{L}$-structures in which the number of isomorphism types of any finite size is finite. Suppose $\mathcal{E}$ is a class of $\mathcal{C}$-embeddings which satisfies JEP and AP. Then there exists a countable $\mathcal{L}$-structure, M (the Fraïssé limit), with the following properties:*

(a) the class of $\mathcal{E}$-embedded substructures of M is equal to $\mathcal{C}$;

(b) M is a union of finite $\mathcal{E}$-embedded substructures;

(c) Suppose $N, P \in \mathcal{C}$. If N is $\mathcal{E}$-embedded in M and $f : N \to P$ is in $\mathcal{E}$, then there exists P^* that is $\mathcal{E}$-embedded in M, containing N and an isomorphism $g : P \to P^*$ such that $gf(x) = x$ for all $x \in N$.

Any two countable structures with these three properties are isomorphic, and any isomorphism between $\mathcal{E}$-embedded finite substructures of M extends to an automorphism of M.

Proof. See Theorem 2.10 of [7].

Finally, we define the central notions in the paper, mentioned in the introduction and taken from [3] and [2].

Definition 2.9 For each k such that $k \geq 2$, we say that a permutation group (G, Ω) *preserves a limit of Steiner k-systems* on Ω, if (G, Ω) is $(k+1)$-transitive but not $(k+1)$-primitive, and there is a linearly ordered set $(J, \leq)$ with no greatest element and an increasing chain $(S_i : i \in J)$ of subsets of Ω such that the following conditions hold.

(i) $\bigcup(S_i : i \in J) = \Omega$.

(ii) For each $i \in J$, $G_{\{S_i\}}$ is k-transitive on S_i and preserves a non-trivial Steiner k-system on S_i.

(iii) For each $i \in J$, $\Omega \backslash S_i$ is a Jordan set for G.

(iv) If $i < j$ then S_i is a subset of the set of points incident with a block of the Steiner system invariant under $G_{\{S_j\}}$.

(v) For all $g \in G$ there is $i_0 \in J$, dependent on g, such that for every $i > i_0$ there is $j \in J$ so that $g(S_i) = S_j$ and g induces a bijection from the block set of S_i to the block set of S_j.

Groups that preserve a limit of Steiner k-systems can arise as the automorphism group of a semilinear tower of Steiner k-systems, which we define below.

Definition 2.10 An (upper) *semilinear tower of Steiner k-systems* is a collection of Steiner k-systems, $((\Omega_\alpha, \mathcal{B}_\alpha, I_\alpha) : \alpha \in \mathcal{T})$ such that

(i) For all $\alpha, \beta \in \mathcal{T}$ there exists $\gamma \in \mathcal{T}$ with $\Omega_\gamma \supseteq \Omega_\alpha \cup \Omega_\beta$.

(ii) $|\Omega_\alpha \cap \Omega_\beta| \geq k \wedge \alpha \neq \beta \rightarrow$
$(((\exists \Sigma_1 \in \mathcal{B}_\alpha)(x \in \Omega_\beta \rightarrow x\mathrm{I}_\alpha \Sigma_1)) \vee ((\exists \Sigma_2 \in \mathcal{B}_\beta)(x \in \Omega_\alpha \rightarrow x\mathrm{I}_\beta \Sigma_2)))$.

The semilinear tower that we shall construct in this paper will have further properties, namely

(iii) For all $\alpha \in \mathcal{T}$ there exists $\beta \in \mathcal{T}$ such that $\Omega_\alpha \subset \Omega_\beta$.

(iv) For all $\alpha, \beta \in \mathcal{T}$ with $|\Omega_\alpha \cap \Omega_\beta| \leq k - 1$ there exists $\gamma \in \mathcal{T}$ such that $\Omega_\gamma \supseteq \Omega_\alpha \cup \Omega_\beta$, and for all $\delta \in \mathcal{T}$ with $\Omega_\delta \supseteq \Omega_\alpha \cup \Omega_\beta$, we have $\Omega_\gamma \subseteq \Omega_\delta$. Further there exists distinct $\Sigma_1, \Sigma_2 \in \mathcal{B}_\gamma$ such that $(x \in \Omega_\alpha \rightarrow x\mathrm{I}_\gamma \Sigma_1) \wedge (x \in \Omega_\beta \rightarrow x\mathrm{I}_\gamma \Sigma_2)$.

Condition (ii) implies in particular that if $|\Omega_\alpha \cap \Omega_\beta| \geq k$ then $\Omega_\alpha \subset \Omega_\beta$ or $\Omega_\beta \subset \Omega_\alpha$, so justifies the term 'semilinear tower'.

We shall construct an upper semilinear tower M^* of Steiner k-systems as a three sorted first order structure. The three sorts will be denoted by $\mathcal{T}, \mathcal{B}$ and Ω. The sort Ω has as its elements, all the points in the various Steiner k-systems, and the sort $\mathcal{B}$ has as its elements, all the blocks of the various Steiner k-systems. The third sort $\mathcal{T}$, corresponds to the index set in Definition 2.10. It will turn out that $\mathrm{Aut}(M^*)$ preserves a limit of Steiner systems, in its action on one of the sorts.

Our language will also have six relations which we interpret informally as follows.

$\leq \subseteq \mathcal{T} \times \mathcal{T} : \alpha \leq \beta$ holds if the Steiner k-system indexed by α has its points contained in the set of points contained in the Steiner k-system indexed by β.

$D \subseteq \mathcal{T} \times \mathcal{T} \times \mathcal{T} : D(\alpha; \beta, \gamma)$ holds if the Steiner k-system indexed by α is the least upper bound (under $\leq$) of the Steiner k-systems indexed by β and γ.

$C \subseteq \mathcal{T} \times \mathcal{B} : C(\alpha, \Sigma)$ holds if Σ is a block of the Steiner k-system indexed by α.

$B \subseteq \mathcal{T} \times \mathcal{B} : B(\alpha, \Sigma)$ holds if the points of the Steiner k-system indexed by α are all incident with Σ. Here Σ must be a block of another Steiner k-system.

$I \subseteq \mathcal{T} \times \Omega \times \mathcal{B} : I(\alpha, x, \Sigma)$ holds if in the Steiner k-system indexed by α, x is incident with the block Σ.

$L \subseteq \mathcal{T} \times \Omega : L(\alpha, x)$ holds if the Steiner k-system indexed by α has x as one of its points.

For clarity, we shall split the construction up into sections. In the next section we shall show how the individual Steiner systems are constructed. Thus we will only use the two sorts, Ω and $\mathcal{B}$. In this construction, the complements of individual blocks form Jordan sets. The Steiner systems constructed in the next section will be isomorphic to those that are built in the tower construction of Section 5. However the automorphism group of the tower will not induce on each Steiner system the full automorphism group of the individual Steiner systems, though the induced group will have strong transitivity properties (Lemma 6.5). The semilinear order is constructed in Section 4, using just the sorts $\mathcal{T}$ and $\mathcal{B}$. In Section 5 we combine these methods to build an upper semilinear tower of Steiner k-systems.

3 Construction of Jordan groups that preserve Steiner k-systems

In this section we build Steiner k-systems for each $k \geq 2$ which have automorphism groups that are Jordan groups in which the complements of sets of points incident with single blocks are Jordan sets. An object with similar properties is built by the Hrushovski construction in [11].The parameter k is fixed throughout the section.

Definition 3.1 We define a two-sorted language $\mathcal{L}$ with a sort Ω for *points* and a sort $\mathcal{B}$ for *blocks*. The language has a single binary relation $I \subseteq \Omega \times \mathcal{B}$, and for $x \in \Omega$ and $\Sigma \in \mathcal{B}$ we write $x I \Sigma$ (rather than $I x \Sigma$) to denote that x is *incident* with Σ (or x *lies on* Σ) and Σ is *incident* with x. We denote by $\mathcal{C}^*$ the class of all $\mathcal{L}$-structures $(\Omega, \mathcal{B}, I)$ satisfying the following axioms:

(i) $\forall \Sigma, \Sigma' \in \mathcal{B}\ (\forall^{\neq} x_1, \ldots, x_k \in \Omega((x_1 I \Sigma \wedge \ldots \wedge x_k I \Sigma \wedge x_1 I \Sigma' \wedge \ldots \wedge x_k I \Sigma') \rightarrow \Sigma = \Sigma')$

(If two blocks are both incident with k common points, they are equal)

(ii) $\forall \Sigma \in \mathcal{B}\ \exists^{\neq} x_1, \ldots, x_k \in \Omega(x_1 I \Sigma \wedge \ldots \wedge x_k I \Sigma)$

(Every block is incident with at least k points)

(iii) $\forall x_1, \ldots, x_k \in \Omega\ \exists \Sigma \in \mathcal{B}\ (x_1 I \Sigma \wedge \ldots \wedge x_k I \Sigma)$

(Every k points are incident with a block)

Thus, $\mathcal{C}^*$ is the class of all partial Steiner k-systems in which every k points lie in a single block. Let $\mathcal{C} = \{M \in \mathcal{C} : |M| < \omega\}$.

Given $M \in \mathcal{C}^*$, denote the elements in the two sorts of M, by Ω_M and $\mathcal{B}_M$. For each $M \in \mathcal{C}^*$, given distinct $x_1, \ldots, x_k \in \Omega_M$, denote the unique block Σ of $\mathcal{B}_M$ such that $x_1 I \Sigma \wedge \ldots \wedge x_k I \Sigma$ by $\Sigma(x_1, \ldots, x_k)$.

Next, define $\widehat{\mathcal{C}}^*$ to be the set of those $M \in \mathcal{C}^*$ such that for any distinct non-incident $x_1, \ldots, x_{k+1} \in \Omega_M$, $\mathrm{cl}_M(x_1, \ldots, x_{k+1})$ is free around $x_1, \ldots, x_{k+1}$. We call the members of $\widehat{\mathcal{C}}^*$, *locally free (partial) Steiner k-systems*. For $M, N \in \widehat{\mathcal{C}}^*$ with $M \leq N$, we write $M \leq_c N$ if $\mathrm{cl}_N(M)$ is free around M. Define $\widehat{\mathcal{C}} := \{M \in \widehat{\mathcal{C}}^* : |M| < \omega\}$. Take $\mathcal{E}_{\widehat{\mathcal{C}}}$ to be the class of all embeddings $f : M \rightarrow N$ of elements of $\widehat{\mathcal{C}}$ such that $f(M) \leq_c N$.

Our first task is to show that $\mathcal{E}_{\widehat{\mathcal{C}}}$ is a class of $\widehat{\mathcal{C}}$-embeddings. For ease of notation we shall identify M with $f(M)$.

Lemma 3.2 *If $M, N, P \in \widehat{\mathcal{C}}$ with $M \leq_c N \leq_c P$, then $\mathrm{cl}_N(M) \leq_c P$.*

Proof. Since $N \leq_c P$ we have a function Parent : $\mathcal{B}_{\mathrm{cl}_P(N)} \to \Omega^{\{k\}}_{\mathrm{cl}_P(N)}$ that satisfies Definition 2.4. Put $S = \mathrm{cl}_P(\mathrm{cl}_N(M))$. In order to show $\mathrm{cl}_N(M) \leq_c P$ we define a function Parent$'$: $\mathcal{B}_S \to \Omega^{\{k\}}_S$ by

- $\forall \Sigma \in \mathcal{B}_{\mathrm{cl}_N(M)}$ Parent$'(\Sigma)$ is undefined

- $\forall \Sigma \in \mathcal{B}_S \backslash \mathcal{B}_{\mathrm{cl}_N(M)}$ Parent$'(\Sigma) = $ Parent(Σ)

Hence for all blocks $\Sigma \in \mathcal{B}_S$, we have Parent$'(\Sigma)$ = Parent(Σ) (where, following Definition 2.4, this means that Parent$'(\Sigma)$ is undefined whenever Parent(Σ) is undefined). We must now show that Parent$'$ is a parent function.

Claim There do not exist any points in $(\Omega_S \backslash \Omega_{\mathrm{cl}_N(M)}) \cap \Omega_N$.
Proof. Suppose towards a contradiction that there is $x \in (\Omega_S \backslash \Omega_{\mathrm{cl}_N(M)}) \cap \Omega_N$. Consider a step by step construction of S around $\mathrm{cl}_N(M)$: at odd steps adjoin all points of S incident with any previously defined blocks and at even steps add any blocks of S incident with at least k existing points. (We are not claiming that this is a free construction.) Call this construction $(*)$.

Since $\mathrm{cl}_N(M)$ is closed in N (in the sense of Definition 2.2), there must be some $x^* \in (\Omega_S \backslash \Omega_{\mathrm{cl}_N(M)}) \cap \Omega_N$ and $\Sigma_1 \in \mathcal{B}_S$ such that $x^* I \Sigma_1$ and $\Sigma_1 \notin \mathcal{B}_N$. By Definition 2.4(iii), $x^* \in$ Parent(Σ_1). Now Σ_1 will have been adjoined at an earlier step than x^* in construction $(*)$, due to the presence of at least k points, $x_1, \ldots, x_k$ say.

Since $x^* \in$ Parent(Σ_1), for some $x' \in \{x_1, \ldots, x_k\}$ we have $x' \notin$ Parent(Σ_1). If $x' \in \Omega_{\mathrm{cl}_N(M)} \subseteq \Omega_N$ this again Definition 2.4(iii) is contradicted, since $\Sigma_1 \notin \mathcal{B}_N$. So assume $x' \notin \Omega_{\mathrm{cl}_N(M)}$. Now x' must have been adjoined incident with at least one block, Σ_2 say. Then by Definition 2.4(iv), we have $x' \in$ Parent(Σ_2). Now if $\Sigma_2 \in \mathcal{B}_{\mathrm{cl}_N(M)} \subseteq \mathcal{B}_N$ then $x' \in$ Parent(Σ_2), contrary 2.4(i) as Parent(Σ_2) is undefined. Alternatively if $\Sigma_2 \notin \mathcal{B}_{\mathrm{cl}_N(M)}$, then Σ_2 was added an earlier step in construction $(*)$, due to the presence of at least k points, one of them not a member of Parent(Σ_2).

Since $\mathcal{B}_S$ is finite, there can only be finitely many steps in construction $(*)$. Hence at some step we find some $x'' \in$ Parent(Σ') for some $\Sigma' \in \mathcal{B}_{\mathrm{cl}_N(M)}$, contrary to 2.4(i). This contradiction proves the claim.

We now show that Parent$'$ satisfies the conditions in Definition 2.4.
(i) By the definition of Parent$'$, Parent$'(\Sigma)$ is undefined for all $\Sigma \in \mathcal{B}_{\mathrm{cl}_N(M)}$.
(ii) First note that $S = \mathrm{cl}_P(\mathrm{cl}_N(M)) \leq \mathrm{cl}_P(N)$. Since $S \backslash \mathrm{cl}_N(M)$ does not contain any points in Ω_N by the claim above, there will be no blocks in $\mathcal{B}_S \cap \mathcal{B}_N$ but not in $\mathrm{cl}_N(M)$. Hence Parent$'$ has been defined for all blocks of S.
(iii) Given $\Sigma \in \mathcal{B}_S \backslash \mathcal{B}_{\mathrm{cl}_N(M)}$ and $x \in \Omega_{\mathrm{cl}_N(M)}$ with $x I \Sigma$, then $x \in$ Parent$'(\Sigma)$ since Parent$'(\Sigma) = $ Parent(Σ) and we have $x \in$ Parent(Σ), since Parent satisfies 2.4.

(iv) Given $x \in \Omega_S \backslash \Omega_{\mathrm{cl}_N(M)}$ with $x\mathrm{I}\Sigma_1$ and $x\mathrm{I}\Sigma_2$ then $x \in \mathrm{Parent}'(\Sigma_1)$ or $x \in \mathrm{Parent}'(\Sigma_2)$, since $\mathrm{Parent}'(\Sigma_1) = \mathrm{Parent}(\Sigma_1)$ and $\mathrm{Parent}'(\Sigma_2) = \mathrm{Parent}(\Sigma_2)$.

(v) We wish to show that for all $x \in \Omega_S \backslash \Omega_{\mathrm{cl}_N(M)}$ there exists a block $\Sigma \in \mathcal{B}_S$ such that $x\mathrm{I}\Sigma$ and $x \notin \mathrm{Parent}'(\Sigma) = \mathrm{Parent}(\Sigma)$. If not, for all $\Sigma \in \mathcal{B}_S$ with $x\mathrm{I}\Sigma$ we have $x \in \mathrm{Parent}'(\Sigma) = \mathrm{Parent}(\Sigma)$. Then argue as in the claim above, with x taking the place of x^*.

Corollary 3.3 *Suppose* $D, M, N \in \widehat{\mathcal{C}}$ *with* $D \leq M \leq_c N$ *and* $\mathrm{cl}_M(D) = D$. *Then* $\mathrm{cl}_N(D)$ *is free around* D *and no blocks in* $\mathcal{B}_M \backslash \mathcal{B}_{\mathrm{cl}_N(D)}$ *have any points in* $\Omega_{\mathrm{cl}_N(D)} \backslash \Omega_M$.

Proof. By Lemma 3.2, $\mathrm{cl}_N(D)$ is free around D since $\mathrm{cl}_M(D)$ is trivially free around D. Thus, there are functions $\mathrm{Parent} : \mathcal{B}_{\mathrm{cl}_N(M)} \to \Omega_{\mathrm{cl}_N(M)}^{\{k\}}$ and $\mathrm{Parent}' : \mathcal{B}_{\mathrm{cl}_N(D)} \to \Omega_{\mathrm{cl}_N(D)}^{\{k\}}$ that satisfy 2.4. By the proof of Lemma 3.2, whenever $\mathrm{Parent}'(\Sigma)$ is defined for some $\Sigma \in \mathcal{B}_{\mathrm{cl}_N(D)}$ we have $\mathrm{Parent}'(\Sigma) = \mathrm{Parent}(\Sigma)$, and whenever $\mathrm{Parent}'(\Sigma)$ is undefined for some $\Sigma \in \mathcal{B}_{\mathrm{cl}_N(D)}$, $\mathrm{Parent}(\Sigma)$ is also undefined.

Suppose that Σ_1 is in $\mathcal{B}_M \backslash \mathcal{B}_{\mathrm{cl}_N(D)}$ and that it is incident with some point $x \in \Omega_{\mathrm{cl}_N(D)} \backslash \Omega_M$. Now by 2.4(iii), $\mathrm{Parent}(\Sigma_1)$ is undefined and so $x \notin \mathrm{Parent}(\Sigma_1)$. Since $\mathrm{cl}_N(D)$ is free around D there exists $\Sigma_2 \in \mathcal{B}_{\mathrm{cl}_N(D)}$ with $x\mathrm{I}\Sigma_2$ and $x \notin \mathrm{Parent}'(\Sigma_2) = \mathrm{Parent}(\Sigma_2)$. Now $\Sigma_2 \notin \mathcal{B}_M \backslash \mathcal{B}_{\mathrm{cl}_N(D)}$ so $\Sigma_1 \neq \Sigma_2$. Hence $\Sigma_1 \neq \Sigma_2 \wedge x\mathrm{I}\Sigma_1 \wedge x\mathrm{I}\Sigma_2$ and $x \notin \mathrm{Parent}(\Sigma_1)$ and $x \notin \mathrm{Parent}(\Sigma_2)$, contrary to 2.4(iv).

Lemma 3.4 *Given* $M, N, P \in \widehat{\mathcal{C}}$ *with* $M \leq P \leq N$, *if* $M \leq_c N$ *then* $M \leq_c P$.

Proof. First we note that clearly $\mathrm{cl}_P(M) \leq \mathrm{cl}_N(M)$ and since $M \leq_c N$ we have a function $\mathrm{Parent} : \mathcal{B}_{\mathrm{cl}_N(M)} \to \Omega_{\mathrm{cl}_N(M)}^{\{k\}}$ that satisfies Definition 2.4. To show $M \leq_c P$ we define a function $\mathrm{Parent}' : \mathcal{B}_{\mathrm{cl}_P(M)} \to \Omega_{\mathrm{cl}_P(M)}^{\{k\}}$ as

- $\mathrm{Parent}'(\Sigma)$ is undefined for all $\Sigma \in \mathcal{B}_M$

- $\mathrm{Parent}'(\Sigma) = \mathrm{Parent}(\Sigma)$ for all $\Sigma \in \mathcal{B}_{\mathrm{cl}_P(M)} \backslash \mathcal{B}_M$

Hence for all blocks $\Sigma \in \mathcal{B}_{\mathrm{cl}_P(M)}$, we have $\mathrm{Parent}'(\Sigma) = \mathrm{Parent}(\Sigma)$.

We show that Parent' satisfies the axioms of Definition 2.4, and hence $M \leq_c P$ as required.

(i) For all $\Sigma \in \mathcal{B}_M$, $\mathrm{Parent}(\Sigma)$ is undefined by the above definition.

(ii) For all $\Sigma \in \mathcal{B}_{\mathrm{cl}_P(M)} \backslash \mathcal{B}_M$, $\mathrm{Parent}(\Sigma)$ is defined since $\mathrm{cl}_P(M) \leq \mathrm{cl}_N(M)$.

(iii) For all $\Sigma \in \mathcal{B}_{\mathrm{cl}_P(M)} \backslash \mathcal{B}_M$ and $x \in \Omega_M$ with $x\mathrm{I}\Sigma$ then $x \in \mathrm{Parent}'(\Sigma)$ since $\mathrm{Parent}'(\Sigma) = \mathrm{Parent}(\Sigma)$ and $x \in \mathrm{Parent}(\Sigma)$.

(iv) For all $x \in \Omega_{\mathrm{cl}_P(M)} \backslash \Omega_M$ with $x\mathrm{I}\Sigma_1 \wedge x\mathrm{I}\Sigma_2$ then $x \in \mathrm{Parent}'(\Sigma_1)$ or $x \in \mathrm{Parent}'(\Sigma_2)$ since $\mathrm{Parent}'(\Sigma_1) = \mathrm{Parent}(\Sigma_1)$ and $\mathrm{Parent}'(\Sigma_2) = \mathrm{Parent}(\Sigma_2)$.

(v) We wish to show that for all $x \in \Omega_{\mathrm{cl}_P(M)} \backslash \Omega_M$ there exists a block $\Sigma \in \mathcal{B}_{\mathrm{cl}_P(M)}$ such that $x\mathrm{I}\Sigma$ and $x \notin \mathrm{Parent}'(\Sigma) = \mathrm{Parent}(\Sigma)$. The argument is similar to that for (v) in the proof of Lemma 3.2 above.

Lemma 3.5 $\mathcal{E}_{\widehat{C}}$ *is a class of $\widehat{C}$-embeddings.*

Proof. We show each condition of Definition 2.6 in turn.

(i) Clearly any isomorphism is in $\mathcal{E}_{\widehat{C}}$.

(ii) Given $M, N, P \in \widehat{C}$ with $M \leq_c N \leq_c P$ we wish to show $M \leq_c P$. By Lemma 3.2, $\mathrm{cl}_N(M) \leq_c P$, and clearly $\mathrm{cl}_P(M) = \mathrm{cl}_P(\mathrm{cl}_N(M))$. Hence we can freely construct $\mathrm{cl}_P(M)$ around M by first freely constructing $\mathrm{cl}_N(M)$ around M as $M \leq_c N$, and then freely constructing $\mathrm{cl}_P(\mathrm{cl}_N(M))$ around $\mathrm{cl}_N(M)$.

(iii) Given $M, N, P \in \widehat{C}$ with $M \leq P \leq N$ and $M \leq_c N$, the fact that $M \leq_c P$ follows from Lemma 3.4.

Lemma 3.6 $\mathcal{E}_{\widehat{C}}$ *satisfies JEP.*

Proof. Given $M, N \in \widehat{C}$, take P to be the element of $\mathcal{C}_k$ with point set $\Omega_P = \Omega_M \cup \Omega_N$ where Ω_M and Ω_N are disjoint, and blocks

$$\mathcal{B}_P = \mathcal{B}_M \cup \mathcal{B}_N \cup \{\Sigma(x_1, \ldots, x_k) : x_i \text{ distinct }, \exists i_1, i_2 \leq k \, (x_{i_1} \in \Omega_M, x_{i_2} \in \Omega_N)\}$$

where $\Sigma(x_1, \ldots, x_k)$ is a new block incident with just the points $x_1, \ldots, x_k$. We shall show that $P \in \widehat{C}$ and then that $M \leq_c P$.

To show $P \in \widehat{C}$, consider a $(k+1)$-set of elements of Ω_P, $\{x_1, \ldots, x_{k+1}\}$ say, with the property that $x_1, \ldots, x_{k+1}$ are non-incident. We wish to show that $\mathrm{cl}_P(x_1, \ldots, x_{k+1})$ is free around $x_1, \ldots, x_{k+1}$. Now without loss of generality assume for some i where $1 \leq i \leq k+1$ that $x_j \in M$ if $j \leq i$ and $x_j \in N$ if $j > i$.

Case 1. $x_1, \ldots, x_{k+1} \in \Omega_M$.

Here $\mathrm{cl}_P(x_1, \ldots, x_{k+1}) = \mathrm{cl}_M(x_1, \ldots, x_{k+1})$, since no blocks in M are incident with any points in $\Omega_P \backslash \Omega_M$. Hence $\mathrm{cl}_P(x_1, \ldots, x_{k+1})$ is free around $x_1, \ldots, x_{k+1}$, since $M \in \widehat{C}$.

Case 2. $x_1, \ldots, x_k \in \Omega_M$, and $x_{k+1} \in \Omega_N$.

Now $\mathrm{cl}_P(x_1, \ldots, x_{k+1})$ consists of the following blocks, $\Sigma(x_1, \ldots, x_k)$ and other blocks incident with just k points, where $k-1$ of these points are incident with $\Sigma(x_1, \ldots, x_k)$ and the remaining one is x_{k+1}. The points of $\mathrm{cl}_P(x_1, \ldots, x_{k+1})$ are x_{k+1} and all those incident with $\Sigma(x_1, \ldots, x_k)$. Again clearly $\mathrm{cl}_P(x_1, \ldots, x_{k+1})$ is free around $x_1, \ldots, x_{k+1}$.

Case 3. $x_1, \ldots x_i \in \Omega_M$, and $x_{i+1}, \ldots, x_{k+1} \in \Omega_N$ $(2 \leq i \leq k-1)$.

In this case $\mathrm{cl}_P(x_1, \ldots, x_{k+1})$ consists of the points $x_1, \ldots, x_{k+1}$ and those blocks incident with exactly k of them, as the $k + 1$ blocks in this subset are each incident with precisely k points. So $\mathrm{cl}_P(x_1, \ldots, x_{k+1})$ is free around $x_1, \ldots, x_{k+1}$.

Case 4. $x_1 \in \Omega_M$, and $x_2, \ldots, x_{k+1} \in \Omega_N$.
Similar to Case 2.

Case 5. $x_1, \ldots, x_{k+1} \in \Omega_N$.
Similar to Case 1.

To show that $M \leq_c P$, consider $\mathrm{cl}_P(M)$. Now since no block in M is incident with any points in N, $\mathrm{cl}_P(M) = M$, so trivially $\mathrm{cl}_P(M)$ is free around M. A similar argument shows $N \leq_c P$, so $\mathcal{E}_{\widehat{c}}$ satisfies JEP.

Lemma 3.7 *Suppose $M, N \in \widehat{C}$ and $M \leq_c N$. If Σ is a block in $\mathcal{B}_M$, and if Γ is the set of all points incident with Σ then for all $\Gamma' \subseteq \Gamma$, we have $M' \leq_c N$ where we define $M' := M \cup \Gamma' \cup \{\Sigma(x_1, \ldots, x_k) : x_1, \ldots, x_k \in \Omega_M \cup \Gamma' \text{ distinct}\}$.*

Proof. We first note that clearly $\mathrm{cl}_N(M') = \mathrm{cl}_N(M)$. Now since $M \leq_c N$ we have a function $\mathrm{Parent} : \mathcal{B}_{\mathrm{cl}_N(M)} \to \Omega_{\mathrm{cl}_N(M)}^{\{k\}}$ that satisfies Definition 2.4. We define $\mathrm{Parent}' : \mathcal{B}_{\mathrm{cl}_N(M')} \to \Omega_{\mathrm{cl}_N(M')}^{\{k\}}$ as follows: for $\Sigma' \in \mathcal{B}_{M'}$, $\mathrm{Parent}'(\Sigma')$ is undefined, and for all $\Sigma' \in \mathcal{B}_{\mathrm{cl}_N(M')} \backslash \mathcal{B}_{M'}$, $\mathrm{Parent}'(\Sigma') := \mathrm{Parent}(\Sigma')$.

Claim If some block Σ^* is in $\mathcal{B}_{M'} \backslash \mathcal{B}_M$, then for all $x^* \in \mathrm{Parent}(\Sigma^*)$ we have $x^* \in \Omega_M \cup \Gamma'$.
Proof. Since $\Sigma^* \in \mathcal{B}_{M'}$ there exists k points, $x_1, \ldots, x_k$ say, with the property that $x_1 \mathrm{I} \Sigma^* \wedge \ldots \wedge x_k \mathrm{I} \Sigma^*$ and $x_1, \ldots, x_k \in \Omega_M \cup \Gamma'$. For each i with $1 \leq i \leq k$, if $x_i \in \Omega_M$ then $x_i \in \mathrm{Parent}(\Sigma^*)$ by Definition 2.4(iii). Alternatively if $x_i \in \Gamma' \backslash \Omega_M$ then $x_i \in \mathrm{Parent}(\Sigma^*)$ since $\mathrm{Birth}_{\mathrm{cl}_N(M)}(x_i) = \Sigma$. Hence for all $x^* \in \mathrm{Parent}(\Sigma^*)$ we have $x^* \in \Omega_M \cup \Gamma'$ as required.

We now show that Parent' satisfies the axioms of 2.4, and hence $M' \leq_c N$ as required.
 (i) By its definition, $\mathrm{Parent}'(\Sigma')$ is undefined for all $\Sigma' \in \mathcal{B}_{M'}$.
 (ii) $\mathrm{Parent}'(\Sigma')$ is defined for all $\Sigma' \in \mathcal{B}_{\mathrm{cl}_N(M')} \backslash \mathcal{B}_{M'}$ since Parent was defined for all blocks in this domain.
 (iii) For all $\Sigma' \in \mathcal{B}_{\mathrm{cl}_N(M')} \backslash \mathcal{B}_{M'}$ and $x \in \Omega_{M'}$ with $x \mathrm{I} \Sigma'$ then $x \in \mathrm{Parent}'(\Sigma')$ since $\mathrm{Parent}'(\Sigma') = \mathrm{Parent}(\Sigma')$ and points in $\Gamma' \backslash \Omega_M$ have Σ as their birth block (in the sense of the function Parent).
 (iv) Suppose $x \in \Omega_{\mathrm{cl}_N(M')} \backslash \Omega_{M'}$ and $\Sigma_1, \Sigma_2 \in \mathcal{B}_{\mathrm{cl}_N(M')}$ with $x \mathrm{I} \Sigma_1$ and $x \mathrm{I} \Sigma_2$. If $\Sigma_1, \Sigma_2 \notin \mathcal{B}_{M'} \backslash \mathcal{B}_M$ then $x \in \mathrm{Parent}'(\Sigma_1)$ or $x \in \mathrm{Parent}'(\Sigma_2)$ as $\mathrm{Parent}'(\Sigma_1) = \mathrm{Parent}(\Sigma_1)$ and $\mathrm{Parent}'(\Sigma_2) = \mathrm{Parent}(\Sigma_2)$. If $\Sigma_1, \Sigma_2 \in \mathcal{B}_{M'} \backslash \mathcal{B}_M$ then $x \mathrm{I} \Sigma_1 \wedge x \mathrm{I} \Sigma_2$ with $x \notin \mathrm{Parent}(\Sigma_1)$ and $x \notin \mathrm{Parent}(\Sigma_2)$ by the claim. This contradicts

2.4(iv). Therefore assume without loss of generality that $\Sigma_1 \in \mathcal{B}_{M'}\backslash\mathcal{B}_M$ and $\Sigma_2 \notin \mathcal{B}_{M'}\backslash\mathcal{B}_M$. Then for all $x^* \in \mathrm{Parent}(\Sigma_1)$ we have $x^* \in \Omega_M \cup \Gamma'$ by the claim above. Therefore by 2.4(iv) $x \in \mathrm{Parent}(\Sigma_2)$, so $x \in \mathrm{Parent}'(\Sigma_2)$ since $\mathrm{Parent}'(\Sigma_2) = \mathrm{Parent}(\Sigma_2)$.

(v) Since $M \leq_c N$, for all $x \in \Omega_{\mathrm{cl}_N(M)}\backslash\Omega_{M'}$ there exists $\Sigma' \in \mathcal{B}_{\mathrm{cl}_N(M)}$ with $x I \Sigma'$ and $x \notin \mathrm{Parent}(\Sigma')$. Now $x \notin \mathrm{Parent}'(\Sigma')$ as if $\Sigma' \notin \mathcal{B}_{M'}\backslash\mathcal{B}_M$ then $\mathrm{Parent}'(\Sigma') = \mathrm{Parent}(\Sigma')$ and if $\Sigma' \in \mathcal{B}_{M'}\backslash\mathcal{B}_M$ then $\mathrm{Parent}'(\Sigma')$ is undefined. Hence, for all $x \in \Omega_{\mathrm{cl}_N(M)}\backslash\Omega_{M'}$, there exists $\Sigma' \in \mathcal{B}_{\mathrm{cl}_N(M')}$ with $x I \Sigma'$ and $x \notin \mathrm{Parent}'(\Sigma')$, as required.

Lemma 3.8 $\mathcal{E}_{\widehat{C}}$ *satisfies AP.*

Proof. Suppose $M, N_1, N_2 \in \widehat{C}$ where $M \leq_c N_1$ and $M \leq_c N_2$. Take P to be the element of $\mathcal{C}_k$ which has point set $\Omega_P = \Omega_{N_1} \cup \Omega_{N_2}$, and blocks $\mathcal{B}_P = \mathcal{B}_{N_1} \cup \mathcal{B}_{N_2} \cup$

$$\{\Sigma(x_1,\ldots,x_k) : x_i \in \Omega_P \text{ distinct}, \exists i_1, i_2 \leq k \, (x_{i_1} \in \Omega_{N_1}\backslash\Omega_M, x_{i_2} \in \Omega_{N_2}\backslash\Omega_M)\}$$

where $\Sigma(x_1,\ldots,x_k)$ is a new block that is incident with just the points $x_1,\ldots,x_k$.

Clearly if a block is incident with points in both $N_1\backslash M$ and $N_2\backslash M$, it either is in M or is incident with just k points. Also, since $M \leq_c N_1$, and $M \leq_c N_2$, no two blocks in M can be incident with a common point in $N_1\backslash M$, or in $N_2\backslash M$.

To show $P \in \widehat{C}$, we consider each $k+1$ set of elements of Ω_P, $x_1,\ldots,x_{k+1}$ say, with the property that $x_1,\ldots,x_{k+1}$ are non-incident. We will show that $\mathrm{cl}_P(x_1,\ldots,x_{k+1})$ is free around $x_1,\ldots,x_{k+1}$. Without loss of generality, assume for some i where $1 \leq i \leq k+1$ that $x_j \in N_1$ if $j \leq i$ and $x_j \in N_2\backslash M$ if $j > i$.

Case 1. $x_1,\ldots,x_{k+1} \in \Omega_{N_1}$.

If $\mathrm{cl}_{N_1}(x_1,\ldots,x_{k+1})$ contains no blocks that are defined in M, then we have that $\mathrm{cl}_P(x_1,\ldots,x_{k+1}) = \mathrm{cl}_{N_1}(x_1,\ldots,x_{k+1})$ and so $\mathrm{cl}_P(x_1,\ldots,x_{k+1})$ is free around $x_1,\ldots,x_{k+1}$.

So suppose $\mathrm{cl}_{N_1}(x_1,\ldots,x_{k+1})$ contains some blocks that are defined in M and in this case set $D := \mathrm{cl}_{N_1}(x_1,\ldots,x_{k+1}) \cap M$. Now clearly $\mathrm{cl}_M(D) = D$, so $\mathrm{cl}_M(D)$ is trivially free around D. Also $\mathrm{cl}_{N_2}(M)$ is free around M, so by Lemma 3.2, $\mathrm{cl}_{N_2}(\mathrm{cl}_M(D)) = \mathrm{cl}_{N_2}(D)$ is free around D. Now $\mathrm{cl}_P(x_1,\ldots,x_{k+1})$ is equal to

$$\mathrm{cl}_{N_1}(x_1,\ldots,x_{k+1}) \cup$$
$$\mathrm{cl}_{N_2}(D) \cup$$
$$\{\Sigma(y_1,\ldots,y_k) : \quad y_i \in \Omega_{\mathrm{cl}_{N_1}(x_1,\ldots,x_{k+1})} \cup \Omega_{\mathrm{cl}_{N_2}(D)} \text{ distinct},$$
$$\exists i_1, i_2 \leq k \, (y_{i_1} \in \Omega_{N_1}\backslash\Omega_M, y_{i_2} \in \Omega_{N_2}\backslash\Omega_M)\}$$

Hence $\mathrm{cl}_P(x_1, \ldots, x_{k+1})$ is free around $x_1, \ldots, x_{k+1}$ by following the free construction of $\mathrm{cl}_{N_1}(x_1, \ldots, x_{k+1})$ around $x_1, \ldots, x_{k+1}$ with the free construction of $\mathrm{cl}_{N_2}(D)$ around D. We note that none of the additional blocks $\Sigma(y_1, \ldots, y_k)$ can be in $\mathcal{B}_M$ as this would contradict Corollary 3.3, as we would be adjoining a block in $\mathcal{B}_M \backslash \mathcal{B}_D$ that is incident with both a point in $\Omega_{N_1} \backslash \Omega_M$ and a point in $\Omega_{N_2} \backslash \Omega_M$. Hence, such blocks are incident with just k points.

Case 2. $x_1, \ldots, x_k \in \Omega_{N_1}$, and $x_{k+1} \in \Omega_{N_2} \backslash \Omega_M$ with at least one element in $\Omega_{N_1} \backslash \Omega_M$.

Subcase (a) $|\{x : x\mathrm{I}\Sigma(x_1, \ldots, x_k)\} \cap M| \leq k - 2$.

In this case we can freely construct $\mathrm{cl}_P(x_1, \ldots, x_{k+1})$ around $x_1, \ldots, x_{k+1}$ as follows. We adjoin all the points that are incident with the block $\Sigma(x_1, \ldots, x_k)$ and then other blocks that are incident with $k - 1$ points from $\Sigma(x_1, \ldots, x_k)$ plus x_{k+1}. These points and blocks can be adjoined by a free construction. Now since these blocks are incident with points in both $\Omega_{N_1} \backslash \Omega_M$ and $\Omega_{N_2} \backslash \Omega_M$, they are either incident with exactly k points or are defined in $\mathcal{B}_M$.

If none of these additional blocks are in $\mathcal{B}_M$ then we have a free construction of $\mathrm{cl}_P(x_1, \ldots, x_{k+1})$ around $x_1, \ldots, x_{k+1}$ as required. It is not possible to have two such blocks in $\mathcal{B}_M$ as this would contradict the fact that $M \leq_c N_2$, as there would be two blocks in $\mathcal{B}_M$ that are both incident with the point $x_{k+1} \in \Omega_{N_2} \backslash \Omega_M$.

Hence, assume that we are adding one such block, $\Sigma \in \mathcal{B}_M$. Next, adjoin one point, $z \in \Omega_M$ that is incident with Σ, but is not incident with $\Sigma(x_1, \ldots, x_k)$. Such z exists, since $\Sigma \in \mathcal{B}_M$ so is incident with at least k points in Ω_M, and $\Sigma(x_1, \ldots, x_k)$ is incident with at most $k - 2$ points in Ω_M. Now Σ and $\Sigma(x_1, \ldots, x_k)$ are incident with exactly $k - 1$ common points so put $\{y_1, \ldots, y_{k-1}\} = \{x : x\mathrm{I}\Sigma \wedge x\mathrm{I}\Sigma(x_1, \ldots, x_k)\}$, and choose some i with $1 \leq i \leq k$ such that $x_i \notin \{y_1, \ldots, y_{k-1}\}$.

Since $N_1 \in \widehat{\mathcal{C}}$ we know that $\mathrm{cl}_{N_1}(y_1, \ldots, y_{k-1}, x_i, z)$ is free around the points $y_1, \ldots, y_{k-1}, x_i, z$. Hence, by Lemma 3.7, $\mathrm{cl}_{N_1}(E)$ is free around E, where E consists of $\{y_1, \ldots, y_{k-1}, x_i, z\}$, the block $\Sigma(x_1, \ldots, x_k)$ and all points of N_1 that are incident with it, and all blocks that are incident with at least k points from $\{y_1, \ldots, y_{k-1}, x_i, z\} \cup \{x : x\mathrm{I}\Sigma(x_1, \ldots, x_k)\}$.

Let $D = \mathrm{cl}_{N_1}(E) \cap M$. Now $\mathrm{cl}_M(D) = D$, and by Lemma 3.2, $\mathrm{cl}_{N_2}(D)$ is free around D. Further $\Sigma \in \mathcal{B}_D$ since $y_1\mathrm{I}\Sigma \wedge \ldots \wedge y_{k-1}\mathrm{I}\Sigma \wedge z\mathrm{I}\Sigma$. Now Lemma 3.7 yields that $\mathrm{cl}_{N_2}(D \cup \Sigma^*)$ is free around $D \cup \Sigma^*$, where we obtain $D \cup \Sigma^*$ from D by adjoining all points of Ω_{N_2} that are incident with Σ, along with all blocks that are incident with k points from $\Omega_D \cup \{x \in \Omega_{N_2} : x\mathrm{I}\Sigma\}$.

Hence, by following the construction below, we can give a free construction of $\mathrm{cl}_P(x_1, \ldots, x_{k+1})$ around $x_1, \ldots, x_{k+1}$, which provides a free construction in the sense of Definition 2.3. Since we have shown that each step is a free construction and $\mathcal{E}_{\widehat{C}}$ is closed under composition (by Lemma 3.5), this construction is free.

The steps of this construction are as follows.

(i) Adjoin all points incident with $\Sigma(x_1, \ldots, x_k)$ and other blocks incident with $k-1$ points from $\Sigma(x_1, \ldots, x_k)$ plus x_{k+1}. This includes the block Σ.

(ii) Adjoin the point z which is incident with the block Σ.

(iii) Follow the free construction of $\mathrm{cl}_{N_1}(E)$ around E.

(iv) Follow the free construction of $\mathrm{cl}_{N_2}(D \cup \Sigma^*)$ around $D \cup \Sigma^*$.

(v) Adjoin any remaining blocks of P that are incident with points in both $\Omega_{N_1} \backslash \Omega_M$ and $\Omega_{N_2} \backslash \Omega_M$. None of these blocks can be in $\mathcal{B}_M$ since this would contradict Corollary 3.3, so they are incident with just k points.

Subcase (b) $|\{x : x\mathrm{I}\Sigma(x_1, \ldots, x_k)\} \cap M| \geq k$.

Let Γ be the set of points incident with $\Sigma(x_1, \ldots, x_k)$. Choose distinct $y_1, \ldots, y_k \in \Gamma \cap \Omega_M$. Then $\mathrm{cl}_{N_2}(y_1, \ldots, y_k, x_{k+1})$ is free around $y_1, \ldots, y_k, x_{k+1}$, since $N_2 \in \widehat{\mathcal{C}}$.

We can freely generate $\mathrm{cl}_P(x_1, \ldots, x_{k+1})$ around $x_1, \ldots, x_{k+1}$ by first adjoining the points of $\Gamma \cap N_1$, and then following the free construction of $\mathrm{cl}_{N_2}(y_1, \ldots, y_k, x_{k+1})$ around $y_1, \ldots, y_k, x_{k+1}$. Then follow the free construction of the structure $\mathrm{cl}_{N_1}(M \cap (\mathrm{cl}_{N_2}(y_1, \ldots, y_k, x_{k+1}) \cup \Gamma^*))$ where

$$\Gamma^* = \Gamma \cup \{\Sigma(z_1, \ldots, z_k) : z_1, \ldots, z_k \in \Gamma \cup \Omega_{\mathrm{cl}_{N_2}(y_1, \ldots, y_k, x_{k+1})} \text{ distinct}\}$$

This last step is possible as $\mathrm{cl}_{N_1}(M \cap (\mathrm{cl}_{N_2}(y_1, \ldots, y_k, x_{k+1}) \cup \Gamma^*))$ is free around $M \cap (\mathrm{cl}_{N_2}(y_1, \ldots, y_k, x_{k+1}) \cup \Gamma^*)$ by the following argument. Apply Lemma 3.2 to obtain that $\mathrm{cl}_{N_1}(M \cap \mathrm{cl}_{N_2}(y_1, \ldots, y_k, x_{k+1}))$ is free around $M \cap \mathrm{cl}_{N_2}(y_1, \ldots, y_k, x_{k+1})$ (for $M \cap \mathrm{cl}_{N_2}(y_1, \ldots, y_k, x_{k+1})$ is clearly closed in M in the sense of Definition 2.2). Then use Lemma 3.7 to get that $\mathrm{cl}_{N_1}(M \cap (\mathrm{cl}_{N_2}(y_1, \ldots, y_k, x_{k+1}) \cup \Gamma^*))$ is free around $M \cap (\mathrm{cl}_{N_2}(y_1, \ldots, y_k, x_{k+1}) \cup \Gamma^*)$.

For the final step of this construction, adjoin any remaining blocks of P that are incident with points in both $\Omega_{N_1} \backslash \Omega_M$ and $\Omega_{N_2} \backslash \Omega_M$. By Corollary 3.3, none of these blocks can be in $\mathcal{B}_M$, so they are incident with just k points.

Subcase (c). $|\{x : x\mathrm{I}\Sigma(x_1, \ldots, x_k)\} \cap M| = k-1$.

In this case, we set $\{y_1, \ldots, y_{k-1}\} := \{x : x\mathrm{I}\Sigma(x_1, \ldots, x_k)\} \cap M$. If $\Sigma(y_1, \ldots, y_{k-1}, x_{k+1})$ is in $\mathcal{B}_M$, we choose $y \neq y_1, \ldots, y_{k-1}$ in Ω_M that is incident with $\Sigma(y_1, \ldots, y_{k-1}, x_{k+1})$. Since $N_1 \in \widehat{\mathcal{C}}$, it follows that $\mathrm{cl}_{N_1}(y_1, \ldots, y_{k-1}, y, x_i)$ is free around $y_1, \ldots, y_{k-1}, y, x_i$, where $x_i \in \Omega_{N_1} \backslash \Omega_M$ for some $1 \leq i \leq k$. We now give a free construction of $\mathrm{cl}_P(x_1, \ldots, x_{k+1})$ around $x_1, \ldots, x_{k+1}$ by the following argument. First, adjoin the points $y_1, \ldots, y_{k-1}$, then add the block $\Sigma(y_1, \ldots, y_{k-1}, x_{k+1})$ and the point y. Now follow the free construction of $\mathrm{cl}_{N_1}(y_1, \ldots, y_{k-1}, y, x_i)$ around $y_1, \ldots, y_{k-1}, y, x_i$. We cannot at this step adjoin a new block that is incident with x_{k+1}, as $x_{k+1}\mathrm{I}\Sigma(y_1, \ldots, y_{k-1}, x_{k+1})$

and $\Sigma(y_1, \ldots, y_{k-1}, x_{k+1}) \in \mathcal{B}_M$. Finally, follow the free construction of $\text{cl}_{N_2}((\text{cl}_{N_1}(y_1, \ldots, y_{k-1}, y, x_i) \cap M) \cup \Gamma^*)$ around $(\text{cl}_{N_1}(y_1, \ldots, y_{k-1}, y, x_i) \cap M) \cup \Gamma^*$ where

$$\Gamma^* = \{x_{k+1}\} \cup$$

$$\{\Sigma(x_{k+1}, y_1', \ldots, y_{k-1}') : y_1', \ldots, y_{k-1}' \in \text{cl}_{N_1}(y_1, \ldots, y_{k-1}, y, x_i) \cap M \text{ distinct}\}.$$

This construction is possible much as in Subcase (b).

Next, we suppose that $\Sigma(y_1, \ldots, y_{k-1}, x_{k+1})$ is not in M, and give a free construction of $\text{cl}_P(x_1, \ldots, x_{k+1})$ around $x_1, \ldots, x_{k+1}$, as follows. First add all the points incident with $\Sigma(x_1, \ldots, x_k)$, then the block $\Sigma(y_1, \ldots, y_{k-1}, x_{k+1})$ and those points incident with $\Sigma(y_1, \ldots, y_{k-1}, x_{k+1})$. The next step of the construction is to adjoin all additional blocks that are incident with at least k of these points. Now these blocks are either incident with exactly k points, or are in $\mathcal{B}_M$, since they are incident with points in both $\Omega_{N_1} \backslash \Omega_M$ and $\Omega_{N_2} \backslash \Omega_M$. If none of these additional blocks are in $\mathcal{B}_M$, we then have a free construction of $\text{cl}_P(x_1, \ldots, x_{k+1})$ around $x_1, \ldots, x_{k+1}$, as required. Also, since $M \leq_c N_2$, it is not possible that two such blocks lie in $\mathcal{B}_M$.

Thus, we may assume that we are adjoining one such block, $\Sigma \in \mathcal{B}_M$. Now as $M \leq_c N_1$ and $M \leq_c N_2$, the sets $\{x \in N_1 \backslash M : xI\Sigma \wedge xI\Sigma(x_1, \ldots, x_k)\}$ and $\{x \in N_2 \backslash M : xI\Sigma \wedge xI\Sigma(y_1, \ldots, y_{k-1}, x_{k+1})\}$ both have size 1. Hence $k - 2$ points that are incident with Σ are in $\{y_1, \ldots, y_{k-1}\}$. Without loss of generality, assume these are $y_1, \ldots, y_{k-2}$. Then $\Sigma = \Sigma(y, z, y_1, \ldots, y_{k-2})$ where y is the unique point in $\Omega_{N_1} \backslash \Omega_M$ that is incident with Σ and $\Sigma(x_1, \ldots, x_k)$, and $z \in \Omega_M$ is incident with Σ but is not incident with $\Sigma(x_1, \ldots, x_k)$. Such a z exists as $\Sigma(x_1, \ldots, x_k)$ is incident with just $k - 1$ points from Ω_M.

Let D denote the structure with point set $\Omega_D = \{x : xI\Sigma(x_1, \ldots, x_k)\} \cup \{z\}$ and all blocks incident with at least k of these points. We shall now show that $\text{cl}_{N_1}(D)$ is free around D. Also, we note that we can follow this step of the free construction, as we have adjoined no other points or blocks in N_1. Since $N_1 \in \hat{\mathcal{C}}$, we know that $\text{cl}_{N_1}(x_1, \ldots, x_k, z)$ is free around D, so by adjoining the $k + 1$ blocks incident with k points from $x_1, \ldots, x_k, z$ and applying Lemma 3.7, we see that $\text{cl}_{N_1}(D)$ is free around D. Also $\Sigma \in \mathcal{B}_{\text{cl}_{N_1}(D)}$ since $\Sigma = \Sigma(y, z, y_1, \ldots, y_{k-2})$.

Let F_1 be the structure consisting of $\text{cl}_{N_1}(D) \cap M$, all points of $\Omega_{N_2} \backslash \Omega_M$ that are incident with Σ, and all blocks incident with at least k points from $(\text{cl}_{N_1}(D) \cap M) \cup \{x \in \Omega_{N_2} \backslash \Omega_M : xI\Sigma\}$. Likewise, let F_2 consist of all points of $\Omega_{N_2} \backslash \Omega_M$ incident with $\Sigma(y_1, \ldots, y_{k-1}, x_{k+1})$ plus any blocks incident with at least k of these points. By Lemma 3.2, $\text{cl}_{N_2}(\text{cl}_{N_1}(D) \cap M)$ is free around $\text{cl}_{N_1}(D) \cap M$, so by Lemma 3.7 twice, $\text{cl}_{N_2}(F_1)$ is free around F_1 and $\text{cl}_{N_2}(F_2)$ is free around F_2.

Thus, for the next step of this construction we follow the free construction of $\text{cl}_{N_2}(F_2)$ around F_2. Finally to complete a free construction of $\text{cl}_P(x_1, \ldots,$

x_{k+1}) around $x_1, \ldots, x_{k+1}$, we adjoin any remaining blocks of P that are incident with points in both $\Omega_{N_1} \backslash \Omega_M$ and $\Omega_{N_2} \backslash \Omega_M$. None of these blocks can be in $\mathcal{B}_M$ since this would contradict Corollary 3.3, so they are incident with just k points.

Case 3. $x_1, \ldots, x_i \in \Omega_{N_1}$, and $x_{i+1}, \ldots, x_{k+1} \in \Omega_{N_2} \backslash \Omega_M$ where $2 \leq i \leq k - 1$ and at least two points are in $\Omega_{N_1} \backslash \Omega_M$.

In this case, we construct $\mathrm{cl}_P(x_1, \ldots, x_{k+1})$ freely around $x_1, \ldots, x_{k+1}$ as follows.

First, adjoin the $k + 1$ blocks that are incident with k points from the set $\{x_1, \ldots, x_{k+1}\}$. Since $M \leq_c N_1$ and $M \leq_c N_2$, at most one of these blocks can be in $\mathcal{B}_M$, and the rest are incident with exactly k points. If no block added is in $\mathcal{B}_M$, we have a free construction of $\mathrm{cl}_P(x_1, \ldots, x_{k+1})$ around $x_1, \ldots, x_{k+1}$ as required. Hence, assume we adjoin some block $\Sigma \in \mathcal{B}_M$. Now for some i, with $1 \leq i \leq k+1$ we have $x_i \not I \Sigma$. Without loss of generality, assume $x_i \in N_1$. Now, using the fact that $N_1 \in \widehat{\mathcal{C}}$ and Lemma 3.7, $\mathrm{cl}_{N_1}(D)$ is free around D where D has point set $\{x_i\} \cup \{x \in N_1 : x I \Sigma\}$ and all blocks incident with at least k of these points.

By Lemma 3.7 and Lemma 3.2, it follows that $\mathrm{cl}_{N_2}(E)$ is free around E, where E is obtained by adjoining to $M \cap \mathrm{cl}_{N_1}(D)$, all points of N_2 that are incident with Σ, along with all blocks incident with at least k points from $(M \cap D) \cup \{x \in \Omega_{N_2} : x I \Sigma\}$. Thus, after the construction of $\mathrm{cl}_{N_1}(D)$ around D, follow that of $\mathrm{cl}_{N_2}(E)$ around E. For the final step, adjoin any remaining blocks of P that are incident with points in both $\Omega_{N_1} \backslash \Omega_M$ and $\Omega_{N_2} \backslash \Omega_M$. None of these blocks can be in $\mathcal{B}_M$ since this would contradict Corollary 3.3, so they are incident with just k points. The steps sketched above give a free construction of $\mathrm{cl}_P(x_1, \ldots, x_{k+1})$ around $x_1, \ldots, x_{k+1}$.

Case 4. $x_1, \ldots, x_i \in \Omega_{N_1}$, and $x_{i+1}, \ldots, x_{k+1} \in \Omega_{N_2} \backslash \Omega_M$ where $1 \leq i \leq k - 1$ and exactly one point is in $\Omega_{N_1} \backslash \Omega_M$.

This is similar to Case 2 (with k points from $x_1, \ldots, x_{k+1}$ in Ω_{N_2}).

Case 5. $x_1, \ldots, x_{k+1} \in \Omega_{N_2}$ and $x_i \in \Omega_{N_2} \backslash \Omega_M$ for some i with $1 \leq i \leq k+1$.
This is like Case 1.

Finally, we verify that $N_1 \leq_c P$. Clearly $\mathrm{cl}_P(N_1) = N_1 \cup \mathrm{cl}_{N_2}(M) \cup$

$$\{\Sigma(x_1, \ldots, x_k) : x_i \text{ distinct } \exists i_1, i_2 \leq k \ (x_{i_1} \in N_1 \backslash M, x_{i_2} \in \mathrm{cl}_{N_2}(M) \backslash M)\}$$

where $\Sigma(x_1, \ldots, x_k)$ is a new block incident with just the points $x_1, \ldots, x_k$. Thus $\mathrm{cl}_P(N_1)$ is free around N_1, by following the free construction of $\mathrm{cl}_{N_2}(M)$ around M. Similarly $N_2 \leq_c P$, so $\mathcal{E}_{\widehat{\mathcal{C}}}$ satisfies AP.

Proposition 3.9 *There exists a countable $\mathcal{L}$-structure S with the following properties*

(i) The class of $\mathcal{E}_{\widehat{C}}$-embedded substructures of S is equal to $\widehat{C}$.

(ii) S is the union of a chain of finite $\mathcal{E}_{\widehat{C}}$-embedded substructures.

(iii) If M is $\mathcal{E}_{\widehat{C}}$-embedded in S and $f : M \to N$ is in $\mathcal{E}_{\widehat{C}}$, then there exists P that is $\mathcal{E}_{\widehat{C}}$-embedded in S containing M and an isomorphism $g : N \to P$ such that $gf(y) = y$ for all $y \in M$.

Furthermore, any two countable structures with these three properties are isomorphic, and any isomorphism between $\mathcal{E}_{\widehat{C}}$-embedded finite substructures of S extends to an automorphism of S. The structure S is a Steiner k-system where each block is incident with ω many points.

Proof. To construct S, apply Theorem 2.8 to $\mathcal{E}_{\widehat{C}}$ and $\widehat{C}$. To see that S is a Steiner k-system, we are just required to check that every block is incident with ω many points, since it is clear from the axioms that every k points are incident with a unique block and that S is non-trivial.

Let M be $\mathcal{E}_{\widehat{C}}$-embedded in S, and let Σ be a block of S that is in M. If we take $N \in \widehat{C}$ to be the structure consisting of M and arbitrary new points $x_1, \ldots, x_n$ (for some $n < \omega$) say, with the additional relations $x_1 I \Sigma \wedge \ldots \wedge x_n I \Sigma$ and additional blocks that are incident with exactly k points, so that any k points are incident with a unique block, then clearly $N \in \widehat{C}$ and $M \leq_c N$. Hence there is an isomorphic (over M) copy P of N which is $\mathcal{E}_{\widehat{C}}$-embedded in S over M. Thus in the limit situation every block will be incident with ω many points.

Lemma 3.10 *The structure consisting of $k + 1$ points that are not incident with a single block is $\mathcal{E}_{\widehat{C}}$-embedded in S. Hence $\mathrm{Aut}(S)$ is transitive on non-incident ordered $k + 1$ sets and indeed is k-transitive.*

Proof. This is immediate, since any such $k + 1$ set is $\mathcal{E}_{\widehat{C}}$-embedded in any finite structure in $\widehat{C}$.

Corollary 3.11 *S is not $\aleph_0$-categorical.*

Proof. By Proposition 3.9, it is clear that for any non-incident $k + 1$ set of points, $x_1, \ldots, x_{k+1}$ say, $\mathrm{cl}_S(x_1, \ldots, x_{k+1})$ is a Steiner k-system that is freely generated around $x_1, \ldots, x_{k+1}$. The corollary now follows from the Ryll-Nardzewski Theorem, for given $n < \omega$, there will always be points $x \in \mathrm{cl}_S(x_1, \ldots, x_{k+1})$ that are not adjoined by step n of a free construction of $\mathrm{cl}_S(x_1, \ldots, x_{k+1})$ around $x_1, \ldots, x_{k+1}$.

Lemma 3.12 *Let Σ be a block of the Steiner k-system S, that was constructed in Proposition 3.9 above. Suppose M is a finite $\mathcal{E}_{\widehat{C}}$-embedded substructure of S and θ is a partial automorphism of M that satisfies the following:*

(i) θ is the identity map on the points incident with Σ in (Domain(θ) $\cup$ Range(θ)).

(ii) Σ is defined in Domain(θ) $\cap$ Range(θ).

(iii) the embeddings f_1 : Domain(θ) $\to M, f_2$: Range(θ) $\to M$ given by inclusion are in $\mathcal{E}_{\widehat{C}}$.

Then there exists a substructure $P \in \widehat{C}$ that is $\mathcal{E}_{\widehat{C}}$-embedded in S and a partial automorphism θ^* of P such that

(a) M is a substructure of P.

(b) $\theta^* \upharpoonright$ Domain(θ) $= \theta$.

(c) Domain(θ^*) $\cap$ Range(θ^*) $\supseteq M$.

(d) θ^* is the identity map on the points incident with Σ in P.

(e) The embeddings g_1 : Domain(θ^*) $\to P, g_2$: Range(θ^*) $\to P$ given by inclusion are in $\mathcal{E}_{\widehat{C}}$.

Proof. Let Γ denote the set of points of S incident with Σ. We first extend θ so that its domain contains M. Now Range(θ) $\leq_c M$ and Σ is defined in Range(θ), so by Lemma 3.7, (Range(θ) $\cup (\Gamma \cap \Omega_M))^* \leq_c M$ (here (Range(θ) $\cup (\Gamma \cap \Omega_M))^*$ denotes Range(θ) $\cup (\Gamma \cap \Omega_M)$ along with all blocks incident with k points from Range(θ) $\cup (\Gamma \cap \Omega_M)$). Similarly (Domain($\theta$) $\cup (\Gamma \cap \Omega_M))^* \leq_c M$.

We can extend θ to θ_1 by putting $\theta_1(y) = \theta(y)$ for all $y \in$ Domain(θ), $\theta_1(x) = x$ for all $x \in \Gamma \cap \Omega_M$ and for each new block that is incident with k points from Domain(θ) $\cup (\Gamma \cap \Omega_M), x_1, \ldots, x_k$ say, we put $\theta_1(\Sigma(x_1, \ldots, x_k)) = \Sigma(\theta_1(x_1), \ldots, \theta_1(x_k))$. We then have an isomorphism from (Domain(θ) $\cup (\Gamma \cap \Omega_M))^*$ to (Range(θ) $\cup (\Gamma \cap \Omega_M))^*$ extending θ that is the identity on $\Gamma \cap \Omega_M$.

There is a copy N_1, of M whose domain consists of (Range(θ) $\cup (\Gamma \cap \Omega_M))^*$ together with elements distinct from M. Choose N_1 so that there is an isomorphism $\Psi_1 : M \to N_1$ with $\Psi_1 \upharpoonright$ Domain(θ) $= \theta$ and $\Psi_1(x) = x$ for all $x \in \Gamma \cap \Omega_M$, hence Ψ_1 extends θ_1. Since (Domain(θ) $\cup (\Gamma \cap \Omega_M))^* \leq_c M$, it follows that (Range(θ) $\cup (\Gamma \cap \Omega_M))^* \leq_c N_1$, so we can amalgamate M and N_1 over (Range(θ) $\cup (\Gamma \cap \Omega_M))^*$ to obtain $N_1^* \in \widehat{C}$ and embeddings $h_1 : M \to N_1^*$, $h_2 : N_1 \to N_1^*$ in $\mathcal{E}_{\widehat{C}}$. We now have a partial automorphism $\theta' : N_1^* \to N_1^*$ given by $\theta'(x) = \Psi_1(x)$ that has Domain(θ') $= M$ and fixes $\Gamma \cap \Omega_M$ pointwise. We must now extend θ' so that its range also contains M.

Find a copy, N_2, of N_1^* whose domain consists of Domain(θ') together with elements distinct from N_1^*. Choose N_2 so that there is an isomorphism $\Psi_2 : N_2 \to N_1^*$ with $\Psi_2(x) = \theta'(x)$ for all $x \in$ Domain(θ'). Now $h_1 : M \to N_1^*$ and $\Psi_2^{-1} h_2 \Psi_2 : M \to N_2$ are in $\mathcal{E}_{\widehat{C}}$, so we can amalgamate N_1^* and N_2 over $M =$ Domain(θ') to get $N_2^* \in \widehat{C}$ and embeddings $h_3 : N_1^* \to N_2^*, h_4 : N_2 \to N_2^*$ in $\mathcal{E}_{\widehat{C}}$. We now have a partial automorphism $\theta^* : N_2^* \to N_2^*$ given by $\theta^*(x) = \Psi_2(x)$ that has Domain(θ^*) $= N_2 \supseteq M$ and Range(θ^*) $= N_1^* \supseteq M$ and fixes $\Gamma \cap \Omega_{N_2^*}$ pointwise. Also, Domain(θ^*) $\leq_c N_2^*$ and Range(θ^*) $\leq_c N_2^*$.

Now M is $\mathcal{E}_{\widehat{C}}$-embedded in S, and the embedding $h_3h_1 : M \to N_2^*$ is in $\mathcal{E}_{\widehat{C}}$. Hence, by Proposition 3.9, there is P which is $\mathcal{E}_{\widehat{C}}$-embedded in S, contains M, and is isomorphic to N_2^*, so that the embeddings $g_1 : \text{Domain}(\theta^*) \to P$, $g_2 : \text{Range}(\theta^*) \to P$ are in $\mathcal{E}_{\widehat{C}}$.

Proposition 3.13 *For each block Σ of S, let Γ denote the points of S incident with Σ. Then $\Omega_S \backslash \Gamma$ is a Jordan set for $\text{Aut}(S)$ acting on Ω_S.*

Proof. Take $y, y' \in \Omega_S \backslash \Gamma$. We need to show that there is $\Psi \in \text{Aut}(S)$ fixing pointwise Γ and mapping y to y'. Take $x_1, \ldots, x_k \in \Gamma$. Since S is a union of a chain of finite $\mathcal{E}_{\widehat{C}}$-embedded structures, there is some M that is $\mathcal{E}_{\widehat{C}}$-embedded in S with $x_1, \ldots, x_k, y, y' \in M$.

Define a partial automorphism θ on M, by $(x_1, \ldots, x_k, y) \mapsto (x_1, \ldots, x_k, y')$. Since $\{x_1, \ldots, x_k, y\}, \{x_1, \ldots, x_k, y'\}$ are $\mathcal{E}_{\widehat{C}}$-embedded in S by Lemma 3.10, we can apply Lemma 3.12, to obtain some N, $\mathcal{E}_{\widehat{C}}$-embedded in S and a partial automorphism θ^* of N with $\text{Domain}(\theta^*) \cap \text{Range}(\theta^*) \supseteq M$, such that θ^* fixes $\Gamma \cap \Omega_N$ pointwise. Also, since S is a union of a chain of finite $\mathcal{E}_{\widehat{C}}$-embedded structures, there is some P in this chain which contains N. Now the embeddings $f_3 : \text{Domain}(\theta^*) \to P$, $f_4 : \text{Range}(\theta^*) \to P$ are in $\mathcal{E}_{\widehat{C}}$ (since $\mathcal{E}_{\widehat{C}}$ is closed under composition by Lemma 3.5) so we can apply Lemma 3.12 again. We repeat this process ω many times to obtain an automorphism of S mapping y to y' fixing Γ pointwise as required.

Theorem 3.14 *For each $k \geq 2$, there exists an infinite primitive Jordan group that is k-transitive, not k-primitive and preserves a Steiner k-system.*

Proof. This follows from Proposition 3.9, Lemma 3.10 and Proposition 3.13.

4 Amalgamation for upper semilinear order

In this section we construct an upper semilinear order, which is essentially an index set for the tower of semilinear orders built in Section 5.

We shall work with a 2-sorted language, $\mathcal{L}$, with sorts $\mathcal{T}$ and $\mathcal{B}$. The sort $\mathcal{T}$ will be the domain of the semilinear order, and in the next section each element of $\mathcal{T}$ will represent a Steiner k-system. We refer to elements of $\mathcal{T}$ as *nodes*. The second sort, $\mathcal{B}$, will in Section 5 represent the different blocks of each Steiner system. We do not here deal with the *points* of the Steiner systems.

Our language will have four relations, which we shall interpret as follows: $\alpha \leq \beta \, (\alpha, \beta \in \mathcal{T})$: *the node α is below or equal to the node β in the semilinear order.*

$D(\alpha; \beta, \gamma) \, (\alpha, \beta, \gamma \in \mathcal{T})$: *the node α is the least upper bound of β and γ (with* *respect to the relation $\leq$).*

$C(\alpha, \Sigma)$ $(\alpha \in \mathcal{T}, \Sigma \in \mathcal{B})$: *the node α has Σ as a block.*

$B(\alpha, \Sigma)$ $(\alpha \in \mathcal{T}, \Sigma \in \mathcal{B})$: *the node α lies in the block Σ.* In the next section we informally interpret $B(\alpha, \Sigma)$ as saying that the points of the Steiner system indexed by α are all incident with Σ.

The relations determine sorts. For example, $B(\alpha, \Sigma) \to \alpha \in \mathcal{T} \wedge \Sigma \in \mathcal{B}$. For ease of notation, if $\alpha \leq \beta$ and $\alpha \neq \beta$ we say $\alpha < \beta$.

Take $\mathcal{S}$ to be the class of all finite $\mathcal{L}$-structures satisfying the following axioms.

A1. $\forall \alpha, \beta, \gamma \in \mathcal{T}$ $(\alpha \leq \beta \wedge \alpha \leq \gamma \to (\beta \leq \gamma \vee \gamma \leq \beta))$
(The set of upper bounds of a node is linearly ordered)

A2. $\forall \alpha, \beta, \gamma \in \mathcal{T}$ $(\alpha \leq \beta \wedge \beta \leq \gamma \to \alpha \leq \gamma)$
($\leq$ is transitive)

A3. $\forall \alpha, \beta \in \mathcal{T}$ $(\alpha = \beta \leftrightarrow (\alpha \leq \beta \wedge \beta \leq \alpha))$

A4. $D(\alpha; \beta, \gamma) \to (\beta \leq \alpha \wedge \gamma \leq \alpha) \wedge D(\alpha; \gamma, \beta)$
(The node α is an upper bound of β and γ whenever $D(\alpha; \beta, \gamma)$ holds)

A5. $\forall \alpha, \beta \in \mathcal{T}$ $\exists \gamma \in \mathcal{T}$ $(D(\gamma; \alpha, \beta))$
(Every two nodes have a least upper bound)

A6. $D(\alpha; \beta, \gamma) \to \forall \sigma \in \mathcal{T}((\beta \leq \sigma \wedge \gamma \leq \sigma) \to \alpha \leq \sigma)$
(The node α is the least upper bound of β and γ)

A7. $\forall \Sigma \in \mathcal{B}$ $\exists \alpha \in \mathcal{T}$ $C(\alpha, \Sigma)$
(Every block is a block of at least one node)

A8. $\forall \alpha, \beta \in \mathcal{T}$ $\forall \Sigma \in \mathcal{B}$ $((C(\alpha, \Sigma) \wedge \alpha \neq \beta) \to \neg C(\beta, \Sigma))$
(Every block is a block of at most one node)

A9. $\forall \alpha, \beta \in \mathcal{T}$ $(\alpha < \beta \to \exists \Sigma \in \mathcal{B}$ $B(\alpha, \Sigma) \wedge C(\beta, \Sigma))$
(If α is below β then α lies in a block of β)

A10. $\alpha < \beta \wedge C(\beta, \Sigma_1) \wedge C(\beta, \Sigma_2) \wedge B(\alpha, \Sigma_1) \wedge B(\alpha, \Sigma_2) \to \Sigma_1 = \Sigma_2$
(A node can only lie in one block of another node)

A11. $B(\alpha, \Sigma) \wedge C(\beta, \Sigma) \wedge \gamma \neq \beta \wedge ((\alpha \leq \gamma < \beta) \vee (\gamma \leq \alpha < \beta)) \to B(\gamma, \Sigma)$
(If α lies in the block Σ of the node β and γ is comparable to α and less than β then γ also lies in Σ)

A12. $D(\alpha; \beta, \gamma) \wedge \beta \not\leq \gamma \wedge \gamma \not\leq \beta \to \exists \Sigma_1, \Sigma_2 \in \mathcal{B}$ $(C(\alpha, \Sigma_1) \wedge C(\alpha, \Sigma_2) \wedge B(\beta, \Sigma_1) \wedge B(\gamma, \Sigma_2) \wedge \Sigma_1 \neq \Sigma_2)$
(Two incomparable nodes lie in different blocks of their least upper bound)

A13. $B(\alpha, \Sigma) \wedge C(\beta, \Sigma) \to \alpha < \beta$
(If α lies in the block Σ of β then α is strictly below β)

Given $M \in \mathcal{S}$, we denote the two sorts of M by $\mathcal{T}_M$ and $\mathcal{B}_M$. If $M \in \mathcal{S}$, then $\leq$ defines an upper semilinear order on $\mathcal{T}_M$. Any two nodes of $\mathcal{T}_M$ will have a least upper bound, and if the two nodes are incomparable then they will lie in different blocks of their least upper bound. Given $M \in \mathcal{S}$, if $\Sigma \in \mathcal{B}_M$ is a block and no nodes of $\mathcal{T}_M$ lie in this block (that is, $\nexists \alpha \in \mathcal{T}_M \, B(\alpha, \Sigma)$), we say that Σ is a *minimal block* of M. Take $\mathcal{E}_\mathcal{S}$ to be the class of all embeddings of elements of $\mathcal{S}$. Then clearly $\mathcal{E}_\mathcal{S}$ is a class of $\mathcal{S}$-embeddings, in the sense of Definition 2.6.

Lemma 4.1 $\mathcal{E}_\mathcal{S}$ *satisfies JEP.*

Proof. Given $M_1, M_2 \in \mathcal{S}$, we take N to be the element of S that has nodes $\mathcal{T}_N = \mathcal{T}_{M_1} \cup \mathcal{T}_{M_2} \cup \{\alpha\}$ and blocks $\mathcal{B}_N = \mathcal{B}_{M_1} \cup \mathcal{B}_{M_2} \cup \{\Sigma_1, \Sigma_2\}$ where $\alpha, \Sigma_1, \Sigma_2 \notin M_1 \cup M_2$. The relations that hold in N are those that hold in M_1 and M_2, along with the following: $\beta \leq \alpha$ (for all $\beta \in \mathcal{T}_N$); $D(\alpha; \beta, \gamma)$ and $D(\alpha; \gamma, \beta)$ (for all $\beta \in \mathcal{T}_N \backslash \mathcal{T}_{M_2}$ and all $\gamma \in \mathcal{T}_N \backslash \mathcal{T}_{M_1}$); $C(\alpha, \Sigma_1), C(\alpha, \Sigma_2)$, $B(\beta, \Sigma_1)$ (for all $\beta \in \mathcal{T}_{M_1}$); and $B(\gamma, \Sigma_2)$ (for all $\gamma \in \mathcal{T}_{M_2}$). There will also be other trivial relations (for D) that are forced to hold via the axioms, such as $D(\alpha; \alpha, \alpha)$. Thus, N consists of a copy of M_1, a copy of M_2 and an additional node α that has two blocks Σ_1 and Σ_2, such that the nodes of $\mathcal{T}_{M_1}$ lie in Σ_1, and the nodes of $\mathcal{T}_{M_2}$ lie in Σ_2. With the usual embeddings $f : M_1 \to N$ and $g : M_2 \to N$, clearly $\mathcal{E}_\mathcal{S}$ satisfies JEP.

Lemma 4.2 *If* $M \in \mathcal{S}$ *with* $|\mathcal{T}_M| = n$, *we can list the nodes of* $\mathcal{T}_M$, *as* $\sigma_1, \ldots, \sigma_n$ *say, so that if* $\sigma_i \leq \sigma_j$ *then* $i \leq j$.

Proof. This holds since any partial order can be extended to a total order.

Lemma 4.3 $\mathcal{E}_\mathcal{S}$ *satisfies AP.*

Proof. Given $M, N_1, N_2 \in \mathcal{S}$ and embeddings $f_i : M \to N_i$ in $\mathcal{E}_\mathcal{S}$, we can construct an element $P \in \mathcal{S}$ and embeddings $g_i : N_i \to P$ in $\mathcal{E}_\mathcal{S}$ such that $g_1 f_1 = g_2 f_2$ as follows. It would be possible to give an explicit description of such P. However, instead we build P around M in stages by successively adding nodes and blocks from N_1 and N_2, as this is similar in style to the amalgamation in the next section. It should be clear that there is some flexibility over the ordering on the added nodes.

First apply Lemma 4.2 to list the nodes of $\mathcal{T}_M$ as $\sigma_1, \ldots, \sigma_n$, so that $i \leq j$ if $\sigma_i \leq \sigma_j$. Now add the elements from $(N_1 \cup N_2) \backslash M$ to M to form P by the following process. We adjoin elements of $N_1 \backslash M$ and $N_2 \backslash M$ so that they are distinct and we make no extra identifications. We first add any new blocks of elements of $\mathcal{T}_M$, i.e. the blocks $\{\Sigma \in \mathcal{B}_{N_1} \cup \mathcal{B}_{N_2} : C(\sigma_i, \Sigma) \text{ for some } i \text{ with } 1 \leq i \leq n\}$. Next, adjoin all nodes $\alpha \in (\mathcal{T}_{N_1} \cup \mathcal{T}_{N_2}) \backslash \mathcal{T}_M$ such that $\alpha < \sigma_1$, then all (not previously adjoined) nodes $\alpha \in (\mathcal{T}_{N_1} \cup \mathcal{T}_{N_2}) \backslash \mathcal{T}_M$ such that $\alpha < \sigma_2$, and so on. Finally, add any 'new' nodes $\alpha \in (\mathcal{T}_{N_1} \cup \mathcal{T}_{N_2}) \backslash \mathcal{T}_M$ such that $\alpha < \sigma_n$, and then those $\alpha \in (\mathcal{T}_{N_1} \cup \mathcal{T}_{N_2}) \backslash \mathcal{T}_M$ such that $\alpha \not< \sigma_n$. Whenever we adjoin a node of $\mathcal{T}_{N_1}$ or $\mathcal{T}_{N_2}$ we also add all of its blocks. The details of this process are covered by the following cases.

Case 1. To add nodes $\alpha \in (\mathcal{T}_{N_1} \cup \mathcal{T}_{N_2}) \backslash \mathcal{T}_M$ such that $\alpha < \sigma_i$ where $\sigma_i \in \mathcal{T}_M$ and there does not exist $\sigma \in \mathcal{T}_M$ with $\sigma < \sigma_i$.

In this case, for each block in $\mathcal{B}_{N_1} \cup \mathcal{B}_{N_2}$ of σ_i we can adjoin all nodes of $(\mathcal{T}_{N_1} \cup \mathcal{T}_{N_2}) \backslash \mathcal{T}_M$ that lie in it. If a block of σ_i has nodes that lie in it from both $\mathcal{T}_{N_1}$ and $\mathcal{T}_{N_2}$, then we are able to put nodes of $\mathcal{T}_{N_1}$ above or below nodes of $\mathcal{T}_{N_2}$ as we desire, as long as we preserve the ordering and the relation B on N_1 and on N_2. If we add $\beta \in \mathcal{T}_{N_1} \backslash \mathcal{T}_M$ and $\gamma \in \mathcal{T}_{N_2} \backslash \mathcal{T}_M$ with $\beta < \gamma$, then for some Σ with $C(\gamma, \Sigma)$ holding we must have $B(\beta, \Sigma)$. When we adjoin nodes of $\mathcal{T}_{N_1}$ and $\mathcal{T}_{N_2}$ we also add all of their blocks.

We will never adjoin $\beta \in \mathcal{T}_{N_1} \backslash \mathcal{T}_M$ and $\gamma \in \mathcal{T}_{N_2} \backslash \mathcal{T}_M$ in such a way that for some $\delta \leq \sigma_i$ and block Σ satisfying $C(\delta, \Sigma)$, we have $B(\beta, \Sigma)$, $B(\gamma, \Sigma)$, β and γ are incomparable, and β is maximal subject to $\beta < \delta$ and γ maximal subject to $\gamma < \delta$. The reason for this is that we do not wish to adjoin two incomparable nodes that lie in the same block of their least upper bound.

Case 2. To add nodes $\alpha \in (\mathcal{T}_{N_1} \cup \mathcal{T}_{N_2}) \backslash \mathcal{T}_M$ such that $\alpha < \sigma_i$ where $\sigma_i \in \mathcal{T}_M$ and there is $\sigma \in \mathcal{T}_M$ with $\sigma < \sigma_i$.

For each block that is a minimal block of M (or a block in $(\mathcal{B}_{N_1} \cup \mathcal{B}_{N_2}) \backslash \mathcal{B}_M$) at σ_i, we can proceed as in Case 1. When we deal with a non-minimal block, Σ say, with $C(\sigma_i, \Sigma)$ holding, we must make the following alteration. Take σ_j to be the maximal element of $\mathcal{T}_M$ such that σ_j lies in Σ (so $j < i$). Let $\beta_1, \ldots, \beta_n$ denote the nodes of $\mathcal{T}_{N_1} \backslash \mathcal{T}_M$ that lie in Σ with $\sigma_j < \beta_1 \wedge \ldots \wedge \sigma_j < \beta_n$. Without loss of generality assume $\sigma_j < \beta_1 < \ldots < \beta_n < \sigma_i$. Similarly denote the nodes of $\mathcal{T}_{N_2}$ in this region by $\gamma_1, \ldots, \gamma_m$ with $\sigma_j < \gamma_1 < \ldots < \gamma_m < \sigma_i$. We can adjoin $\beta_1, \ldots, \beta_n, \gamma_1, \ldots, \gamma_m$ in any order (as long as we respect the structure of N_1 and N_2), and when we add these nodes we add all their blocks and all other nodes that lie in these blocks, ie

$$\{\Sigma' \in \mathcal{B}_{N_1} \cup \mathcal{B}_{N_2} : \text{There exists } \delta \in \mathcal{T}_{N_1} \cup \mathcal{T}_{N_2} \text{ with } \sigma_j < \delta < \sigma_i, C(\delta, \Sigma')\} \cup$$

$$\{\delta, \Sigma' : \delta \in \mathcal{T}_{N_1} \cup \mathcal{T}_{N_2}, \delta < \sigma_i \wedge \sigma_j \not< \delta \wedge B(\delta, \Sigma), \Sigma' \in \mathcal{B}_{N_1} \cup \mathcal{B}_{N_2}, C(\delta, \Sigma')\}$$

We adjoin the β_i and γ_i so that they are distinct.

Case 3. To add nodes $\alpha \in (\mathcal{T}_{N_1} \cup \mathcal{T}_{N_2}) \backslash \mathcal{T}_M$ such that $\alpha \not< \sigma_n$.

We adjoin all nodes $\alpha \in (\mathcal{T}_{N_1} \cup \mathcal{T}_{N_2}) \backslash \mathcal{T}_M$ such that $\sigma_n \leq \alpha$ in any order as long as we preserve the order and the relation B on N_1 and on N_2. Again, when we add these nodes we add all their blocks and all other nodes that lie in their blocks, ie

$$\{\Sigma \in \mathcal{B}_{N_1} \cup \mathcal{B}_{N_2} : \text{There exists } \delta \in \mathcal{T}_{N_1} \cup \mathcal{T}_{N_2} \text{ with } \sigma_n < \delta, C(\delta, \Sigma)\} \cup$$

$$\{\delta, \Sigma : \delta \in \mathcal{T}_{N_1} \cup \mathcal{T}_{N_2}, \delta \not\leq \sigma_n \wedge \sigma_n \not\leq \delta, \Sigma \in \mathcal{B}_{N_1} \cup \mathcal{B}_{N_2}, C(\delta, \Sigma)\}$$

The only other relations in the language $\mathcal{L}$ that hold are those that held in N_1 and N_2 and those forced to hold via the axioms. For every two nodes in $\mathcal{T}_P$ it is clear which node is their common least upper bound. Also for each α and β with $\alpha < \beta$, α must lie in some block Σ where $C(\beta, \Sigma)$ holds. In this case $B(\alpha, \Sigma)$ must also hold.

By this method we obtain an element $P \in \mathcal{S}$ such that the embeddings $g_i : N_i \to P$ given by inclusion are in $\mathcal{E}_{\mathcal{S}}$.

Remark 4.4 The Fraïssé limit of this class will be a dense semilinear order on the sort $\mathcal{T}$ as required. In the language of Adeleke and Neumann [5], it is the doubly homogeneous semilinear order of positive type and ramification order $\aleph_0$.

5 Construction of the semilinear tower

In this section we fix an integer $k \geq 2$ and combine the last two constructions to produce an example of an upper semilinear tower of Steiner k-systems that has an automorphism group that is a $(k+1)$-transitive, not $(k+1)$-primitive Jordan group.

We shall work with a 3-sorted language, $\mathcal{L}$, where the sort $\mathcal{T}$ will be interpreted by the set of Steiner systems, the sort Ω will be a set of points (of the various Steiner systems), and finally the elements of the sort $\mathcal{B}$ will be the blocks of the Steiner systems. In order to keep parts of this construction as similar as possible to the preceding sections, our language will have six relations which we shall interpret informally as follows.

$\alpha \leq \beta$ $(\alpha, \beta \in \mathcal{T})$: *the Steiner system α is below or equal to the Steiner system β in the semilinear order.*

$D(\alpha; \beta, \gamma)$ $(\alpha, \beta, \gamma \in \mathcal{T})$: *the Steiner system α is the least upper bound of β and γ (with respect to the relation $\leq$).*

$C(\alpha, \Sigma)$ $(\alpha \in \mathcal{T}, \Sigma \in \mathcal{B})$: *the Steiner system α has Σ as one of its blocks.*

$B(\alpha, \Sigma)$ $(\alpha \in \mathcal{T}, \Sigma \in \mathcal{B})$: *the Steiner system α lies in the block Σ.*

$I(\alpha, x, \Sigma)$ $(\alpha \in \mathcal{T}, x \in \Omega, \Sigma \in \mathcal{B})$: *the point x is incident with the block Σ in the Steiner system α.*

$L(\alpha, x)$ $(\alpha \in \mathcal{T}, x \in \Omega)$: *the Steiner system α has x as one of its points.*

The first four relations correspond to those used to construct the semilinear order in the previous section and the relation I corresponds to the incidence relation I, that was used in the construction of single Steiner systems. As in the previous section, the relations here also determine sorts: for example, we assume $L(\alpha, x) \to \alpha \in \mathcal{T} \wedge x \in \Omega$. For ease of notation if $\alpha \leq \beta$ and $\alpha \neq \beta$ we say $\alpha < \beta$.

Take $\mathcal{D}^*$ to be the class of all $\mathcal{L}$-structures satisfying the axioms A1–A13 and the following additional axioms, and $\mathcal{D}$ to be the class of all finite members of $\mathcal{D}^*$.

A14. $\forall \alpha \in \mathcal{T} \ \forall x \in \Omega \ \forall \Sigma \in \mathcal{B} \ (I(\alpha, x, \Sigma) \rightarrow C(\alpha, \Sigma) \wedge L(\alpha, x))$

(If x is incident with Σ in the Steiner system α then x and Σ are both in the Steiner system α)

A15. $\forall \alpha \in \mathcal{T} \ \forall \Sigma, \Sigma' \in \mathcal{B} \ (\forall^{\neq} x_1, \ldots, x_k \in \Omega((I(\alpha, x_1, \Sigma) \wedge \ldots \wedge I(\alpha, x_k, \Sigma) \wedge I(\alpha, x_1, \Sigma') \wedge \ldots \wedge I(\alpha, x_k, \Sigma')) \rightarrow (\Sigma = \Sigma')))$

(In each Steiner system, if two blocks are both incident with k points then they are equal)

A16. $\forall \alpha \in \mathcal{T} \ \forall \Sigma \in \mathcal{B}(C(\alpha, \Sigma) \rightarrow \exists^{\neq} x_1, \ldots, x_k \in \Omega(I(\alpha, x_1, \Sigma) \wedge \ldots \wedge I(\alpha, x_k, \Sigma)))$

(Every block of a Steiner system is incident with at least k points)

A17. $\forall \alpha \in \mathcal{T} \ \forall x_1, \ldots, x_k \in \Omega \ (L(\alpha, x_1) \wedge \ldots \wedge L(\alpha, x_k)) \rightarrow \exists \Sigma \in \mathcal{B} \ (I(\alpha, x_1, \Sigma) \wedge \ldots \wedge I(\alpha, x_k, \Sigma))$

(If k points belong to a Steiner system, then they are all incident with some block of that Steiner system)

A18. $\forall \alpha, \beta \in \mathcal{T} \ \forall x \in \Omega \ (L(\alpha, x) \wedge \alpha \leq \beta \rightarrow L(\beta, x))$

(If the Steiner system α is below the Steiner system β and x is a point of α, then x is also a point of β)

A19. $\forall \alpha, \beta \in \mathcal{T} \ \forall^{\neq} x_1, \ldots, x_k \in \Omega(L(\alpha, x_1) \wedge \ldots \wedge L(\alpha, x_k) \wedge L(\beta, x_1) \wedge \ldots \wedge L(\beta, x_k) \rightarrow (\alpha \leq \beta \vee \beta \leq \alpha))$

(If two Steiner systems have k points in common, then they are comparable)

A20. $\forall \alpha, \beta \in \mathcal{T} \ \forall x \in \Omega \ \forall \Sigma \in \mathcal{B} \ (L(\alpha, x) \wedge B(\alpha, \Sigma) \wedge C(\beta, \Sigma) \rightarrow I(\beta, x, \Sigma))$

(If the Steiner system α lies in the block Σ of the Steiner system β and x is a point of α, then x is incident with Σ in the Steiner system β)

A21. $\forall \alpha \in \mathcal{T} \ \exists^{\neq} x_1, \ldots, x_k \in \Omega(L(\alpha, x_1) \wedge \ldots \wedge L(\alpha, x_k))$

(Every Steiner system has at least k points)

A22. $\forall x \in \Omega \ \exists \alpha \in \mathcal{T} \ L(\alpha, x)$

(Every point belongs to some Steiner system)

The axioms tell us that the sorts $\mathcal{T}$ and $\mathcal{B}$ will define a semilinear order as before. The additional axioms tell us that each element of $\mathcal{T}$ corresponds to a (partial) Steiner k-system and explains the interaction between points in more than one Steiner system. Each (partial) Steiner k-system must contain at least k points and hence one block. Also, any two incomparable Steiner systems will lie in different blocks of the Steiner system that is their least common upper bound. For $M \in \mathcal{D}^*$ we denote the elements in the three sorts by $\Omega_M, \mathcal{B}_M$ and $\mathcal{T}_M$.

We wish to restrict the class of structures that we are interested in. If $M \in \mathcal{D}^*$ then it will be of the form $\{\Omega_M, \mathcal{T}_M, \mathcal{B}_M; \leq, D, C, B, I, L\}$. We define a new class as follows.

$$\widehat{\mathcal{D}}^* := \{M \in \mathcal{D}^* : \forall \alpha \in \mathcal{T}_M \text{ the structure } \{x, \Sigma : \ L(\alpha, x), \ C(\alpha, \Sigma)\} \in \widehat{\mathcal{C}}^*$$
$$\text{under the relation induced by } I(\alpha, \ldots)\}$$

So if $M \in \widehat{\mathcal{D}}^*$, then every partial Steiner system in the tower is a locally free (partial) Steiner k-system. Let $\widehat{\mathcal{D}} := \{M \in \widehat{\mathcal{D}}^* : |M| < \omega\}$ – the collection of finite elements of $\widehat{\mathcal{D}}^*$. The parameter k implicit in the definition of $\widehat{\mathcal{D}}$ will be the same as that in the definition of $\widehat{\mathcal{C}}$. If $M \in \widehat{\mathcal{D}}^*$ and $\alpha \in \mathcal{T}_M$, let M_α denote the structure consisting of the points $\{x \in \Omega_M : L(\alpha, x)\}$, and the blocks $\{\Sigma \in \mathcal{B}_M : C(\alpha, \Sigma)\}$. Also let $\Sigma_{M_\alpha}(x_1, \ldots, x_k)$ denote the unique block of M_α that is incident with $x_1, \ldots, x_k$ in M.

Definition 5.1 Let $\mathcal{E}_{\widehat{\mathcal{D}}}$ denote the class of all embeddings $f : M \to N$ (where $M, N \in \widehat{\mathcal{D}}$) which satisfy the following conditions. For ease of notation we shall identify M with $f(M)$.

(1) For all $\alpha \in \mathcal{T}_M$, the embedding $M_\alpha \to N_\alpha$ with the relation induced by $I(\alpha, \ldots)$ is in $\mathcal{E}_{\widehat{\mathcal{C}}}$.

This says essentially that if the embedding extends a Steiner system of M, then it extends it with an embedding from $\mathcal{E}_{\widehat{\mathcal{C}}}$.

(2) Suppose Σ is a minimal block of M, and Σ is a block of the Steiner system $\alpha \in M$ (so $C(\alpha, \Sigma)$). Let $x_1, \ldots, x_n$ be an enumeration of the set $S := \{x \in \Omega_M : I(\alpha, x, \Sigma)\}$. By A16, we must have $n \geq k$. Also, by A20, if $\sigma \in \mathcal{T}_N$ with $B(\sigma, \Sigma)$ and $x \in \Omega_M$ with $L(\sigma, x)$ then $I(\alpha, x, \Sigma)$. Hence if σ has any points in Ω_M then they must belong to S. For each $\sigma \in \mathcal{T}_N$ with $B(\sigma, \Sigma)$, we require that one of the following three conditions holds.

(2a) $|\{x : L(\sigma, x) \wedge x \in \Omega_M\}| \leq k$

The Steiner system σ contains at most k points from Ω_M.

(2b) $\{x : L(\sigma, x) \wedge x \in \Omega_M\} = \{x_1, \ldots, x_n\}$ with $n > k$ and there exists $\Sigma' \in \mathcal{B}_N$ such that $C(\sigma, \Sigma') \wedge I(\sigma, x_1, \Sigma') \wedge \ldots \wedge I(\sigma, x_n, \Sigma')$. Also, $\{x_1, \ldots, x_n\} \cup \{\Sigma'\}$ is $\mathcal{E}_{\widehat{\mathcal{C}}}$-embedded in N_σ with the relation induced by $I(\sigma, \ldots)$. The latter is in fact forced since $\{x_1, \ldots, x_n\} \cup \{\Sigma'\}$ forms a trivial partial Steiner system.

The Steiner system σ contains all the (at least $k+1$) points of S, and they are all incident with the same block of σ.

(2c) $\{x : L(\sigma, x) \wedge x \in \Omega_M\} = \{x_1, \ldots, x_n\}$ with $n > k$ and for any distinct $i_1, \ldots, i_{k+1} \in \{1, \ldots, n\}$, we have $\neg I(\sigma, x_{i_{k+1}}, \Sigma_{N_\sigma}(x_{i_1}, \ldots, x_{i_k}))$. Moreover $\{x_1, \ldots, x_n\} \cup \{\Sigma^* \in \mathcal{B}_N : C(\sigma, \Sigma^*) \wedge I(\sigma, x_{i_1}, \Sigma^*) \wedge \ldots \wedge I(\sigma, x_{i_k}, \Sigma^*) \ (1 \leq i_1 < \ldots < i_k \leq n)\}$ is $\mathcal{E}_{\widehat{\mathcal{C}}}$-embedded in N_σ with the relation induced by $I(\sigma, \ldots)$.

The Steiner system σ contains all the (at least $k+1$) points of S, and no $k+1$ of them are all incident with the same block of σ. Given M, N and Σ, by A20 there can only be one Steiner system satisfying condition (c). If it exists we say that σ is the *specified Steiner system relating to M, N and Σ*, which we denote by $SS(M, N, \Sigma)$.

(3) Suppose Σ is a non-minimal block of M, Σ is a block of the Steiner system $\alpha \in M$ and $\beta \in \mathcal{T}_M$ is maximal with respect to $\beta < \alpha$ and $B(\beta, \Sigma)$. Note that by A12, there is at most one such β, since incomparable maximal such Steiner systems would lie in different blocks of the Steiner system α. Let $x_1, \ldots, x_n$ be an enumeration of the set $S := \{x \in \Omega_M : I(\alpha, x, \Sigma)\}$, so that for some $m \leq n$, $L(\beta, x_i)$ holds if $i \leq m$ and $\neg L(\beta, x_i)$ holds whenever $i > m$. By A21, we have $m \geq k$.

Now by A18, for each $\sigma \in \mathcal{T}_N$ with $\beta < \sigma < \alpha$ (and therefore also $B(\sigma, \Sigma)$) holding, $L(\sigma, x_i)$ holds for $1 \leq i \leq m$, and by A20, $x_1, \ldots, x_m$ will all be incident with the same block of σ. By A20, σ has only those points from Ω_M which are incident with Σ in α. Now for each $\sigma \in \mathcal{T}_N$ with $\beta < \sigma < \alpha$ and $B(\sigma, \Sigma)$, one of the following three conditions holds.

(3a) $\{x \in \Omega_N : L(\sigma, x) \wedge x \in \Omega_M\} = \{x_1, \ldots, x_m\}$

The Steiner system σ contains $x_1, \ldots, x_m$ and no other points from Ω_M. By A20, $x_1, \ldots, x_m$ will all be incident with the same block of σ.

(3b) $\{x \in \Omega_N : L(\sigma, x) \wedge x \in \Omega_M\} = \{x_1, \ldots, x_n\}$ with $n > m$ and there exists $\Sigma' \in \mathcal{B}_N$ such that $C(\sigma, \Sigma') \wedge I(\sigma, x_1, \Sigma') \wedge \ldots \wedge I(\sigma, x_n, \Sigma')$. Furthermore, $\{x_1, \ldots, x_n\} \cup \{\Sigma'\}$ is $\mathcal{E}_{\widehat{C}}$-embedded in N_σ with the relation induced by $I(\sigma, \ldots)$. Again, this property is forced since $\{x_1, \ldots, x_n\} \cup \{\Sigma'\}$ forms a trivial partial Steiner system.

The Steiner system σ contains all the points of S, and they are all incident with the same block of σ.

(3c) $\{x \in \Omega_N : L(\sigma, x) \wedge x \in \Omega_M\} = \{x_1, \ldots, x_n\}$ with $n > m$ and for each set $i_1, \ldots, i_k \in \{1, \ldots, n\}$ distinct, $\{x \in \Omega_M : I(\sigma, x, \Sigma_{N_\sigma}(x_{i_1}, \ldots, x_{i_k})\}$ equals either $\{x_{i_1}, \ldots, x_{i_k}\}$ or $\{x_1, \ldots, x_m\}$. Also, $\{x_1, \ldots, x_n\} \cup \{\Sigma^* \in \mathcal{B}_N : C(\sigma, \Sigma^*) \wedge I(\sigma, x_{i_1}, \Sigma^*) \wedge \ldots \wedge I(\sigma, x_{i_k}, \Sigma^*) \ (1 \leq i_1 < \ldots < i_k \leq n)\}$ is $\mathcal{E}_{\widehat{C}}$-embedded in N_σ with the relation induced by $I(\sigma, \ldots)$.

This says essentially that the Steiner system σ contains all the points of S, and $x_1, \ldots, x_m$ will all be incident with a single block and each x_i $(i > m)$ will not be incident with this block. All other blocks of σ will be incident with at most k points from Ω_M. Again, given M, N and Σ, by A20 there can only be one Steiner system satisfying (c). If it exists, we say that the σ in this case is the *specified Steiner system relating to M, N and Σ*, which we denote by $SS(M, N, \Sigma)$.

Example 5.2 This example shows the relevance of specified Steiner systems. Suppose $M \in \widehat{\mathcal{D}}$ consists of a single Steiner system α, a single block Σ, and n points (where $n \geq k + 2$) $x_1, \ldots, x_n$. Also, suppose $L(\alpha, x_1) \wedge \ldots \wedge L(\alpha, x_n) \wedge C(\alpha, \Sigma)$ and $I(\alpha, x_1, \Sigma) \wedge \ldots \wedge I(\alpha, x_n, \Sigma)$ holds in M. Now there is $N_1 \in \widehat{\mathcal{D}}$ containing M and an additional Steiner system β such that $\beta < \alpha$ and $L(\beta, x_1) \wedge \ldots \wedge L(\beta, x_n)$. We assume that in the Steiner system β, no $k + 1$ points from $\{x_1, \ldots, x_n\}$ are all incident with a single block (so $\beta = SS(M, N_1, \Sigma)$). There is also $N_2 \in \widehat{\mathcal{D}}$ containing M, with an additional Steiner system γ so that $\gamma < \alpha$ and $L(\gamma, x_1) \wedge \ldots \wedge L(\gamma, x_n)$. We may suppose there is $\Sigma^* \in \mathcal{B}_{N_2}$ such that $I(\gamma, x_1, \Sigma^*) \wedge \ldots \wedge I(\gamma, x_{n-1}, \Sigma^*) \wedge \neg I(\gamma, x_n, \Sigma^*)$. Here $\gamma \neq SS(M, N_2, \Sigma)$ and in fact γ is not one of the types of Steiner systems described in Definition 5.1.

Now by the following argument, we cannot amalgamate N_1 and N_2 over M. Since $L(\beta, x_1) \wedge \ldots \wedge L(\beta, x_n) \wedge L(\gamma, x_1) \wedge \ldots \wedge L(\gamma, x_n)$ and $n \geq k+2$, by A19 we have $\gamma < \beta$ or $\beta < \gamma$ or $\gamma = \beta$. Now $\gamma \not< \beta$ since otherwise β would have to contain $x_1, \ldots, x_n$ and they must all be incident with a single block, a contradiction. By the same argument, $\beta \not< \gamma$. Also, we cannot have $\gamma = \beta$ since γ contains a block that is incident with $n - 1$ points from $x_1, \ldots, x_n$ but β has no such block. Hence we cannot amalgamate N_1 and N_2 over M.

The use of the specified Steiner systems avoids the above problem as the embedding from M into N_2 would not lie in $\mathcal{E}_{\widehat{\mathcal{D}}}$, because of γ. If the embedding from M into N_2 did adjoin a Steiner system with $x_1, \ldots, x_n$ as points that are not all incident with a single block, then no $k + 1$ of them would be incident with a single block, so the Steiner system would be $SS(M, N_2, \Sigma)$. We would then be able to identify together specified Steiner systems such as $SS(M, N_1, \Sigma)$ and $SS(M, N_2, \Sigma)$ if they exist.

Lemma 5.3 $\mathcal{E}_{\widehat{\mathcal{D}}}$ *is a class of $\widehat{\mathcal{D}}$-embeddings.*

Proof. We show each condition (i)–(iii)of Definition 2.6 in turn. For convenience of notation, we shall assume embeddings are the identity, that is, identify structures with their images. We think of an embedding $f : M \to N$ as adjoining additional elements (in $N \setminus M$) to M.

(i) Clearly any isomorphism is in $\mathcal{E}_{\widehat{\mathcal{D}}}$.

(ii) Suppose we have embeddings $f : M \to N, g : N \to P$ in $\mathcal{E}_{\widehat{\mathcal{D}}}$. We require that the embedding $gf : M \to P$ is in $\mathcal{E}_{\widehat{\mathcal{D}}}$. By Lemma 3.5, $\mathcal{E}_{\widehat{\mathcal{C}}}$ is closed under composition so gf satisfies Definition 5.1(1).

We now verify that 5.1(2) also holds for gf. Suppose Σ is a minimal block of M, and belongs to the Steiner system α. We shall show that each Steiner system in $P \setminus gf(M)$ which lies in Σ is of one of the forms in Definition 5.1 (2).

First, suppose the embedding $f : M \to N$ introduces a specified Steiner system $\sigma = SS(M, N, \Sigma)$. Each block of σ will be incident with at most k points from Ω_M. We suppose the embedding $g : N \to P$ also adjoins some new Steiner systems that lie in Σ (since otherwise it is trivial that gf satisfies Definition 5.1(2) for the block Σ.) Suppose that $\delta \in \mathcal{T}_P \setminus \mathcal{T}_N$ is one such. Then either $\delta < \sigma$, $\sigma < \delta$ or δ and σ are incomparable. Observe that by A20, any points from Ω_M incident with δ must be incident with Σ (in M_α). If $\delta < \sigma$, then since the points $\{x \in \Omega_M : L(\delta, x)\}$ must all be incident with a single block of σ by A20, we have $|\{x \in \Omega_M : L(\delta, x)\}| \leq k$. If $\sigma < \delta$, then by A18, we have $\{x \in \Omega_M : L(\delta, x)\} = \{x \in \Omega_M : L(\sigma, x)\} = \{x \in \Omega_M : I(\alpha, x, \Sigma)\}$, and by A20 these points must all be incident with a single block, Σ' say, of δ. Also, $\{x : L(\delta, x) \wedge x \in \Omega_M\} \cup \{\Sigma'\}$ is $\mathcal{E}_{\hat{c}}$-embedded in P_δ with the relation induced by $I(\delta, \ldots)$. If σ and δ are incomparable, then $|\{x \in \Omega_P : L(\delta, x)\} \cap \{x \in \Omega_P : L(\sigma, x)\}| \leq k - 1$. Hence $|\{x \in \Omega_M : L(\delta, x)\}| \leq k - 1$, since if $x \in \Omega_M$ and $L(\delta, x)$ holds then $L(\sigma, x)$ also holds. Thus in this case, the embedding $gf : M \to P$ satisfies Definition 5.1(2), for the block Σ.

Next, suppose the embedding $f : M \to N$ does not adjoin a specified Steiner system $SS(M, N, \Sigma)$. Now each Steiner system γ of N that lies in Σ will have one of the following properties.

(a) $\{x \in \Omega_M : L(\gamma, x)\} = \{x \in \Omega_M : I(\alpha, x, \Sigma)\}$ and there is $\Sigma' \in \mathcal{B}_N$ with $C(\gamma, \Sigma')$ and $\{x \in \Omega_M : I(\gamma, x, \Sigma')\} = \{x \in \Omega_M : L(\gamma, x)\}$. If there is such a γ, there is a unique minimal one, β say, with corresponding block Σ_β. If such a γ does not exist, we set β equal to α and $\Sigma_\beta = \Sigma$.

(b) There is $\gamma^* \in \mathcal{T}_N$ that satisfies (a) above and some $\Sigma^* \in \mathcal{B}_N$ with $C(\gamma^*, \Sigma^*)$ and $|\{x \in \Omega_M : I(\gamma^*, x, \Sigma^*)\}| < k$, and we have $B(\gamma, \Sigma^*)$. In this case, we have $|\{x \in \Omega_M : L(\gamma, x)\}| < k$ by A20.

(c) $|\{x \in \Omega_M : L(\gamma, x)\} \cap \{x \in \Omega_M : I(\alpha, x, \Sigma)\}| \leq k$ and $B(\gamma, \Sigma_\beta)$ holds. If there is such a γ then there must be a unique maximal such γ, denoted by $\hat{\gamma}$ say, since two incomparable maximal such elements would lie in different blocks of β.

We show that the embedding $gf : M \to P$ satisfies 5.1(2), for the block Σ. Assume that the embedding $g : N \to P$ adds some new Steiner system δ which lies in Σ (since otherwise it is trivial that gf satisfies 5.1(2) for Σ.) So $\delta \in \mathcal{T}_P \backslash \mathcal{T}_N$ and $B(\delta, \Sigma)$ holds. Now either $\delta < \beta, \beta < \delta$ or δ and β are incomparable.

If $\delta < \beta$ and $\neg B(\delta, \Sigma_\beta)$ then by A20 we have $|\{x \in \Omega_M : L(\delta, x)\}| \leq k - 1$. Now suppose $\delta < \beta$ and $B(\delta, \Sigma_\beta)$ holds. If $\hat{\gamma}$ does not exist, then since $\{x \in \Omega_M : I(\beta, x, \Sigma_\beta)\} = \{x \in \Omega_M : I(\alpha, x, \Sigma)\}$ and Σ_β is a minimal block of N, it is clear that δ will be of one of the forms described in Definition 5.1(2) with β in place of α and δ in place of γ.

Thus, we may suppose $\hat{\gamma}$ exists. If $\hat{\gamma}$ does exist then δ will be of one of the following forms by 5.1.

(I) $\hat{\gamma} < \delta$ and $|\{x \in \Omega_M : L(\delta, x)\}| = |\{x \in \Omega_M : L(\hat{\gamma}, x)\}| \leq k$.

(II) $\hat{\gamma} < \delta$ and $\{x \in \Omega_M : L(\delta, x)\} = \{x \in \Omega_M : I(\beta, x, \Sigma_\beta)\} = \{x \in \Omega_M : I(\alpha, x, \Sigma)\}$ and these points are all incident with the same block, Σ' say of δ. Also, $\{x : L(\delta, x) \wedge x \in \Omega_M\} \cup \Sigma'$ is trivially $\mathcal{E}_{\hat{c}}$-embedded in P_δ with the relation induced by $I(\delta, \ldots)$.

(III) $\hat{\gamma} < \delta$ and $\{x \in \Omega_M : L(\delta, x)\} = \{x \in \Omega_M : I(\beta, x, \Sigma_\beta)\} = \{x \in \Omega_M : I(\alpha, x, \Sigma)\}$ and no $k + 1$ of these points are incident with the same block of

δ. The points $\{x \in \Omega_N : L(\hat{\gamma}, x)\}$ will all be incident with a single block, but $|\{x \in \Omega_M : L(\hat{\gamma}, x)\}| \leq k$. By Lemma 3.2,

$$\{x \in \Omega_M : L(\delta, x)\} \cup \{\hat{\Sigma} \in \mathcal{B}_P : \quad \exists^{\neq} y_1, \ldots, y_k \in \Omega_M \text{ with} \\ I(\delta, y_1, \hat{\Sigma}) \wedge \ldots \wedge I(\delta, y_k, \hat{\Sigma})\}$$

is $\mathcal{E}_{\hat{\mathcal{C}}}$-embedded in P_δ with the relation induced by $I(\delta, \ldots)$, because it is is easily seen to be $\mathcal{E}_{\hat{\mathcal{C}}}$-embedded in

$$\{x \in \Omega_N : L(\delta, x)\} \cup \{\hat{\Sigma} \in \mathcal{B}_P : \quad \exists^{\neq} y_1, \ldots y_k \in \Omega_N \text{ with} \\ I(\delta, y_1, \hat{\Sigma}) \wedge \ldots \wedge I(\delta, y_k, \hat{\Sigma})\}$$

with the relation induced by $I(\delta, \ldots)$ which by Definition 5.1 is $\mathcal{E}_{\hat{\mathcal{C}}}$-embedded in P_δ with the relation induced by $I(\delta, \ldots)$.

(IV) $\delta < \hat{\gamma}$ and then by A5, $|\{x \in \Omega_M : L(\delta, x)\}| \leq |\{x \in \Omega_M : L(\hat{\gamma}, x)\}| \leq k$.

(V) δ and $\hat{\gamma}$ are incomparable. Then since δ and $\hat{\gamma}$ lie in different blocks of their least upper bound (which will be either β or one of the forms described in (I),(II),(III) or (IV) above), we have $|\{x \in \Omega_M : L(\delta, x)\}| \leq k$.

If $\beta < \delta$ then by A20, $\{x \in \Omega_M : L(\delta, x)\} = \{x \in \Omega_M : L(\beta, x)\}$ and these points are all incident with a unique block, Σ' say of δ. Also, $\{x : L(\delta, x) \wedge x \in \Omega_M\} \cup \{\Sigma'\}$ is trivially $\mathcal{E}_{\hat{\mathcal{C}}}$-embedded in P_δ with the relation induced by $I(\delta, \ldots)$. If β and δ are incomparable, then since δ lies in a different block to β of their least upper bound, we have $|\{x \in \Omega_M : L(\delta, x)\}| \leq k - 1$.

We have now shown that each Steiner system that is adjoined by the embedding gf and that lies in Σ is of one of the forms described in Definition 5.1(2), so the embedding $gf : M \to P$ satisfies (ii). By a similar argument, it also satisfies 5.1(3) and hence $\mathcal{E}_{\hat{\mathcal{D}}}$ is closed under composition.

(iii) That this condition holds for $\mathcal{E}_{\hat{\mathcal{C}}}$ is shown in Lemma 3.5, so Definition 5.1(1) is satisfied. Conditions (2) and (3) are satisfied follows since they depend on the *description* of certain Steiner systems (in fact only the points that are in Ω_M) and not their existence.

Lemma 5.4 $\mathcal{E}_{\hat{\mathcal{D}}}$ *satisfies JEP.*

Proof. Given $M, N \in \hat{\mathcal{D}}$, take P to be the structure in $\mathcal{D}_k$ consisting of a copy of M and a disjoint copy of N along with an additional element, $\beta \in \mathcal{T}_P$ such that $\forall \alpha \in \mathcal{T}_P$ ($\alpha \leq \beta$) and the Steiner system β contains two blocks, one incident with all the points of M, and the other incident with all the points of N. This Steiner system must also have other blocks incident with just k points. This Steiner system is in $\hat{\mathcal{C}}$ by Lemma 3.6 and so $P \in \hat{\mathcal{D}}$. The embeddings $f : M \to P$ and $g : N \to P$ are in $\mathcal{E}_{\hat{\mathcal{D}}}$ so $\mathcal{E}_{\hat{\mathcal{D}}}$ satisfies JEP.

Lemma 5.5 $\mathcal{E}_{\hat{\mathcal{D}}}$ *satisfies AP.*

Proof. Suppose we have $M, N_1, N_2 \in \widehat{\mathcal{D}}$ and embeddings $f_i : M \rightarrow N_i$ in $\mathcal{E}_{\widehat{\mathcal{D}}}$. Then we construct an element $P \in \widehat{\mathcal{D}}$ and embeddings $g_i : N_i \rightarrow P$ in $\mathcal{E}_{\widehat{\mathcal{D}}}$ such that $g_1 f_1 = g_2 f_2$ as follows. We construct P around M by a step by step process as it would be difficult to give a clear explicit definition of P. As usual, we identify M with its images in the N_i.

We first amalgamate $(\mathcal{T}_{N_1}, \mathcal{B}_{N_1})$ and $(\mathcal{T}_{N_2}, \mathcal{B}_{N_2})$ over $(\mathcal{T}_M, \mathcal{B}_M)$ by the same method as Lemma 4.3. However we make two changes to this technique as follows, adding certain elements to M (and corresponding elements to N_1 and N_2).

Firstly for each minimal block of $\mathcal{T}_M$ that is incident with at least $k+1$ points, and each non-minimal block of $\mathcal{T}_M$ that is incident with more points than are in the unique maximal Steiner system that lies in it, we shall adjoin an additional Steiner system σ that lies in this block. We call σ a *specified Steiner system*. We also add certain blocks which we describe as follows. At present, we are only deal with the sorts $\mathcal{T}$ and $\mathcal{B}$, so we do not add any points. However we describe what points from Ω_M will be incident with these blocks when we adjoin the points later on.

If Σ is a minimal block of some Steiner system, α say, and $|\{x \in \Omega_M : I(\alpha, x, \Sigma)\}| = n$ then we adjoin $\binom{n}{k}$ additional blocks, all as blocks of σ. Later on, when we consider points, we will require σ to contain all the points $\{x \in \Omega_M : I(\alpha, x, \Sigma)\}$ and for each of the $\binom{n}{k}$ new blocks, exactly k points from $\{x \in \Omega_M : I(\alpha, x, \Sigma)\}$ will be incident with it. We have $\sigma < \alpha$ and $B(\sigma, \Sigma)$.

Suppose Σ is a non-minimal block of some Steiner system, α say, and $\beta \in \mathcal{T}_M$ is maximal with respect to $B(\beta, \Sigma)$ holding. Put $m := |\{x \in \Omega_M : L(\beta, x)\}|$ and $n := |\{x \in \Omega_M : I(\alpha, x, \Sigma)\}|$ and suppose $m < n$. Now we adjoin $\binom{n}{k} - \binom{m}{k} + 1$ additional blocks, all as blocks of σ. Later, we will require σ to contain all the points $\{x \in \Omega_M : I(\alpha, x, \Sigma)\}$. However, the points $\{x \in \Omega_M : L(\beta, x)\}$ will all be incident with the unique block, $\widehat{\Sigma}$ say, that is adjoined satisfying with $B(\beta, \widehat{\Sigma}) \wedge C(\sigma, \widehat{\Sigma})$, and each other block will be incident with exactly k points from $\{x \in \Omega_M : I(\alpha, x, \Sigma)\}$. Again, $\beta < \sigma < \alpha$ and $B(\sigma, \Sigma)$.

If either $SS(M, N_1, \Sigma)$ or $SS(M, N_2, \Sigma)$ exists, we identify it with σ, and for each block of $SS(M, N_1, \Sigma)$ (respectively $SS(M, N_2, \Sigma)$) that is incident with at least k points from Ω_M, we identify this block with a block of σ. Clearly if $SS(M, N_1, \Sigma)$ and $SS(M, N_2, \Sigma)$ both exist and both have a block incident with the same k points from Ω_M, we must identify them both together with the same block of σ.

If neither $SS(M, N_1, \Sigma)$ or $SS(M, N_2, \Sigma)$ exists, we still must adjoin σ to M and the N_i. We add σ so that it is below any Steiner systems τ in the N_i which contain all the points $\{x \in \Omega_M : I(\alpha, x, \Sigma)\}$, all int he same block of τ.

Also we must add σ so it is above any Steiner systems in the N_i that lie in Σ but contain at most k points from $\{x \in \Omega_M : I(\alpha, x, \Sigma)\}$. Further suppose $N_1 \backslash M$ and $N_2 \backslash M$ both contain a Steiner system that lies in Σ and is incident with at most k points from $\{x \in \Omega_M : I(\alpha, x, \Sigma)\}$. Let $\sigma_1 \in \mathcal{T}_{N_1} \backslash \mathcal{T}_M$ and $\sigma_2 \in \mathcal{T}_{N_2} \backslash \mathcal{T}_M$ be maximal with respect to this condition holding. Now unless σ_1 and σ_2 are both incident with exactly k points from $\{x \in \Omega_M : I(\alpha, x, \Sigma)\}$, and these points are the same, they must lie in different blocks of σ. Otherwise (when σ_1 and σ_2 are incident with the same k points from $\{x \in \Omega_M : I(\alpha, x, \Sigma)\}$), σ_1 and σ_2 must both lie in the same block of σ.

If only $SS(M, N_1, \Sigma)$ exists then it is clear where σ lies in the semilinear ordering on the amalgam $\mathcal{T}_P$ in relation to the Steiner systems in $\mathcal{T}_{N_2}$, much as in the last paragraph; likewise if only $SS(M, N_2, \Sigma)$ exists.

There is a further issue if Σ is a minimal block of M and there are $\beta \in \mathcal{T}_{N_1}, \gamma \in \mathcal{T}_{N_2}$ with $B(\beta, \Sigma)$ and $B(\gamma, \Sigma)$. Now each $\beta \in \mathcal{T}_{N_1}$ with $B(\beta, \Sigma)$ holding will contain at most k points from Ω_M, by the definition of *specified Steiner system*. Suppose $\beta_1, \ldots, \beta_l$ denote the Steiner systems in $\mathcal{T}_{N_1}$ that lie in Σ and have exactly k points from Ω_M. Then any two of the β_i are comparable, so without loss of generality, assume $\beta_1 < \ldots < \beta_l < \sigma$. Similarly each $\gamma \in \mathcal{T}_{N_2}$ with $B(\gamma, \Sigma)$ holding will contain at most k points from Ω_M. Let $\gamma_1, \ldots, \gamma_p$ denote the Steiner systems in $\mathcal{T}_{N_2}$ that lie in Σ with exactly k points from Ω_M, and assume $\gamma_1 < \ldots < \gamma_p < \sigma$.

We then adjoin $\beta_1, \ldots, \beta_l, \gamma_1, \ldots, \gamma_p$ as in Case I of Lemma 4.3. This can be done since $\beta_1, \ldots, \beta_l, \gamma_1, \ldots, \gamma_p$ all have the same k points from Ω_M in common. Now let β' denote the maximal Steiner system of N_1 that lies in the block of β_1 that is incident with k points from Ω_M (or that lies in Σ if β_1 does not exist), provided there such β' exists. Then β' is unique since incomparable Steiner systems lie in different blocks of their least upper bound. Similarly let γ' denote the maximal Steiner system of N_2 that lies in the block of γ_1 that is incident with k points from Ω_M (or that lies in Σ if γ_1 does not exist), ifsuch γ' exists.

If neither β' or γ' exist then no further adjoining is needed here. If β' exists, but γ' does not, then we can add β' and all Steiner systems that are below it, as in Case I of Lemma 4.3. Similarly if γ' exists, but β' does not. Finally, suppose β' and γ' both exist. Then we adjoin a new element, δ say, to $\mathcal{T}_P$ with $B(\delta, \Sigma)$ holding, so that δ contains two blocks and β' and γ' lie below δ in different δ-blocks. We also require that $\delta < \beta_1$ and $\delta < \gamma_1$ (if they exist) and δ lies in the blocks of β_1 and γ_1 that have k points from Ω_M. We shall call δ a *splitting element*. We may be forced to adjoin several such splitting elements, corresponding to different minimal blocks of M. The use of splitting elements ensures that two incomparable Steiner systems lie in different blocks of the Steiner system that is their least upper bound.

We can now complete the semilinear amalgamation as in the previous section, and must next adjoin the points of each Steiner system. The sort

$\mathcal{T}_P$ has now been completely defined. However we may still be forced to add more blocks so we have not yet defined the sort $\mathcal{B}_P$.

Next, apply Lemma 4.2 to order the elements of $\mathcal{T}_P$ so that for $\sigma_i, \sigma_j \in \mathcal{T}_P$, if we have $\sigma_i \leq \sigma_j$ then $i \leq j$. We consider each Steiner system in this order in turn, and adjoin all the points of each Steiner system. It is convenient to work in this fashion, as points of a Steiner system belong to all Steiner systems above it. There are several cases. In each case, we consider a Steiner system $\alpha \in \mathcal{T}_P$. When the case has been handled, we will have defined the points of α ($\{x \in \Omega_P : L(\alpha, x)\}$), its blocks ($\{\Sigma \in \mathcal{B}_P : C(\alpha, \Sigma)\}$), and incidence between them.

Case A. $\alpha \in \mathcal{T}_{N_1} \backslash \mathcal{T}_M$ or $\alpha \in \mathcal{T}_{N_2} \backslash \mathcal{T}_M$ and α not a specified Steiner system.

Without loss of generality assume $\alpha \in \mathcal{T}_{N_1} \backslash \mathcal{T}_M$. For each point $x \in \Omega_{N_1}$ such that $L(\alpha, x)$ holds in Ω_{N_1}, $L(\alpha, x)$ must also hold in P. For these points we define $I(\alpha, x, \Sigma)$ to hold for some $\Sigma \in \mathcal{B}_P$ with $C(\alpha, \Sigma)$ holding in P, whenever $I(\alpha, x, \Sigma)$ holds in N_1.

Now consider each block of α that has some Steiner systems in $\mathcal{T}_P$ that lie in it, in turn. Let Σ be such a block. Now we have already specified the full incidence structure in P of each Steiner system that lies in Σ. Let $\beta \in \mathcal{T}_P$ be maximal with respect to $B(\beta, \Sigma^*)$ holding: a unique such β exists since incomparable Steiner systems lie in different blocks of their least upper bound. Now if $\{x \in \Omega_P : L(\beta, x)\} = \{x_1, \ldots, x_m\}$. then we require that $x_1, \ldots, x_m$ lie in α and $I(\alpha, x_1, \Sigma^*) \wedge \ldots \wedge I(\alpha, x_m, \Sigma^*)$ holds. This will ensure that A20 holds in P. Since the maximal Steiner systems that lie in different blocks of α are incomparable, there will not be two such maximal Steiner systems that have k points in common by A19. Hence no two blocks of α will have k points in common, so A15 holds.

After adjoining to α these new points for each block of α, we add new blocks to $\mathcal{B}_P$ as blocks of α, so that every k points of α are incident with a unique block of α.

If some point of $\Omega_{N_2} \backslash \Omega_M$ was in two different Steiner systems that lay in different blocks of α, then since α would be the least upper bound of the two Steiner systems, it would belong to $\mathcal{T}_{N_2}$, which is a contradiction. Hence, when we adjoined to α a new point in the previous two paragraphs, it was incident with just a single existing block of α. It is therefore clear that the identity embedding from N_{1_α} into P_α is in $\mathcal{E}_{\widehat{\mathcal{C}}}$, wih respect to $I(\alpha, \ldots)$.

Case B. $\alpha \in \mathcal{T}_M$.

Since $M_\alpha, N_{1_\alpha}, N_{2_\alpha} \in \widehat{\mathcal{C}}$ under the relation induced by $I(\alpha, \ldots)$ we are able to apply Lemma 3.8 to amalgamate N_{1_α} and N_{2_α} over M_α. Since we have already adjoined the blocks of $\mathcal{B}_{N_1}$ and $\mathcal{B}_{N_2}$, we must identify any of these blocks that are blocks of α, with the corresponding blocks that are produced in the amalgamation.

Now similarly to the previous case, for each maximal Steiner system that lies in a block of α, we must adjoin all the points of this Steiner system to α, so that A20 holds in P. Further, we must add new blocks to α so that every k points of α are incident with a unique block of α. Again, we find that N_{1_α} and N_{2_α} are $\mathcal{E}_{\hat{C}}$-embedded into P_α (with respect to $I(\alpha, \ldots)$).

Case C. α is a specified Steiner system.

Suppose α is a specified Steiner system, which was adjoined because of a block, Σ say, of the Steiner system β, and $\{x \in \Omega_M : I(\beta, x, \Sigma)\} = \{x_1, \ldots, x_n\}$. Then we adjoin these n points, as points of α. If there exists $\gamma \in \mathcal{T}_M$ with $\gamma < \beta$ and γ is maximal with respect to this condition, let $\{x \in \Omega_M : L(\gamma, x)\} = \{x_1, \ldots, x_m\}$ (so $m < n$). We now define incidence for the Steiner system α between these n points and the blocks we adjoined earlier. If γ exists, we define $x_1, \ldots, x_m$ to all be incident with a single block of α, and otherwise we define incidence so that every k points are incident with a single block of α. If any of the blocks of α have Steiner systems from $\mathcal{T}_P$ that lie in them, we must ensure that we define incidence on these blocks so that all the points from Ω_M of the Steiner systems that lie in them, are incident with the block. This is possible since these points will either be $x_1, \ldots, x_m$ (in the case where γ lies in this block), or number at most k. We now consider the following four cases.

1. If $\alpha \in \mathcal{T}_{N_1} \wedge \alpha \in \mathcal{T}_{N_2}$, we amalgamate N_{1_α} and N_{2_α} over the points $x_1, \ldots, x_n$ and the blocks of α that are described in the preceding paragraph, as in case II above. This is possible since the structure consisting of these points and blocks along with the incidence relation induced by $I(\alpha, \ldots)$ is $\mathcal{E}_{\hat{C}}$-embedded in N_{1_α} and in N_{2_α} by Definition 5.1.

2. Suppose $\alpha \in \mathcal{T}_{N_1} \backslash \mathcal{T}_{N_2}$. Then adjoin to α the points and blocks of N_{1_α}, and define incidence as in N_1. Now as before adjoin all points of all maximal Steiner systems that lie in α, and define incidence for α so that these points are incident with just the block containing the corresponding maximal Steiner system. Also add new blocks so that every k points are incident with a block.

Now suppose β is minimal with respect to $\beta \in \mathcal{T}_{N_2}$ and $\alpha < \beta$. Further suppose $\hat{\Sigma} \in \mathcal{B}_{N_2}$ with $C(\beta, \hat{\Sigma})$ and $B(\alpha, \hat{\Sigma})$ holding. Let $\{x \in \Omega_{N_2} : I(\beta, x, \hat{\Sigma})\} = \{y_1, \ldots, y_l\}$. Add all of these l points (if not already in α) to α. For any such point y, put $\neg I(\alpha, y, \Sigma')$ for any existing block Σ' of α. This helps ensure that the embedding $g_2 : N_2 \to P$ is in $\mathcal{E}_{\hat{D}}$. We shall give details of this later on, in the last paragraph of this proof.

3. The case when $\alpha \in \mathcal{T}_{N_2} \backslash \mathcal{T}_{N_1}$ is like Case 2.

4. If $\alpha \notin \mathcal{T}_{N_1} \wedge \alpha \notin \mathcal{T}_{N_2}$, we adjoin all points of all maximal Steiner systems that lie in α, and define incidence for α so that these points are incident with just the block in which they lie. Add new blocks so that every k points are

incident with a block.

Now suppose β_1 is minimal with respect to $\beta_1 \in \mathcal{T}_{N_1}$ and $\alpha < \beta_1$. Suppose $\hat{\Sigma} \in \mathcal{B}_{N_1}$ with $C(\beta_1, \hat{\Sigma})$ and $B(\alpha, \hat{\Sigma})$, and let $\{x \in \Omega_{N_1} : I(\beta_1, x, \hat{\Sigma})\} = \{y_1, \ldots, y_l\}$. Similarly suppose β_2 is minimal with respect to $\beta_2 \in \mathcal{T}_{N_2}$ and $\alpha < \beta_2$, and that $\Sigma' \in \mathcal{B}_{N_2}$ with $C(\beta_2, \Sigma')$ and $B(\alpha, \Sigma')$, and let $\{x \in \Omega_{N_2} : I(\beta_2, x, \Sigma')\} = \{z_1, \ldots, z_t\}$. Adjoin all of these $l + t$ points, if not already in α, to α. For any such point x, put $\neg I(\alpha, x, \Sigma'')$ for any existing block, Σ'', of α.

At the end, we once more add blocks (each with k points) to α so that every k points of α are incident with a unique block of α.

Case D. α is a splitting element.

We have defined this Steiner system to have just two blocks. For each block, we adjoin the points of the maximal Steiner system that lies in it, and define incidence so that A20 holds.

Now suppose β_1 is minimal with respect to $\beta_1 \in \mathcal{T}_{N_1}$ and $\alpha < \beta_1$ and that $\hat{\Sigma} \in \mathcal{B}_{N_1}$ with $C(\beta_1, \hat{\Sigma})$ and $B(\alpha, \hat{\Sigma})$. Let $\{x \in \Omega_{N_1} : I(\beta_1, x, \hat{\Sigma})\} = \{x_1, \ldots, x_m\}$. Similarly suppose β_2 is minimal with respect to $\beta_2 \in \mathcal{T}_{N_2}$ and $\alpha < \beta_2$ and that $\Sigma' \in \mathcal{B}_{N_2}$ with $C(\beta_2, \Sigma')$ and $B(\alpha, \Sigma')$ holding. Let $\{x \in \Omega_{N_2} : I(\beta_2, x, \Sigma')\} = \{y_1, \ldots, y_n\}$. We now adjoin to α any of these $m+n$ points which are not already in α. For any such point x, put $\neg I(\alpha, x, \Sigma'')$ for any existing block Σ'' of α. Finally we add new blocks to α so that every k points of α are incident with a unique block of α.

Once the above construction is complete, we interpret the relations of $\mathcal{L}$ on P in the obvious way. It just remains to show that $P \in \hat{\mathcal{D}}$ and the embeddings $g_i : N_i \to P$ $(i = 1, 2)$ are in $\mathcal{E}_{\hat{\mathcal{D}}}$. Each individual Steiner system in P is in $\hat{\mathcal{C}}$ under the incidence relation induced by I, hence clearly $P \in \hat{\mathcal{D}}$.

Consider the embeddings $g_i : N_i \to P$. Then Lemma 3.8 and our construction above ensures that the g_i satisfy Definition 5.1(1) (we always worked with embeddings from $\mathcal{E}_{\hat{\mathcal{C}}}$). We verify that conditions (2) and (3) also hold for the embedding $g_1 : N_1 \to P$. For this to be the case we require that every Steiner system in $\mathcal{T}_P \backslash \mathcal{T}_{N_1}$ is of one of the forms described in Definition 5.1.

Suppose $\alpha, \beta \in \mathcal{T}_{N_1}$ and $\sigma \notin \mathcal{T}_{N_1}$ is not a specified Steiner system or splitting element. If $\alpha < \sigma < \beta$ (and α is maximal in $\mathcal{T}_P \backslash \mathcal{T}_{N_2}$, and β minimal in $\mathcal{T}_{N_1}$ with respect to this condition), then

$$\{x \in N_1 : L(\sigma, x)\} = \{x \in N_1 : L(\alpha, x)\}$$

and further there exists $\Sigma \in \mathcal{B}_P$ such that

$$\{x \in N_1 : L(\sigma, x)\} = \{x \in N_1 : I(\sigma, x, \Sigma)\}.$$

Any points of σ that are not incident with Σ will be in $\Omega_{N_2} \backslash \Omega_M$ by case A above. Hence σ is of one of the forms described in Definition 5.1 for the embedding $g_1 : N_1 \to P$.

Suppose $\beta \in \mathcal{T}_{N_1}$ and $\sigma \notin \mathcal{T}_{N_1}$ is not a specified Steiner system or splitting element, and $\sigma < \beta$ (with β minimal with respect to this condition) but there does not exist $\alpha \in \mathcal{T}_{N_1}$ with $\alpha < \sigma$. In this case it follows from Case A above, and Definition 5.1, that $|\{x \in N_1 : L(\sigma, x)\}| \le k$. Hence again, σ is of one of the forms described in Definition 5.1 for the embedding $g_1 : N_1 \to P$.

Now suppose $\beta \in \mathcal{T}_{N_1}$ and $\sigma \notin \mathcal{T}_{N_1}$ and α is a specified Steiner system or splitting element with $\alpha < \sigma < \beta$ (with α maximal in $\mathcal{T}_P \backslash \mathcal{T}_{N_2}$ and β minimal in $\mathcal{T}_{N_1}$ with respect to this condition). By inspection of Cases C and D above, where we adjoin additional points to specified Steiner systems and splitting elements, we know that

$$\{x \in N_1 : L(\alpha, x)\} = \{x \in N_1 : I(\beta, x, \Sigma)\}$$

where Σ is the block of the Steiner system β with $B(\sigma, \Sigma)$ holding. Therefore it follows that

$$\{x \in N_1 : L(\sigma, x)\} = \{x \in N_1 : I(\beta, x, \Sigma)\}$$

and these points are all incident with a single block of σ. Hence again, σ is one of the forms described in Definition 5.1 for the embedding $g_1 : N_1 \to P$.

Therefore, a problem could only arise for σ a specified Steiner system or splitting element with $\alpha < \sigma < \beta$ (where $\alpha, \beta \in \mathcal{T}_{N_1}$ and $\sigma \notin \mathcal{T}_{N_1}$ with α maximal and β minimal such) or if σ is a specified Steiner system with $\sigma < \beta$ (where $\beta \in \mathcal{T}_{N_1}$ and $\sigma \notin \mathcal{T}_{N_1}$ with β minimal such) and there does not exist $\alpha \in \mathcal{T}_{N_1}$ with $\alpha < \sigma$.

Again, we verify that (2) and (3) of Definition 5.1 hold by inspection of cases C and D above. We shall highlight how this works in the case where σ is a specified Steiner system (adjoined because of some block of the Steiner system $\delta \in \mathcal{T}_M$) that is not in N_1, but there exists $\alpha \in \mathcal{T}_M, \beta \in \mathcal{T}_{N_1}$ with $\alpha < \sigma < \beta \le \delta$ holding in P. Let Σ denote the unique block of $\mathcal{B}_{N_1}$ such that $C(\beta, \Sigma) \wedge B(\alpha, \Sigma)$ holds. Without loss of generality we assume that α is maximal in $\mathcal{T}_M$ with respect to $\alpha < \sigma$ and β is minimal in $\mathcal{T}_{N_1}$ with respect to $\sigma < \beta$. Since σ is a specified Steiner system, all points $\{x \in N_1 : L(\sigma, x)\}$ will be incident with the block Σ in β, but will not all be incident with a single block of σ. For Definition 5.1(3) to hold, we require that

$$\{x \in N_1 : L(\sigma, x)\} = \{x \in N_1 : I(\beta, x, \Sigma)\},$$

and by Case C this would be the case. Without the adjoining in Case C it is possible that

$$\{x \in N_1 : L(\sigma, x)\} \subset \{x \in N_1 : I(\beta, x, \Sigma)\},$$

and this situation would arise if the block Σ contains any points in $N_1 \backslash M$.

By similar arguments to that in the previous paragraph, it can be shown that 5.1(3) also holds when σ is a splitting element, and (2) holds when σ is a specified Steiner system that is not above any Steiner systems of N_1. Hence the embedding $g_1 : N_1 \to P$ is in $\mathcal{E}_{\widehat{\mathcal{D}}}$ as required. Similarly $g_2 : N_2 \to P$ is in $\mathcal{E}_{\widehat{\mathcal{D}}}$.

Proposition 5.6 *There exists a countable $\mathcal{L}_k$-structure M^* with the following properties.*

1. The class of $\mathcal{E}_{\widehat{\mathcal{D}}}$-embedded substructures of M^* is equal to $\widehat{\mathcal{D}}$.

2. M^* is the union of a chain of finite $\mathcal{E}_{\widehat{\mathcal{D}}}$-embedded substructures.

3. If N is $\mathcal{E}_{\widehat{\mathcal{D}}}$-embedded in M^* and $f : N \to N'$ is in $\mathcal{E}_{\widehat{\mathcal{D}}}$, then there exists P that is $\mathcal{E}_{\widehat{\mathcal{D}}}$-embedded in M^* containing N and an isomorphism $g : N' \to P$ such that $gf(a) = a$ for all $a \in N$.

Any two countable structures with these three properties are isomorphic, and any isomorphism between $\mathcal{E}_{\widehat{\mathcal{D}}}$-embedded finite substructures of M^*, extends to an automorphism of M^*. Also, M^* is an upper semilinear tower of Steiner k-systems and each individual Steiner system is isomorphic to that constructed in Proposition 3.9.

Proof. Apply Theorem 2.8 to $\widehat{\mathcal{D}}$ and $\mathcal{E}_{\widehat{\mathcal{D}}}$. The fact that M^* is an upper semilinear tower of Steiner k-systems follows easily from the fact that M^* is the Fraïssé limit of $\widehat{\mathcal{D}}$, and the axioms for $\widehat{\mathcal{D}}^*$. Each individual Steiner system is isomorphic to that constructed in Proposition 3.9, since the conditions of Proposition 3.9 are satisfied.

6   Properties of the semilinear tower

In this final section we show that the automorphism group of the tower satisfies the properties required to prove the following theorem (which is just a restatement of Theorem 1.2). Recall that a permutation group G on Ω is k-*primitive* if it is k-transitive and for any distsinct $x_1, \ldots, x_{k-1} \in \Omega$, $G_{x_1 \ldots x_{k-1}}$ is primitive on $\Omega \setminus \{x_1, \ldots, x_{k-1}\}$.

Theorem 6.1 *For each k with $2 \leq k < w$ there is an infinite primitive Jordan group (G, Ω) such that*

(i) (G, Ω) is $(k+1)$-transitive, not $(k+1)$-primitive,

(ii) G preserves a limit of Steiner k-systems on Ω,

(iii) G preserves no familiar relation on Ω; that is, no linear or circular order, linear or general betweenness relation, separation relation, C-relation, D-relation or semilinear order,

(iv) G does not preserve a non-trivial Steiner $(k+1)$-system in which the complements of sets of points incident with individual blocks form Jordan sets,

(v) $G_{x_1,\ldots,x_{k-1}}$ *preserves a C-relation (defined below) on* $\Omega\backslash\{x_1,\ldots,x_{k-1}\}$ *for distinct* $x_1,\ldots,x_{k-1}\in\Omega$.

See Remark 6.14 regarding a possible strengthening of (iv).

In the proof of Theorem 6.1, we let M^* be the structure built in Proposition 5.6, and Ω will be its point set Ω_{M^*} and G the group induced by $\mathrm{Aut}(M^*)$ on Ω_{M^*}. From now on we shall refer to this group as (G,Ω_{M^*}). We shall sometimes still refer to elements of $\mathrm{Aut}(M^*)$ We use the notation $\Sigma_\alpha(x_1,\ldots,x_k)$ (instead of $\Sigma_{M^*_\alpha}(x_1,\ldots,x_k)$) to denote the unique block of the Steiner system α of M^* that is incident with $x_1,\ldots,x_k$.

Lemma 6.2 *Given distinct* $x_1,\ldots,x_{k+1}\in\Omega_{M^*}$, *there exists a unique* $\alpha\in$ $\mathcal{T}_{M^*}$ *such that* $L(\alpha,x_1)\wedge\ldots\wedge L(\alpha,x_{k+1})\wedge\neg I(\alpha,x_{k+1},\Sigma_\alpha(x_1,\ldots,x_k))$.

Proof. Since M^* is the union of a chain of finite $\mathcal{E}_{\widehat{\mathcal{D}}}$-embedded structures, there is some N that is $\mathcal{E}_{\widehat{\mathcal{D}}}$-embedded in M^* that contains $x_1,\ldots,x_{k+1}$. If N contains a Steiner system that contains $x_1,\ldots,x_{k+1}$ and they are non-incident, then such an α exists and is unique by A20.

Otherwise, in every Steiner system of $\mathcal{T}_N$ that contains $x_1,\ldots,x_{k+1}$, they are all incident with a single block. Now there exists some Steiner system $\beta\in\mathcal{T}_N$ with $L(\beta,x_1)\wedge\ldots\wedge L(\beta,x_{k+1})$ holding by A18 and the fact that $\leq$ forms an upper semilinear order. Now without loss of generality, assume that β is minimal with this condition holding (a unique such β exists by A19). Now by our assumption, there exists $\Sigma\in\mathcal{B}_N$ with $I(\beta,x_1,\Sigma)\wedge\ldots\wedge$ $I(\beta,x_{k+1},\Sigma)$. By the use of specified Steiner systems in Definition 5.1, it is clear that there exists $N'\in\widehat{\mathcal{D}}$ extending N and a mapping $f:N\to N'$ in $\mathcal{E}_{\widehat{\mathcal{D}}}$, where N' contains a Steiner system α such that $B(\alpha,\Sigma)$ holds and $L(\alpha,x_1)\wedge\ldots\wedge L(\alpha,x_{k+1})\wedge\neg I(\alpha,x_{k+1},\Sigma_\alpha(x_1,\ldots,x_k))$ as required. A gain α is unique by A20. Now apply Proposition 5.6(3).

Given $x_1,\ldots,x_{k+1}\in\Omega_{M^*}$, denote the unique α in the sense of Lemma 6.2 above by $S(x_1,\ldots,x_{k+1})$.

Lemma 6.3 G *is* $(k+1)$-*transitive on* Ω_{M^*}. *Also,* $\mathrm{Aut}(M^*)$ *is transitive on the set of all Steiner k-systems.*

Proof. By Lemma 3.10, any non-incident $(k+1)$-set of points of a Steiner system α, is $\mathcal{E}_{\widehat{C}}$-embedded in α. Hence it follows that if $N\in\widehat{\mathcal{D}}$ with $\mathcal{T}_N=\{\alpha\}$, $\Omega_N=\{x_1,\ldots,x_{k+1}\}$ and

$$L(\alpha,x_1)\wedge\ldots\wedge L(\alpha,x_{k+1})\wedge\neg I(\alpha,x_{k+1},\Sigma_\alpha(x_1,\ldots,x_k)),$$

then N is $\mathcal{E}_{\widehat{\mathcal{D}}}$-embedded in M^*. Thus, since α is unique by Lemma 6.2, we can apply Proposition 5.6 to get that G is $(k+1)$-transitive on Ω_{M^*} and that $\mathrm{Aut}(M^*)$ is transitive on the set of all Steiner k-systems.

Corollary 6.4 M^* *is not $\aleph_0$-categorical.*

Proof. This follows from Corollary 3.11 since each individual Steiner system is isomorphic to that constructed in Proposition 3.9, by Proposition 5.6.

For $\alpha \in \mathcal{T}_{M^*}$, let $M_\alpha^* := \{x \in \Omega_{M^*} : L(\alpha, x)\}$. This notation is different to that in previous sections, as we no longer include blocks.

Lemma 6.5 *For each $\alpha \in \mathcal{T}_{M^*}$, $G_{\{M_\alpha^*\}}$ is transitive on non-incident ordered $(k+1)$-sets and indeed is k-transitive.*

Proof. This follows from Lemma 6.3, since G is $(k+1)$-transitive, and the group G maps Steiner systems to Steiner systems.

Lemma 6.6 *Let $\alpha \in \mathcal{T}_{M^*}$, and suppose that N is an $\mathcal{E}_{\widehat{\mathcal{D}}}$-embedded substructure of M^* and θ is a partial automorphism of N that satisfies the following.*
  (i) θ is the identity map on $M_\alpha^* \cap (\mathrm{Domain}(\theta) \cup \mathrm{Range}(\theta))$.
  (ii) M_α^* contains at least k points in $\mathrm{Domain}(\theta) \cap \mathrm{Range}(\theta)$.
  (iii) The embeddings $f_1 : \mathrm{Domain}(\theta) \to N, f_2 : \mathrm{Range}(\theta) \to N$ given by inclusion are in $\mathcal{E}_{\widehat{\mathcal{D}}}$.
  Then there exists a substructure N^* that is $\mathcal{E}_{\widehat{\mathcal{D}}}$-embedded in M^* and a partial automorphism θ^* of N^* such that
  (a) N is a substructure of N^*.
  (b) $\theta^* \upharpoonright \mathrm{Domain}(\theta) = \theta$.
  (c) $\mathrm{Domain}(\theta^*) \cap \mathrm{Range}(\theta^*) \supseteq N$.
  (d) θ^* is the identity map on $M_\alpha^* \cap N^*$.
  (e) The embeddings $g_1 : \mathrm{Domain}(\theta^*) \to N^*, g_2 : \mathrm{Range}(\theta^*) \to N^*$ given by inclusion are in $\mathcal{E}_{\widehat{\mathcal{D}}}$.

Proof. This lemma is proved is the same manner as Lemma 3.12, after we show the following claim, where $R := \mathrm{Range}(\theta) \cup (M_\alpha^* \cap N) \cup \{\sigma, \Sigma : \sigma \in \mathcal{T}_N, \sigma \le \alpha, \Sigma \in \mathcal{B}_N, C(\sigma, \Sigma)\} \cup \{\Sigma' \in \mathcal{B}_N : \exists \gamma \in \mathrm{Range}(\theta), x_1, \ldots, x_k \in (M_\alpha^* \cap N) \cup \mathrm{Range}(\theta) \ (\alpha < \gamma \wedge I(\gamma, x_1, \Sigma') \wedge \ldots \wedge I(\gamma, x_k, \Sigma'))\}$.

Claim R is $\mathcal{E}_{\widehat{\mathcal{D}}}$-embedded in N.
Proof. We show that the conditions of Definition 5.1 are satisfied. We consider each Steiner system $\beta \in \mathcal{T}_R$ in turn.

First suppose $\beta \in \mathcal{T}_R$ with $\beta \le \alpha$. Then $M_\beta^* \cap R = M_\beta^* \cap N$ and there can be no specified elements that lie below β, since R contains all Steiner systems in N that lie below β.

If $\beta \in \mathcal{T}_R$ with $\alpha < \beta$ then $M_\beta^* \cap R = (M_\beta^* \cap \mathrm{Range}(\theta)) \cup \Gamma$ where Γ is the set of points in M_α^*. By A20 the points in Γ must all be incident with the same block of β (which is also in R), so Definition 5.1(1) is satisfied for β by Lemma 3.7. Now Definition 5.1(2) and (3) also hold since the points of

M_α^* must all be incident with the same block of each Steiner system γ with $\alpha < \gamma \le \beta$, hence specified Steiner systems exist in the embedding $R \to N$ if and only if they existed in the embedding $f_2 : \text{Range}(\theta) \to N$. We note that the additional blocks adjoined to β in going from $\text{Range}(\theta)$ to R are incident with just k points from N, so no specified Steiner systems can lie in them.

Finally if $\beta \in T_R$ with $\alpha \not\le \beta$ and $\beta \not\le \alpha$ then $M_\beta^* \cap R = M_\beta^* \cap \text{Range}(\theta)$, since we do not alter any Steiner systems that are incomparable with α. Hence the conditions of Definition 5.1 are satisfied, since $f_2 : \text{Range}(\theta) \to N$ is in $\mathcal{E}_{\widehat{D}}$.

This proves the claim, and the lemma follows.

Lemma 6.7 *For each $\alpha \in T_{M^*}$, the set $\Omega_{M^*} \backslash M_\alpha^*$ is a Jordan set for G.*

Proof. Take $y, y' \in \Omega_{M^*} \backslash M_\alpha^*$. We will show there is an automorphism of M^* mapping y to y' and fixing M_α^* pointwise. Then, clearly the induced action of this automorphism on Ω_{M^*} will be in G. Choose $x_1, \ldots, x_k \in M_\alpha^*$ and let β, β' denote the Steiner systems $S(x_1, \ldots, x_k, y)$ and $S(x_1, \ldots, x_k, y')$ respectively. Then $\alpha < \beta, \beta'$. Since M^* is a union of a chain of finite $\mathcal{E}_{\widehat{D}}$-embedded structures, there exists N that is $\mathcal{E}_{\widehat{D}}$-embedded in M^* such that $x_1, \ldots, x_k, y, y', \alpha, \beta, \beta' \in N$.

By initially mapping $(x_1, \ldots, x_k, y, \alpha, \beta)$ to $(x_1, \ldots, x_k, y', \alpha, \beta')$, define a partial automorphism θ on N. Now $\text{Domain}(\theta)$ and $\text{Range}(\theta)$ are $\mathcal{E}_{\widehat{D}}$-embedded in M^*, so we can apply Lemma 6.6 to extend to some P, $\mathcal{E}_{\widehat{D}}$-embedded in M^* and a partial automorphism θ^* of P with $\text{Domain}(\theta^*) \cap \text{Range}(\theta^*) \supseteq N$, fixing pointwise $M_\alpha^* \cap P$ such that the two embeddings $\text{Domain}(\theta) \to P$ and $\text{Range}(\theta) \to P$ given by inclusion are in $\mathcal{E}_{\widehat{D}}$.

Since M^* is a union of a chain of finite $\mathcal{E}_{\widehat{D}}$-embedded structures, we take some P^* in this chain such that P^* contains P. Now the embeddings $f_3 : \text{Domain}(\theta^*) \to P^*$, $f_4 : \text{Range}(\theta^*) \to P^*$ are in $\mathcal{E}_{\widehat{D}}$ (since $\mathcal{E}_{\widehat{D}}$ is closed under composition) so we can apply Lemma 6.6 again. We repeat this process ω many times to build $g \in \text{Aut}(M^*)$ such that $g(y) = y'$ and $g \upharpoonright M_\alpha^*$ is the identity.

Recall that a ternary relation C on a set Δ is called a *C-relation* of the following universal axioms hold [5]

C1 $C(\alpha; \beta, \gamma) \Rightarrow C(\alpha; \gamma, \beta)$;

C2 $C(\alpha; \beta, \gamma) \Rightarrow \neg C(\beta; \alpha, \gamma)$;

C3 $C(\alpha; \beta, \gamma) \wedge \neg C(\delta; \beta, \gamma) \Rightarrow C(\alpha; \delta, \gamma))$;

C4 $(\alpha \ne \beta) \Rightarrow (\exists \gamma) (\beta \ne \gamma \wedge C(\alpha; \beta, \gamma))$;

C5 $(\exists \alpha) \, C(\alpha; \beta, \gamma)$.

Lemma 6.8 *Let* $x_1, \ldots, x_{k-1} \in \Omega_{M^*}$ *be distinct. Then* $G_{x_1,\ldots,x_{k-1}}$ *preserves a* C-*relation on* $\Omega_{M^*} \backslash \{x_1, \ldots, x_{k-1}\}$.

Proof. Given $x, y, z \in \Omega_{M^*} \backslash \{x_1, \ldots, x_{k-1}\}$, we can define

$$K_{x_1,\ldots,x_{k-1}}(x; y, z) \Leftrightarrow S(x_1, \ldots, x_{k-1}, x, y) > S(x_1, \ldots, x_{k-1}, y, z)$$

(i.e. $S(x_1, \ldots, x_{k-1}, y, z) \leq S(x_1, \ldots, x_{k-1}, x, y)$ and $S(x_1, \ldots, x_{k-1}, y, z) \neq S(x_1, \ldots, x_{k-1}, x, y)$). Note that if $S(x_1, \ldots, x_{k-1}, x, y) > S(x_1, \ldots, x_{k-1}, y, z)$ then $x_1, \ldots, x_{k-1}, y, z$ are incident with a single block of $S(x_1, \ldots, x_{k-1}, x, y)$ that is not incident with x.

We now show that $K_{x_1,\ldots,x_{k-1}}$ defines a C-relation by verifying the axioms of the definition above, and that this is invariant under $\mathrm{Aut}(M^*)$.

(C1) holds as $S(x_1, \ldots, x_{k-1}, x, y) > S(x_1, \ldots, x_{k-1}, y, z) \Rightarrow S(x_1, \ldots, x_{k-1}, x, z) > S(x_1, \ldots, x_{k-1}, z, y)$, as $x_1, \ldots, x_{k-1}, y$ and z are all incident with the same block of $S(x_1, \ldots, x_{k-1}, x, y)$, which is not incident with x.

(C2) holds as $S(x_1, \ldots, x_{k-1}, x, y) > S(x_1, \ldots, x_{k-1}, y, z) \Rightarrow S(x_1, \ldots, x_{k-1}, z, x) = S(x_1, \ldots, x_{k-1}, x, y)$ since $x_1, \ldots, x_{k-1}, y$ and z are all incident with the same block of $S(x_1, \ldots, x_{k-1}, x, y)$ which is not incident with x.

(C3) holds as if we have $S(x_1, \ldots, x_{k-1}, x, y) > S(x_1, \ldots, x_{k-1}, y, z)$ and $S(x_1, \ldots, x_{k-1}, t, y) \leq S(x_1, \ldots, x_{k-1}, y, z)$, then we have $S(x_1, \ldots, x_{k-1}, x, t) > S(x_1, \ldots, x_{k-1}, t, z)$, since $S(x_1, \ldots, x_{k-1}, t, z) \leq S(x_1, \ldots, x_{k-1}, y, z)$.

(C4) holds since the semilinear order has no minimal elements.

(C5) holds since the semilinear order has no maximal elements.

Corollary 6.9 G *is not* $(k+1)$-*primitive.*

Proof. By the definition of multiply primitive groups, if $G_{x_1,\ldots,x_{k-1}}$ is not 2-primitive on $\Omega_{M^*} \backslash \{x_1, \ldots, x_{k-1}\}$ then G is not $(k+1)$-primitive. Hence this result follows from the fact that a permutation group H that preserves a C-relation on a set Δ is not 2-primitive. To see the latter, we use the fact (see Theorem 12.4 of [5]) that Δ can be identified with a dense set of maximal chains in a semilinear order, with $C(\alpha; \beta, \gamma)$ holding precisely if $\alpha \cap \beta \subseteq \beta \cap \gamma$ and $\alpha \notin \{\beta, \gamma\}$. Choose $\alpha \in \Delta$. Then there is an H_α-invariant equivalence relation $\sim$ on $\delta \backslash \{\alpha\}$ given by $\beta \sim \gamma$ if and only if $\beta \cap \alpha = \gamma \cap \alpha$.

Lemma 6.10 (G, Ω_{M^*}) *preserves a limit of Steiner* k-*systems.*

Proof. By Lemma 6.3, G is $(k+1)$-transitive and by Corollary 6.9 is not $(k+1)$-primitive. Let $\mathcal{T}^*$ be a maximal linearly ordered subset of $\mathcal{T}_{M^*}$. Then $(\mathcal{T}^*, \leq)$ has no maximal elements. We now verify the remaining five conditions of Definition 2.9

(i) $\bigcup\{M_\alpha^* : \alpha \in T^*\} = \Omega_{M^*}$, since every point in Ω_M belongs to some Steiner system by A9, and hence to some Steiner system in T^* by A5.

(ii) For each $\alpha \in T^*$, $G_{\{M_\alpha^*\}}$ is k-transitive on M_α^* and preserves a non-trivial Steiner k-system on M_α^* by Lemma 6.5.

(iii) For each $\alpha \in T$, the set $\Omega_{M^*}\backslash M_\alpha^*$ is a Jordan set for G by Lemma 6.7.

(iv) If $\alpha < \beta$ then M_α^* is a subset of the points incident with a block of the Steiner system β by A7.

(v) The automorphism group of the semilinear tower of Steiner k-systems, must preserve the structure of the individual Steiner k-systems, and the ordering $\leq$. Therefore for all $g \in G$ there is $\alpha_0 \in T^*$ dependent on g, such that for every $\alpha > \alpha_0$ there is $\beta \in T^*$ so that $g(M_\alpha^*) = M_\beta^*$ and the image under g of every block of M_α^* is a block of M_β^*.

Hence $\mathrm{Aut}(M^*)$ preserves a limit of Steiner k-systems on Ω_{M^*}.

Lemma 6.11 *G preserves no invariant Steiner $(k+1)$-systems, in which the complements of sets of points incident with individual blocks are Jordan sets.*

Proof. If G preserves a Steiner $(k + 1)$-system and a semilinear tower of Steiner k-systems, then the pointwise stabiliser of $k - 2$ points preserves a Steiner 3-system and a limit of Steiner 2-systems, by [3] (Theorem 5.8.4). Hence we are only required to show this result for the case when $k = 2$.

Assume G preserves a non-trivial Steiner 3-system, S, on Ω_{M^*} where the complements of sets of points incident with individual blocks are Jordan sets. For each $x_1, x_2, x_3 \in \Omega_{M^*}$, let $\Phi(x_1, x_2, x_3)$ denote the set of points incident with the unique block of S, which is incident with x_1, x_2 and x_3.

Choose $x_1, x_2, x_3 \in \Omega_{M^*}$ and let $\alpha = S(x_1, x_2, x_3)$. Trivially we have $x_1, x_2, x_3 \in \Phi(x_1, x_2, x_3)$ and we now describe the remaining points that are incident with this block.

Suppose towards a contradiction that $(\Omega_{M^*}\backslash M_\alpha^*) \cap \Phi(x_1, x_2, x_3) \neq \emptyset$ and choose $x \in (\Omega_{M^*}\backslash M_\alpha^*) \cap \Phi(x_1, x_2, x_3)$. Now for all $x' \in (\Omega_{M^*}\backslash M_\alpha^*)$, there is $g \in G$ with $g(x) = x'$ such that g fixes M_α^* pointwise, by Lemma 6.7. Since $g(x_1, x_2, x_3, x) = (x_1, x_2, x_3, x')$ we have $g(\Phi(x_1, x_2, x_3)) = \Phi(x_1, x_2, x_3)$ and so $x' \in \Phi(x_1, x_2, x_3)$. Hence it follows that $(\Omega_{M^*}\backslash M_\alpha^*) \subseteq \Phi(x_1, x_2, x_3)$.

Now consider $y_1, y_2, y_3 \notin \Phi(x_1, x_2, x_3)$ and put $\beta = S(y_1, y_2, y_3)$. Now since M^* is an upper semilinear tower of Steiner 2-systems, there exists $z_1, z_2, z_3 \in (\Omega_{M^*}\backslash M_\alpha^*) \cap (\Omega_{M^*}\backslash M_\beta^*)$. Hence z_1, z_2, z_3 are incident with both the distinct blocks that have point sets $\Phi(x_1, x_2, x_3)$ and $\Phi(y_1, y_2, y_3)$. This is a contradiction, so $(\Omega_{M^*}\backslash M_\alpha^*) \cap \Phi(x_1, x_2, x_3) = \emptyset$.

Now $\Phi(x_1, x_2, x_3) \subseteq M_\alpha^*$. We now assume $x_1', x_2', x_3' \in M_\alpha^* \cap \Phi(x_1, x_2, x_3)$ and $I(\alpha, x_3', \Sigma_\alpha(x_1', x_2'))$. Now $S(x_1', x_2', x_3') < S(x_1, x_2, x_3)$, so by the above argument we have $\Phi(x_1', x_2', x_3') \subseteq M_{S(x_1', x_2', x_3')}^*$ and so $\Phi(x_1', x_2', x_3') \neq \Phi(x_1, x_2, x_3)$. This is a contradiction.

Hence $\Phi(x_1, x_2, x_3)$ consists of the points x_1, x_2, x_3 and all other points belong to $S(x_1, x_2, x_3)$. However, no three points are incident with a single block of $S(x_1, x_2, x_3)$.

Choose $x', x'' \in M_\alpha^*$ such that $I(\alpha, x', \Sigma_\alpha(x_1, x_2))$ and $I(\alpha, x'', \Sigma_\alpha(x_1, x_3))$. Now by the above paragraph $x', x'' \notin \Phi(x_1, x_2, x_3)$. Since the complements of blocks are Jordan sets for G, there exists $g \in G$ such that $g(x_1, x_2, x_3, x') = (x_1, x_2, x_3, x'')$. But this map does not preserve the structure of $S(x_1, x_2, x_3)$, thus we have a contradiction. Hence G does not preserve a non-trivial Steiner 3-system in which the complements of sets of points incident with individual blocks form Jordan sets.

Lemma 6.12 (G, Ω_{M^*}) *does not preserve a linear or circular order, a linear or general betweenness relation, C-relation or a semilinear order.*

Proof. These relations are all binary or ternary hence this result follows from the fact that G is 3-transitive.

Lemma 6.13 (G, Ω_{M^*}) *does not preserve a separation relation or D-relation.*

Proof. Since G is $(k + 1)$-transitive, we need only verify this result for the case when $k = 2$. In this case however, the reader is referred to the proof given by Adeleke [2] as this result follows by a similar argument.

We have now shown that G has the desired properties, proving Theorem 6.1.

Remark 6.14 We have only shown in Theorem 6.1 that G does not preserve an individual Steiner $(k+1)$-system in which the complements of sets of points incident with individual blocks form Jordan sets. The intention was to show that G does not preserve any Steiner $(k+1)$-system on Ω, but this has not been achieved. However, in all the known examples of $(k + 1)$-transitive Jordan groups preserving a Steiner $(k+1)$-system for some $k \geq 2$, the complements of blocks form Jordan sets. Thus, if our group does preserve a limit of Steiner k-systems and a single Steiner $(k+1)$-system then it is rather unusual. However, in Johnson [9] there is an account of a similar unusual example of a Jordan group preserving a non-trivial C-relation and a single Steiner 3-system.

References

[1] S. A. Adeleke, 'Examples of irregular infinite Jordan groups', preprint, Western Illinois University, 1994.

[2] S. A. Adeleke, 'Semilinear tower of Steiner systems', *Journal of Combinatorial Theory*, Series A, 72 (1995) 243-255.

[3] S. A. Adeleke, H. D. Macpherson, 'Classification of infinite primitive Jordan permutation groups', *Proc. London. Math. Soc.* (3) 72 (1996) 63-123.

[4] S. A. Adeleke, P. M. Neumann, 'Primitive permutation groups with primitive Jordan sets', *Journal of the London Mathematical Society, Series 2,* 53 (1996) 209-229.

[5] S. A. Adeleke, P. M. Neumann, 'Relations related to betweenness: their structure and automorphisms', *Memoirs of the American Mathematical Society* Vol 131, No 623 (1998).

[6] M. Bhattacharjee, H. D. Macpherson, R. G. Möller, P. M. Neumann, *Notes on infinite permutation groups,* Lecture Notes in Mathematics 1698, Springer-Verlag, Berlin, 1998.

[7] D. M. Evans, 'Examples of $\aleph_0$-categorical structures', *Automorphisms of first-order Structures* (eds R.W. Kaye and H.D. Macpherson, Oxford University Press, 1994), pp 33-72.

[8] E. Hrushovski, 'A new strongly minimal set', *Annals of Pure and Applied Logic* 62 (1993) 147-166

[9] K. Johnson, 'Model-theoretic constructions of infinite permutation groups', PhD thesis, University of Leeds, 1999.

[10] W. M. Kantor, 'Homogeneous designs and geometric lattices', J. Combin. Theory Ser. A 8 (1985), 64–77.

[11] H. D. Macpherson, 'A survey of Jordan groups', *Automorphisms of first-order Structures* (eds R. W. Kaye and H. D. Macpherson, Oxford University Press, 1994), pp 73-110.

[12] P. M. Neumann, 'Some primitive permutation groups', Proc. London Math. Soc. (3), 50 91985), 265–281.

Introduction to the Lascar Group

Martin Ziegler*

1 Introduction

The aim of this article is to give a short introduction to the Lascar Galois group $\mathrm{Gal_L}(T)$ of a complete first order theory T. We prove that $\mathrm{Gal_L}(T)$ is a quasicompact topological group in section 5. $\mathrm{Gal_L}(T)$ has two canonical normal closed subgroups: $\Gamma_1(T)$, the topological closure of the identity, and $\mathrm{Gal_L^0}(T)$, the connected component. In section 6 we characterize these two groups by the way they act on bounded hyperimaginaries. In the last section we give examples which show that every compact group occurs as a Lascar Galois group and an example in which $\Gamma_1(T)$ is non–trivial.

None of the results, except possibly Corollary 26, are new, but some technical lemmas and proofs are. In particular, the treatment of the topology of $\mathrm{Gal_L}(T)$ in sections 4 and 5 avoids ultraproducts, by which the topology was originally defined in [6]. Most of the theory expounded here was taken from that article, and the more recent [7], [4] and [2].

I thank Katrin Tent for reading the manuscript carefully, Markus Tressl, who found a serious mistake in an earlier version, and Anand Pillay, who helped me with the proof of Theorem 23.

2 The group

We fix a complete theory T. Let $\mathbb{C}$ be a saturated[1] model of T, of cardinality larger than as $2^{|T|}$, and let $\mathrm{Aut}(\mathbb{C})$ its automorphism group. The subgroup $\mathrm{Autf_L}(\mathbb{C})$ generated by all point–wise stabilizers $\mathrm{Aut}_M(\mathbb{C})$ of elementary[2]

---

*Partially supported by the Mittag-Leffler Institute, Stockholm

[1]T may not have saturated models. In this case we take for $\mathbb{C}$ a *special* model (see [3] Chapter 10.4) of T and use the $\mathrm{cf}|\mathbb{C}|$ instead of $|\mathbb{C}|$. Especially we assume that $\mathrm{cf}|\mathbb{C}| > 2^{|T|}$.

[2]In the sequel *submodel* will always mean *elementary submodel*.

submodels M is called the group of *Lascar strong* automorphisms. $\mathrm{Autf}_L(\mathbb{C})$ is a normal subgroup of $\mathrm{Aut}(\mathbb{C})$. The quotient is the *Lascar (Galois) group* of $\mathbb{C}$:

$$\mathrm{Gal}_L(\mathbb{C}) = \mathrm{Aut}(\mathbb{C})/\mathrm{Autf}_L(\mathbb{C}).$$

We will show that $\mathrm{Gal}_L(\mathbb{C})$ does not depend on the choice of $\mathbb{C}$.

Lemma 1 *Let M and N be two small [3] submodels of $\mathbb{C}$ and f an automorphism. Then the class of f in $\mathrm{Gal}_L(\mathbb{C})$ is determined by the type of $f(M)$ over N.*

PROOF: Let $(m_i)_{i \in I}$ be an enumeration of M. By the type of $f(M)$ over N we mean the type of the infinite tuple $(f(m_i))_{i \in I}$ over N. This is a type in variables $(x_i)_{i \in I}$. We denote by $S_I(N)$ the set of all such types over N.

Let $g(M)$ have the same type over N as $f(M)$. Choose an automorphism s which fixes N and maps $f(M)$ to $g(M)$. Then s is a Lascar strong automorphism, as is $t = (sf)^{-1}g$, which fixes M. Now we see that $g = sft$ and f have the same class in $\mathrm{Gal}_L(\mathbb{C})$. □

Two possibly infinite tuples a and b from $\mathbb{C}$ are said to have the same *Lascar strong type* iff $f(a) = b$ for a Lascar strong automorphism f.

Lemma 2 *a and b have the same Lascar strong type iff there is a sequence of tuples $a = a_0, \dots, a_n = b$ and a sequence of small submodels $N_1, \dots, N_n$ such that, for each i, a_{i-1} and a_i have the same type over N_i.*

PROOF: Clear □

Corollary 3 *a and b have the same Lascar strong type in $\mathbb{C}$ if they have the same Lascar strong type in an elementary extension of $\mathbb{C}$.*

PROOF: If $a_0, \dots, N_n$ exist in an elementary extension of $\mathbb{C}$, we find by saturation in $\mathbb{C}$ a sequence $a'_0, \dots, N'_n$ which has the same type over ab as $a_0, \dots, N_n$. This sequence shows that a and b have the same Lascar strong type in $\mathbb{C}$. □

Theorem 4 ([6]) *$\mathrm{Gal}_L(\mathbb{C})$ depends only on T and not on the choice of $\mathbb{C}$.*

PROOF: If $\mathbb{C}'$ is another big saturated model of T we can assume that $\mathbb{C}'$ is an elementary extension of $\mathbb{C}$ and of larger cardinality. We can extend every

---

[3]of smaller cardinality than $\mathbb{C}$

automorphism f of $\mathbb{C}$ to an automorphism f' of $\mathbb{C}'$. Since all such f' differ only by elements of $\mathrm{Aut}_{\mathbb{C}}(\mathbb{C}')$, this defines a homomorphism $\mathrm{Aut}(\mathbb{C}) \to \mathrm{Gal}_{\mathrm{L}}(\mathbb{C}')$. If f Lascar strong, f' is Lascar strong as well. Whence we have a well defined natural map

$$\mathrm{Gal}_{\mathrm{L}}(\mathbb{C}) \to \mathrm{Gal}_{\mathrm{L}}(\mathbb{C}'),$$

which will turn out to be an isomorphism.

To prove surjectivity, fix an automorphism g of $\mathbb{C}'$. Choose two small submodels M and N of $\mathbb{C}$. By saturation we find a submodel M' of $\mathbb{C}$ which has the same type over N as $g(M)$. There is an automorphism f of $\mathbb{C}$ which maps M to M'. Extend f to an automorphism f' of $\mathbb{C}'$. Then $f'(M)$ and $g(M)$ have the same type over N. Whence, by the last lemma f' and g represent the same element of $\mathrm{Gal}_{\mathrm{L}}(\mathbb{C}')$.

Now assume that $f \in \mathrm{Aut}(\mathbb{C})$ extends to a Lascar strong automorphism f' of $\mathbb{C}'$. Fix a small submodel M of $\mathbb{C}$. Then M and $f(M)$ have the same Lascar strong type in $\mathbb{C}'$, whence also in $\mathbb{C}$ by Corollary 3. So M can be mapped to $f(M)$ by a Lascar strong automorphism of $\mathbb{C}$. Such an automorphism agrees with f on M, whence f is also strong. This shows that $\mathrm{Gal}_{\mathrm{L}}(\mathbb{C}) \to \mathrm{Gal}_{\mathrm{L}}(\mathbb{C}')$ is injective. $\qquad\square$

Definition *The Lascar group of T is the quotient*

$$\mathrm{Gal}_{\mathrm{L}}(T) = \mathrm{Aut}(\mathbb{C})/\mathrm{Autf}_{\mathrm{L}}(\mathbb{C}),$$

where $\mathbb{C}$ is any big saturated model of T.

Corollary 5 *The cardinality of $\mathrm{Gal}_{\mathrm{L}}(T)$ is bounded by $2^{|T|}$.*

PROOF: The class of f in $\mathrm{Gal}_{\mathrm{L}}(T)$ is determined by the type of $f(M)$ over N. If M and N are chosen to be of cardinality T, there are at most $2^{|T|}$ possible types. $\qquad\square$

3 Digression: Lascar strong types and thick formulas

Definition *Let $\theta(x,y)$ be a formula in two tuples of variables x and y having the same length. $\theta(x,y)$ is thick, if it has no infinite antichain, that is a sequence of tuples $a_0, a_1, \ldots$ such that $\mathbb{C} \models \neg\,\theta(a_i, a_j)$ for all $i < j$.*

Clearly $\theta(x,y)$ is thick iff there is no indiscernible sequence $a_0, a_1, \ldots$ such that $\mathbb{C} \models \neg\theta(a_0, a_1)$. With this description it is easy to see that the intersection of two thick formulas is thick again and that a formulas remains thick if one interchanges the role of x and y.

Lemma 6 *Let $\Theta(x,y)$ be the set of all thick formulas in x and y and let a and b two tuples of the same length. Then the following are equivalent:*

a) $\mathbb{C} \models \Theta(a,b)$

b) a *and* b *belong to an infinite indiscernible sequence.*

PROOF: Assume $\mathbb{C} \models \Theta(a,b)$. Then, if $\psi(x,y)$ is satisfied by ab, $\neg\psi$ is not thick, so there is an infinite sequence of indiscernibles $a_0, a_1, \ldots$ such that $\psi(a_0, a_1)$ is true. Whence, by compactness, there is one infinite sequence of indiscernibles such that $a_0 a_1$ has the same type as ab.

If conversely a, b are the first two elements of an infinite indiscernible sequence they have to satisfy all thick formulas   □

Lemma 7
1. *If $\mathbb{C} \models \Theta(a,b)$, there is a model over which a and b have the same type.*

2. *If a and b have the same type over some model, the pair ab satisfies the relational product $\Theta \circ \Theta$. I.e. there is a tuple a' such that $\mathbb{C} \models \Theta(a,a')$ and $\mathbb{C} \models \Theta(a',b)$.*

PROOF:
1. Let I be an infinite sequence of indiscernibles and M any small model. Then there are indiscernibles I' over M of the same type as I. Whence there is a model M' of the same type as M over which I is indiscernible. Therefore, if a, b are the first elements of some I, they have the same type over some model M'. Now apply Lemma 6.

A more direct proof, which avoids Lemma 6, uses the observation that two sequences a and b of the same length have the same type over a model iff ab satisfies all formulas of the form

$$\exists z\, \varphi(z) \to \exists z \left(\varphi(z) \wedge \bigwedge_{i=1}^{n} \psi_i(x,z) \leftrightarrow \psi_i(y,z) \right) \tag{1}$$

for all finite variable tuples z and formulas $\varphi(z), \psi_1(x,z), \ldots, \psi_n(x,z)$. All formulas (1) are thick, antichains have length at most 2^n.

2. Assume that a and b have the same type over M. If θ is a thick formula, consider a maximal antichain $a_1, \ldots, a_n$ for θ in M. Then, since M is an

elementary substructure, $a_1, \ldots, a_n$ is also a maximal antichain in $\mathbb{C}$. Whence $\mathbb{C} \models \theta(a_i, a)$ for some i. Since b has the same type over M, we have $\mathbb{C} \models \theta(a_i, b)$. This proves that for every finite subset Θ_0 of Θ there is an a' such that $\mathbb{C} \models \Theta_0(a', a)$ and $\mathbb{C} \models \Theta_0(a', b)$. This proves the claim using compactness and the observation that Θ defines a symmetric relation. $\qquad\square$

Corollary 8 *The relation of having the same Lascar strong type is the transitive closure of the relation defined by Θ.* $\qquad\square$

Let π be a type defined over the empty set. A formula $\theta(x, y)$ is *thick on π* if θ has no infinite antichain in $\pi(\mathbb{C})$. Let Θ_π be the set of all formulas which are thick over π.

Corollary 9 *Two realizations of π, a and b, have the same Lascar strong type if the pair (a, b) is in the transitive closure of the relation defined by Θ_π.*

PROOF: Assume that a and b have the same type over a model M. The proof of Lemma 7 (1) shows that we can assume that M is ω–saturated. If θ is thick on π, let $a_1, \ldots, a_n$ be a maximal antichain for θ in $\pi(M)$. Then, since is ω–saturated, $a_1, \ldots, a_n$ is also maximal in $\pi(\mathbb{C})$. Now proceed as in Lemma 7 (2). $\qquad\square$

4 The topology

Let M and N be two small submodels of $\mathbb{C}$. Assign to every automorphism f of $\mathbb{C}$ the type of $f(M)$ over N. This defines a surjective map μ from $\mathrm{Aut}(\mathbb{C})$ to $\mathrm{S}_M(N)$, the set all types over N of conjugates of M. By Lemma 1 the projection $\mathrm{Aut}(\mathbb{C}) \to \mathrm{Gal}_L(T)$ factors through μ:

$$\mathrm{Aut}(\mathbb{C}) \xrightarrow{\mu} \mathrm{S}_M(N) \xrightarrow{\nu} \mathrm{Gal}_L(T).$$

$\mathrm{S}_M(N)$, as a closed subspace of $\mathrm{S}_I(N)$, is a boolean space. We give $\mathrm{Gal}_L(T)$ the quotient topology with respect to ν.

To show that this does not depend on the choice of M and N we consider another pair M' and N'. We may assume that $M \subset M'$ and $N \subset N'$. The map $\mathrm{S}_{M'}(N') \longrightarrow \mathrm{Gal}_L(T)$ then factors as

$$\mathrm{S}_{M'}(N') \longrightarrow \mathrm{S}_M(N) \xrightarrow{\nu} \mathrm{Gal}_L(T),$$

where the first map is restriction of types. Since restriction is continuous and the spaces are compact, $\mathrm{S}_M(N)$ carries the quotient topology of $\mathrm{S}_{M'}(N')$, which implies that on $\mathrm{Gal}_\mathrm{L}(T)$ the two topologies, coming from $\mathrm{S}_{M'}(N')$ and $\mathrm{S}_M(N)$, are the same.

A quotient of a quasicompact space remains quasicompact. So we have

Lemma 10 $\mathrm{Gal}_\mathrm{L}(T)$ *is quasicompact.* □

Let p and q be types in $\mathrm{S}_M(N)$. Two realizations M' and M'' of p and q have the same Lascar strong type iff $\nu(p) = \nu(q)$. Whence, by Corollary 8, the equivalence relation
$$p \approx q \iff \nu(p) = \nu(q)$$
is the transitive closure of the relation D, where $D(p,q)$ holds if p and q have realizations M' and M'' with $\mathbb{C} \models \Theta(M', M'')$.

Lemma 11

1. D *is a closed subset of* $\mathrm{S}_M(N) \times \mathrm{S}_M(N)$

2. $\approx$ *is a* F_σ*-set, i.e. a countable union of closed sets.*

PROOF:
1. This is clear, because
$$D(p,q) \iff p(x) \cup q(y) \cup \Theta(x,y) \text{ consistent.}$$

2. $\approx$ is the union of all powers
$$\mathrm{D}^n = \underbrace{\mathrm{D} \circ \cdots \circ \mathrm{D}}_{n \text{ times}}.$$

So, is suffices to show that all D^i are closed. This follows from the fact that, in compact spaces, the product of two closed relations is closed again. To see this, note that, for binary relations R and S, $R \circ S$ is the projection of $\{(p,q,r) | R(p,q) \wedge S(q,r)\}$ onto the first and third variable. □

In general the map $\mathrm{S}_M(N) \xrightarrow{\nu} \mathrm{Gal}_\mathrm{L}(T)$ is not open.[4] But it has a property that comes close to openness. Define for $p \in \mathrm{S}_M(N)$
$$D[p] = \{q \in \mathrm{S}_M(N) \mid D(p,q)\}.$$

---

[4] If $\mathrm{Aut}(\mathbb{C})$ is endowed with the topology of point–wise convergence, μ becomes continuous (see Lemma 29). If ν were always open, $\mathrm{Aut}(\mathbb{C}) \to \mathrm{Gal}_\mathrm{L}(T)$ would be open too: If a, b are two (finite) tuples, choose N, M in such a way that $a, b \in M = N$. Then the basic open set $\{f \in \mathrm{Aut}(\mathbb{C}) | f(a) = b\}$ will be mapped onto an open subset of $\mathrm{S}_M(N)$ and whence, by assumption, onto an open subset of $\mathrm{Gal}_\mathrm{L}(T)$. Whence, the closedness of $\mathrm{Autf}_\mathrm{L}(\mathbb{C})$ would imply that $\mathrm{Gal}_\mathrm{L}(T)$ is hausdorff. That this is not true shows one of the examples in [2] ($Th(M^*)$ in Proposition 4.5).

Lemma 12 *If* $D[p]$ *is contained in the interior of some subset* $O \subset S_M(N)$, *then* $\nu(p)$ *is an inner point of* $\nu(O)$.

PROOF: $D[p]$ is the intersection of all

$$D_\delta[p] = \{q \in S_M(N) \mid p(x) \cup q(y) \cup \{\delta(x,y)\} \text{ consistent}\}, \quad (\delta \in \Theta).$$

By compactness some $D_\delta[p]$ is contained in (the interior of) O.

Claim 1: p is an inner point of $D_\delta[p]$.

Proof: Since δ is thick, there is a finite set $\{H_1, \dots, H_n\}$ of realizations of p such that for every other realization H we have $\mathbb{C} \models \delta(H_i, H)$ for some i. By compactness this is true for every realization H of any p' contained in a small enough neighborhood C of p, which implies that C is contained in $D_\delta[p]$.

After replacing O by $\nu^{-1}(\nu O)$ we can assume that O is closed under $\approx$ (i.e. is a union of $\approx$–classes.) We set

$$U = \{q \in S_M(N) \mid D_\delta[q] \subset O \text{ for some } \delta \in \Theta\}.$$

U contains p.

Claim 2: U is closed under $\approx$.

Proof: Let q be in U, witnessed by $D_\delta[q] \subset O$, and $q \approx r$. Then a realization H of q is mapped by a Lascar strong automorphism f to a realization $f(H) = K$ of r. In order to show that r belongs to U we fix an element r' of $D_\delta[r]$. We have then a realization K' of r' such that $\mathbb{C} \models \delta(K, K')$. Let q' be the type of $H' = f^{-1}(K')$ over M. Since $\mathbb{C} \models \delta(H, H')$, q' belongs to $D_\delta[q]$ and therefore to O. Since $q \approx q'$ and O is closed under $\approx$, we have $q' \in O$. It follows $D_\delta[r] \subset O$.

Claim 3: U is open.

Proof: U is a subset of the interior of O by Claim 1. Since U is closed under $\approx$, it is contained in the open set

$$U' = \{q \in S_M(N) \mid D[q] \subset \text{interior}(O)\},$$

which, by compactness, equals

$$U'' = \{q \in S_M(N) \mid D_\delta[q] \subset \text{interior}(O) \text{ for some } \delta \in \Theta\}.$$

But U'' is contained in U, which shows that $U = U'$.

By Claims 2 and 3 the projection of U is an open subset of $\nu(O)$ and contains $\nu(p)$. This completes the proof of Lemma. $\square$

Corollary 13 *If L is countable, $\mathrm{Gal}_L(T)$ has a countable basis.*

PROOF: If L is countable we can choose countable M and N. $S_M(N)$ has then a countable base, $\mathcal{B}$. We can assume that $\mathcal{B}$ is closed under finite unions. Let us show that the set of all $\nu(B)^\circ$, $(B \in \mathcal{B})$, is a basis of $\mathrm{Gal}_L(T)$. Let Ω be open and $\alpha \in \Omega$. Choose a preimage p of α and a basic open set B, such that $\mathrm{D}[p] \subset B \subset \nu^{-1}(\Omega)$. This is possible, since B is compact and $\mathcal{B}$ closed under finite unions. Then $\nu(B)^\circ \subset \Omega$ is an open neighborhood of p. $\square$

The following corollary is a reformulation of Corollary 3.5 in [2].

Corollary 14 *Let X be a subset of $\mathrm{Gal}_L(T)$. Then*

$$\overline{X} = \nu(\overline{\nu^{-1}(X)}).$$

PROOF: Since ν is continuous the right hand side lies inside $\overline{X}$. Let $\nu(p)$ be an element of $\mathrm{Gal}_L(T)$ which does not belong to $\nu(\overline{\nu^{-1}(X)})$. Then the whole $\approx$–class of p, which contains $\mathrm{D}[p]$, is disjoint from $\overline{\nu^{-1}(X)}$. By Lemma 12 the complement of $\overline{\nu^{-1}(X)}$ is mapped to a neighborhood of $\nu(p)$, which is disjoint from X. This shows $\nu(p) \notin \overline{X}$. $\square$

Corollary 15 *$\mathrm{Gal}_L(T)$ is hausdorff iff $\approx$ is closed.*

PROOF: "$\mathrm{Gal}_L(T)$ hausdorff $\Rightarrow$ $\approx$ closed" is an easy consequence of the continuity of ν.

Now assume that $\approx$ is closed. Consider two different elements x, y of $\mathrm{Gal}_L(T)$. Since $\approx$ is closed, we can separate each element of $\nu^{-1}(x)$ from each element of $\nu^{-1}(y)$ by a pair of neighborhoods which projects onto disjoint subsets of $\mathrm{Gal}_L(T)$. But $\nu^{-1}(x)$ and $\nu^{-1}(y)$ are compact. This implies that there is one pair of open sets, O and U, which separate $\nu^{-1}(x)$ and $\nu^{-1}(y)$ and have disjoint projections $\nu(O)$ and $\nu(U)$, which are, by the lemma, neighborhoods of x and y. $\square$

We will see in section 7 (Theorem 28) that $\mathrm{Gal}_L(T)$ need not to be hausdorff.

5 The topological group

Theorem 16 (Lascar) $\mathrm{Gal}_L(T)$ *is a topological group.*

For the proof we fix again two small submodels M and N and consider the natural mappings

$$\mathrm{Aut}(\mathbb{C}) \xrightarrow{\mu} S_M(N) \xrightarrow{\nu} \mathrm{Gal}_L(T).$$

Lemma 17 *The projections of multiplication*

$$\mathcal{M} = \big\{ \big(\mu(f), \mu(g), \mu(fg)\big) \mid f, g \in \mathrm{Aut}(\mathbb{C}) \big\}$$

and of inversion

$$\mathcal{I} = \big\{ \big(\mu(f), \mu(f^{-1})\big) \mid f \in \mathrm{Aut}(\mathbb{C}) \big\}$$

are closed subset of $S_M(N) \times S_M(N) \times S_M(N)$ and of $S_M(N) \times S_M(N)$, respectively.

PROOF: We introduce two unary function symbols F and G and express the fact that F are G automorphisms by the $L \cup \{F, G\}$–theory $A(F, G)$. Then (p, q, r) belongs to $\mathcal{M}$ iff there are are functions $f, g : \mathbb{C} \to \mathbb{C}$ which satisfy the theory

$$B(F, G, p, q, r) = A(F, G) \cup p(F(M)) \cup q(G(M)) \cup r(F(G(M))).$$

Since $\mathbb{C}$ is saturated, $B(F, G, p, q, r)$ can be satisfied in $\mathbb{C}$ if it is consistent with the theory of $\mathbb{C}_{M,N}$. This is a closed condition on p, q, r.

The closedness of $\mathcal{I}$ is similar. □

The graphs of the multiplication and inversion in $\mathrm{Gal}_L(T)$ are the projections of $\mathcal{M}$ and $\mathcal{I}$. If $\mathrm{Gal}_L(T)$ is hausdorff, the projections are closed, which, by compactness, implies that multiplication and inversion are continuous in $\mathrm{Gal}_L(T)$.

For the general case we need the following notation: For two subsets of A and B of $S_M(N)$ define

$$A * B = \big\{ r \in S_M(N) \mid (p, q, r) \in \mathcal{M} \text{ for a pair } (p, q) \in A \times B \big\}.$$

Lemma 18 *If A and B are closed and $A * B$ is contained in the open set W, there are neighborhoods U and V of A and B such that $U * V \subset W$.*

PROOF: Let W' be the complement of W. $A \times B$ is disjoint from the projection C of

$$\mathcal{M} \cap \left(S_M(N) \times S_M(N) \times W' \right)$$

on the first two coordinates. Since C is closed (and A and B are compact) there are neighborhoods U and V of A and B such that $U \times V$ is disjoint from C. It follows that $U * V \subset W$. $\qquad\square$

We can now prove that multiplication in $\mathrm{Gal}_L(T)$ is continuous. Let $\alpha = \nu(p)$ and $\beta = \nu(q)$ be elements of $\mathrm{Gal}_L(T)$ and Ω an open neighborhood of $\alpha\beta$. Then

$$\mathrm{D}[p] * \mathrm{D}[q] \subset \nu^{-1}(\alpha) * \nu^{-1}(\beta) \subset \nu^{-1}(\alpha\beta) \subset \nu^{-1}(\Omega).$$

By the last lemma there neighborhoods U and V of $\mathrm{D}[p]$ and $\mathrm{D}[q]$, respectively, such that $U * V \subset \nu^{-1}(\Omega)$. This implies $\nu(U)\nu(V) \subset \Omega$. Finally, we remark that, by Lemma 12, $\nu(U)$ and $\nu(V)$ are neighborhoods of α and β.

The continuity of inversion is proved in the same manner, which completes the proof of the theorem.

6   Two subgroups

$\mathrm{Gal}_L(T)$ has two canonical normal subgroups:

- $\Gamma_1(T)$, the closure of $\{1\}$.

- $\mathrm{Gal}_L^0(T)$, the connected component of 1.

Since $\mathrm{Gal}_L(T)$ is quasicompact, we have

Lemma 19
1. *The quotient* $\mathrm{Gal}_L^c(T) = \mathrm{Gal}_L(T)/\Gamma_1(T)$ *is a compact group, the closed Galois group of* T.

2. $\mathrm{Gal}_L^0(T)$ *is the intersection of all closed (normal) subgroups of finite index.*

PROOF: $\mathrm{Gal}_L^c(T)$ is quasicompact and hausdorff, i.e. compact. For the second part, note that the quotient $\mathrm{Gal}_L(T)/\mathrm{Gal}_L^0(T)$ is totally disconnected ([12, §2]) and compact, whence a profinite group. In a profinite group the intersection of all normal closed subgroups of finite index is the identity. $\qquad\square$

An *imaginary* element of $\mathbb{C}$ is a class of a $\emptyset$–definable equivalence relation on a cartesian power $\mathbb{C}^n$. Automorphisms of $\mathbb{C}$ act in a natural way on imaginaries. An imaginary with only finitely many conjugates under $\mathrm{Aut}(\mathbb{C})$ is called *algebraic*.

Let us prove that algebraic imaginaries are fixed by Lascar strong automorphisms: Let a/E be an algebraic imaginary with k conjugates. This means that E partitions the set of all conjugates of a into k classes. It follows that the type of a contains a formula $\varphi(x)$ whose realization set meets exactly k equivalence classes. Let f fix the model M. Then $\varphi(M)$ meets the same classes as $\varphi(\mathbb{C})$, which implies that a/E contains an element b of M, which must also belong to $f(a)/E$. It follows that $a/E = f(a)/E$.

This result extends easily to *hyperimaginaries*. Hyperimaginaries are equivalence classes of type–definable equivalence relations E, which are defined by a set of formulas Φ without parameters:

$$E(a,b) \Leftrightarrow \mathbb{C} \models \Phi(a,b).$$

a and b are, possible infinite, tuples of elements of $\mathbb{C}$, of length smaller than $|\mathbb{C}|$. A hyperimaginary is *bounded* if it has less than $|\mathbb{C}|$ conjugates.

Lemma 20 *Bounded hyperimaginaries are fixed by Lascar strong automorphisms.*

PROOF: Let a/E be a bounded hyperimaginary and E defined by $\Phi(x,y)$. Then $\Phi \subset \Theta_\pi$, where $\pi = \mathrm{tp}(a)$, since otherwise some $\theta \in \Phi$ would have antichains in $\pi(\mathbb{C})$ of arbitrary length, contradicting the assumption that a/E is bounded. If f is Lascar strong, a and $f(a)$ have the same Lascar strong type. By Corollary 9, $E(a, f(a))$. $\qquad\qquad\square$

If, conversely, a hyperimaginary h is fixed by all Lascar strong automorphisms, $f(h)$ is determined by the class of f in $\mathrm{Gal}_L(T)$. Whence h has no more than $2^{|T|}$–many conjugates and is bounded.

We conclude that $\mathrm{Gal}_L(T)$ acts on bounded hyperimaginaries in a well defined way.

Theorem 21
1. $\Gamma_1(T)$ *is the set of all elements of* $\mathrm{Gal}_L(T)$ *which fix all bounded hyperimaginaries.*

2. $\mathrm{Gal}_L^0(T)$ *is the set of all elements of* $\mathrm{Gal}_L(T)$ *which fix all algebraic imaginaries.*

PROOF:

1. Let a/E be a bounded hyperimaginary and $\Gamma \leq \mathrm{Gal_L}(T)$ the stabilizer of a/E. The preimage of Γ in $\mathrm{S}_M(N)$ is

$$\nu^{-1}(\Gamma) = \{\mathrm{tp}(f(M)/N) \mid f \in \mathrm{Aut}(\mathbb{C}),\ E(f(a),a)\}.$$

Choose M containing a, let $N = M$ and E be axiomatized by Φ. Then

$$\nu^{-1}(\Gamma) = \{p(x) \in \mathrm{S}_M(N) \mid \Phi(x',a) \subset p(x)\},$$

where the variables x' are a subtuple of x, as a is a subtuple of (m_i), the enumeration of M. Whence Γ is closed and we conclude $\Gamma_1(T) \subset \Gamma$. This shows that the elements of $\Gamma_1(T)$ fix all bounded imaginaries.

For the converse consider the inverse image G_1 of $\Gamma_1(T)$ in $\mathrm{Aut}(\mathbb{C})$. For $|T|$–tuples a, b let $E(a,b)$ denote the equivalence relation of being in the same G_1–orbit. Since the index of G_1 is bounded by $2^{|T|}$, E has at most $2^{|T|}$ classes. Since $\Gamma_1(T)$ is closed, E is type–definable. To see this, write the closed set $\nu^{-1}(\Gamma_1(T))$ as $\{p(x) \in \mathrm{S}_M(N) \mid \Psi(x) \subset p(x)\}$ for a set $\Psi(x)$ of $L(N)$–formulas. Then

$$E(a,b) \ \Leftrightarrow \ \text{for some } f \in \mathrm{Aut}(\mathbb{C})\ \ \mathbb{C} \models f(a) = b \land \Psi(f(M)).$$

This shows, by an argument similar to that in the proof of Lemma 17, that E is can be defined by a set of formulas with parameters from M and N. Since $\Gamma_1(T)$ is a normal subgroup, G_1 is a normal subgroup of $\mathrm{Aut}(\mathbb{C})$. This implies that E is invariant under automorphisms, and whence can be defined by a set of formulas without parameters.

Now assume that $\alpha \in \mathrm{Gal_L}(T)$ fixes all bounded hyperimaginaries. Take a model K of cardinality $|T|$ and consider it as a $|T|$–tuple. Then K/E is a bounded hyperimaginary and fixed by α. This means that α is represented by an automorphism which agrees on K with an automorphism f from G_1. Since K is a model, this implies that α is represented by f and belongs to $\Gamma_1(T)$.

2. Let i be an algebraic imaginary and Γ the stabilizer of i in $\mathrm{Gal_L}(T)$. Γ is closed and has finite index, since the index equals the number of conjugates of i. It follows that $\mathrm{Gal_L^0}(T) \subset \Gamma$. Thus the elements of $\mathrm{Gal_L^0}(T)$ fix all algebraic imaginaries.

For the converse it suffices to show that every normal closed $\Gamma \leq \mathrm{Gal_L}(T)$ of finite index is the stabilizer of an algebraic imaginary. The first part of the proof shows that Γ, being a normal[5] closed subgroup, is the stabilizer of a

---

[5] A slight variation of the argument shows that normality is not necessary: Let G be the preimage of Γ, and K a model of size $|T|$. Define $E(a,b)$ to be true if $a = b$ or, for some

bounded hyperimaginary a/E. Since Γ has finite index, a/E has only a finite number of conjugates. We will show that a/E has the same stabilizer as an algebraic imaginary a/F. (If a is an infinite tuple, we can replace it by the finite subtuple of elements which occur in F.)

Let E be defined by Φ and let $a_1/E, \ldots, a_n/E$ be the different conjugates of a/E. By compactness there is a symmetric formula[6] $\theta \in \Phi$ such that no pair (a_i, a_j) $(i \neq j)$ satisfies $\theta^2 = \theta \circ \theta$.[7] This means that the sets $\theta(a_i, \mathbb{C})$ are disjoint. Since they cover the set of conjugates of a, there is a formula $\varphi(x)$ satisfied by a such that the intersections

$$D_i = \varphi(\mathbb{C}) \cap \theta(a_i, \mathbb{C})$$

form a partition of $\varphi(\mathbb{C})$. In order to ensure that this partition is invariant under automorphisms, we choose $\theta \in \Phi$ so small that no pair (a_i, a_j) satisfies θ^4. This implies that $\theta^2(c, d)$ is never true for $c \in D_i$ and $d \in D_j$ and, therefore, that

$$F(x, y) \;=\; (\neg\varphi(x) \wedge \neg\varphi(y)) \vee (\varphi(x) \wedge \varphi(y) \wedge \theta^2(x, y))$$

defines an equivalence relation, with classes $\neg\varphi(\mathbb{C}), D_1, \ldots, D_n$. Thus a/F is an algebraic imaginary. Since a/E and a/F contain the same conjugates of a, they have the same stabilizer. $\square$

Corollary 22

1. $\mathrm{Gal}_{\mathrm{L}}^{\mathrm{c}}(T)$ is the automorphism group of the set of all bounded hyperimaginaries of length $|T|$.

2. $\mathrm{Gal}_{\mathrm{L}}(T)/\mathrm{Gal}_{\mathrm{L}}^0(T)$ is the automorphism group of the set of all algebraic imaginaries.

$\square$

It was shown in [7] that every bounded hyperimaginary has the same (point-wise) stabilizer as a set of bounded hyperimaginaries of finite length. So $\mathrm{Gal}_{\mathrm{L}}^{\mathrm{c}}(T)$ is the automorphism group of the set of all bounded hyperimaginaries of finite length.

The set of algebraic imaginaries is often called $\mathrm{acl}^{\mathrm{eq}}(\emptyset)$. The group

$$\mathrm{Gal}_{\mathrm{L}}(T)/\mathrm{Gal}_{\mathrm{L}}^0(T) = \mathrm{Aut}(\mathrm{acl}^{\mathrm{eq}}(\emptyset))$$

$f \in \mathrm{Aut}(\mathbb{C})$ and $g \in G$, $f(K) = a$ and $fg(K) = b$. Then K/E is a bounded hyperimaginary with Γ as its stabilizer. See [7, 4.12].

[6]Assume Φ closed under conjunction.

[7]Recall that $\theta(x, y)$ is the formula $\exists z\, \theta(x, z) \wedge \theta(z, y)$.

is the Galois group introduced by Poizat in [9].

For stable T two tuples a and b which have the same *strong* type (i.e. the same type over $\mathrm{acl}^{\mathrm{eq}}(\emptyset)$) have the same type over any model which is independent from ab. It follows that $\mathrm{Gal}_{\mathrm{L}}^{0}(T) = 1$. This was extended to supersimple theories in [1]. Whether this is true for all simple[8] theories is an open problem. All we know is Kim's result ([5]) that $\Gamma_1(T) = 1$ for simple T.

7 Two Examples

The first part of this section is concerned with the proof of the following unpublished result of E. Bouscaren, D. Lascar and A. Pillay:

Theorem 23 *Any compact Lie group is the Galois group of a countable complete theory.*

First we need a lemma on O-minimal structures. Recall that a structure M with a distinguished linear order $<$ is O-minimal if every definable subset of M is a union of finitely many points and intervals with endpoints in M. Note that every structure elementarily equivalent to an O-minimal structure is itself O-minimal.

Lemma 24 *Every automorphism of a big saturated O-minimal structure is Lascar strong.*

PROOF: Let $\mathbb{C}$ be a big saturated O-minimal structure. We prove that any two small submodels M, N of the same type have the same type over some model K. This implies, as in the proof of Lemma 1, that every automorphism which maps M to N is the product of an automorphism which fixes M and an automorphism which fixes K.

It is enough (and equivalent, see the proof of Lemma 7 (1)) to show the following : Every consistent formula $\varphi(z)$ has a realization c over which M and N have the same type.

We prove this by induction on the length of z. Assume that z consists of a tuple z_1 and a single variable z_2. By induction there is a realization c_1 of $\exists z_2 \varphi(z_1, z_2)$ over which M and N have the same type. Let $\psi(m, c_1, z_2)$ be any formula over Mc_1, and let the tuple $n \in N$ correspond to m. By O-minimality, and since m and n have the same type over c_1, either both $\psi(m, c_1, \mathbb{C})$ and $\psi(n, c_1, \mathbb{C})$ contain a non–empty final segment of $\varphi(c_1, \mathbb{C})$ or $\neg\psi(m, c_1, \mathbb{C})$ and $\neg\psi(n, c_1, \mathbb{C})$ contain a non–empty segment. If we choose c_2

---

[8]See [11] for an introduction to simple theories.

in the intersection of all these segments, $c = c_1 c_2$ realizes $\varphi(z)$ and M and N have the same type over c. □

Now fix a compact Lie group G. The group G together with its structure of a real analytic manifold can be defined inside an expansion $\mathcal{R}$ of the field $\mathbb{R}$ by a finite number of analytic functions which are defined on bounded rectangles. By a result of van den Dries $\mathcal{R}$ is O–minimal[9] (see [10]).

Let $\mathcal{R}^*$ a big saturated extension of $\mathcal{R}$ and G^* the resulting extension of G. The intersection μ of all $\emptyset$–definable neighborhoods of the unit element of G^* is the normal subgroup of *infinitesimal* elements. The compactness of G implies that every element of G^* differs by an infinitesimal from some element of G. Whence G^* is the semi–direct product of G and μ.

Lemma 25 μ *is the set of all commutators* $[\varphi, h] = h^{-1}\varphi(h)$, *where* $h \in G^*$ *and* $\varphi \in \mathrm{Aut}(\mathcal{R}^*)$.

PROOF: Let φ be an automorphism of $\mathcal{R}^*$ and let h differ from $h_0 \in G$ by an infinitesimal ε. Since φ fixes $\mathbb{R}$, it fixes h_0. Whence $h^{-1}\varphi(h) = (h_0\varepsilon)^{-1}\varphi(h_0\varepsilon) = \varepsilon^{-1}\varphi(\varepsilon)$ is infinitesimal.

Let conversely $\varepsilon \in \mu$ be given. Consider a *generic* type $p \in \mathrm{S}(\emptyset)$ of G (cf. [8]). This means that p can be axiomatized by formulas which define (non–empty) open subsets $O(G)$ of G. Each $O(G^*)$ contains two elements h and $h\varepsilon$ (pick any $h \in O(G)$). Whence, by saturation, p has two realizations h and $h\varepsilon$. Choose an automorphism φ with $\varphi(h) = h\varepsilon$. Then $\varepsilon = h^{-1}\varphi(h)$.[10] □

Consider the two–sorted structure

$$\mathcal{M} = (\mathcal{R}, X, \cdot)$$

where $\cdot$ is a regular action of G on the set X. We will show that G is the Galois group of the complete theory of $\mathcal{M}$.

Let $\mathcal{M}^* = (\mathcal{R}^*, X^*)$ be a big saturated elementary extension of $\mathcal{M}$. To describe the automorphisms of $\mathcal{M}^*$ we fix a base point $x_0 \in X^*$. Any element of X^* can then uniquely be written as

$$x = h \cdot x_0$$

---

[9] As A. Pillay has told me, compact Lie groups are semi–algebraic. This means that here (and in the proof of Corollary 26) one can actually assume that $\mathcal{R}$ is the field of reals with a finite tuple of named parameters.

[10] A variant of the proof shows that one can find a φ which fixes an elementary submodel of $\mathcal{R}^*$.

for some $h \in G^*$. We extend each automorphism φ of $\mathcal{R}^*$ to $\mathcal{M}^*$ by

$$\overline{\varphi}(x) = \varphi(h) \cdot x_0.$$

The automorphisms which leave $\mathcal{R}^*$ fixed have the form $\overline{g}$, where

$$\overline{g}(x) = hg^{-1} \cdot x_0.$$

This implies that every automorphism of $\mathcal{M}^*$ is a product

$$\Phi = \overline{g}\,\overline{\varphi}.$$

Note the commutation rule $\overline{\varphi}\,\overline{g} = \overline{\varphi(g)}\,\overline{\varphi}$.

Elementary substructures of $\mathcal{M}^*$ have the form $(\mathcal{R}', G' \cdot x)$, where $\mathcal{R}'$ is an elementary substructure of $\mathcal{R}^*$ and $x = h \cdot x_0$ is any element of X^*. Therefore an automorphism fixes a submodel iff it can be written as $\overline{h}^{-1}\overline{\varphi}\,\overline{h}$, for some φ which fixes an elementary submodel of $\mathcal{R}^*$. It follows that an automorphism is Lascar strong iff it is a product of conjugates of automorphisms of the form $\overline{\varphi}$, for Lascar strong φ.

By Lemma 24 all φ are Lascar strong. The formula

$$\overline{h}^{-1}\overline{\varphi}\,\overline{h} = \overline{[\varphi, h]}\,\overline{\varphi},$$

together with the last Lemma, implies that $\Phi = \overline{g}\,\overline{\varphi}$ is Lascar strong iff g is infinitesimal. We conclude that

$$g \mapsto \text{class of } \overline{g}$$

defines an isomorphism $\iota : G \to \text{Gal}_{\text{L}}(\mathcal{M}^*)$.[11]

Finally we have to prove that ι is a homeomorphism. Let $U(G)$ be a $\emptyset$–definable neighborhood of $1 \in G$. Consider the map $\nu : S_\mathcal{M}(\mathcal{M}) \to \text{Gal}_{\text{L}}(\mathcal{M}^*)$. Then $\nu^{-1}\iota(U(G))$ consists of those $\text{tp}(f(\mathcal{M})/\mathcal{M})$ for which $\mathcal{M}^* \models U(f(1))$. Whence, if 1 has index 1 in the enumeration of $\mathcal{M}$,

$$\nu^{-1}\iota(U(G)) = \{p \in S_\mathcal{M}(\mathcal{M}) \mid U(x_1) \in p\}.$$

This proves that $\iota(U_n)$ is open. So ι is an open map. Since $\text{Gal}_{\text{L}}(\mathcal{M}^*)$ is quasicompact and G is hausdorff, ε must also be continuous. *This completes the proof of Theorem 23.*

Corollary 26 *Every compact group is the Galois group of a complete theory.*

---

[11]The proof shows that two elements of X^* differ by an infinitesimal if they have the same Lascar strong type. It is easy to verify that this happens iff they have the same type over a submodel of $\mathcal{M}^*$.

PROOF: Let G be a compact group. G is the direct limit of a directed system $(G_i, f_{i,j})_{i \leq j \in I}$ of compact Lie groups ([12, §25]). Again let $\mathcal{R}$ be an expansion of the reals by bounded analytic functions, in which all the G_i and the maps f_{ij} can be defined. The elements of G are then given by certain infinite tuples $g = (g_i)_{i \in I}$ from the direct product of the G_i.

G will be the Galois group (of the complete theory) of the many–sorted structure

$$\mathcal{M} = (\mathcal{R}, X_i, f'_{ij})_{i \leq j \in I},$$

where the directed system of sets $(X_i, f'_{i,j})_{i \leq j \in I}$ is a copy of $(G_i, f_{i,j})_{i \leq j \in I}$ and each G_i operates (regularly) on X_i as it operates on itself by left multiplication.

Let again $\mathcal{M}^*$ be a big saturated elementary extension of $\mathcal{M}$ and G^* the inverse limit of the G_i^*. We call an element $\varepsilon = (\varepsilon_i)$ of G^* infinitesimal if all its components are infinitesimal. Let μ the subgroup of all infinitesimals. It is easy to see that G is isomorphic to the quotient G^*/μ.

Fix a base point $x_0 = (x_{0i})_{i \in I}$ in the (non–empty) inverse limit of the X_i^*. Then every automorphism of $\mathcal{M}^*$ has the form $\Phi = \overline{g}\,\overline{\varphi}$ for $\varphi \in \mathrm{Aut}(\mathcal{R}^*)$ and $g \in G^*$, where $\overline{g}$ and $\overline{\varphi}$ are defined as in the proof of the theorem. Thus, it suffices to show that Φ is Lascar strong iff g is infinitesimal.

Assume first that Φ is Lascar strong. Then each g_i is infinitesimal, since Φ restricted to $(\mathcal{R}^*, X_i)$ is Lascar strong. Conversely, if g is infinitesimal, we find for every i an $h_i \in G_i$ such that h_i and $h_i g_i$ have the same type. A compactness argument shows that we can find the sequence (h_i) in G^*. Then h and hg have the same type. Let ψ be an automorphism of $\mathcal{R}^*$ with $\psi(h) = hg$. As in the proof of the theorem, it is easy to see that $\overline{g}\overline{\psi} = \overline{h}^{-1}\overline{\psi}\,\overline{h}$ is Lascar strong. Whence also $\Phi = (\overline{h}^{-1}\overline{\psi}\,\overline{h})\psi^{-1}\overline{\varphi}$ is Lascar strong. $\qquad\square$

We construct our second example from the circle group S, the unit circle in the complex number plane. Let us fix some notation: λ_s denotes multiplication by s. R is the cyclic ordering on S, where $R(r, s, t)$ holds if s comes before t in the counter–clockwise ordering of $S \setminus \{r\}$.

Fix a natural number N, write σ_N for $\lambda_{\frac{2\pi i}{N}}$ and consider the structure

$$\mathcal{S}_N = (S, R, \sigma_N).$$

Let $\mathbb{C}_N$ a big saturated elementary extension and f an automorphism of $\mathbb{C}_N$. If f is Lascar strong, let $|f|$ be the smallest n such that f is the product of n automorphisms which fix elementary submodels. If f is not Lascar strong, write $|f| = \infty$.

We will make use of the following lemma, which can be proved from Lemma 24 (see [2] for details).

Lemma 27
1. *Every automorphism of $\mathbb{C}_N$ is the product of some σ_N^n and some f with $|f| \leq 2$.*

2. *$|\sigma_N^n| = |n| + 2$, whenever $0 < |n| \leq \frac{N}{2}$.*

Let $\mathcal{S}_\infty$ be the disjoint union of the $\mathcal{S}_1, \mathcal{S}_2, \ldots$ viewed as a many–sorted structure[12] with saturated extension $\mathbb{C}_\infty = (\mathbb{C}_1, \mathbb{C}_2, \ldots)$.

Theorem 28 ([2]) *For each N let C_N be the N–element cyclic group with generator c_N. Let B be the group of all sequences $(\mathrm{c}_N^{e_N})$ with a bounded sequence (e_N) of exponents. Then*

$$\mathrm{Gal}_\mathrm{L}(\mathbb{C}_\infty) \cong \prod_N \mathrm{C}_N / B.$$

$\mathrm{Gal}_\mathrm{L}(\mathbb{C}_\infty)$ carries the indiscrete topology.

PROOF: The map $(\mathrm{c}_N^{e_N}) \mapsto (\sigma_N^{e_N})$ defines a map from $\prod_N \mathrm{C}_N$ to $\mathrm{Aut}(\mathbb{C}_\infty)$, which yields a homomorphism

$$\mu : \prod_N \mathrm{C}_N \to \mathrm{Gal}_\mathrm{L}(\mathbb{C}_\infty).$$

Let (f_N) be any automorphism of $\mathbb{C}_\infty$. If we apply part 1 of the Lemma to each component we see that we can write (f_N) as a product of some $(\sigma_N^{e_N})$ and two automorphisms which fix a model. This shows that μ is surjective.

Let $(\mathrm{c}_N^{e_N})$ be an arbitrary element of $\prod_N \mathrm{C}_N$. We can assume that $|e_N| \leq \frac{N}{2}$. Then by part 2 of the lemma it is immediate that $(\sigma_N^{e_N})$ is Lascar strong iff (e_N) is bounded, which means that B is the kernel of μ.

It remains to show that the topology of $\mathrm{Gal}_\mathrm{L}(\mathbb{C}_\infty)$ is indiscrete, or

$$\mathrm{Gal}_\mathrm{L}(\mathbb{C}_\infty) = \Gamma_1(\mathbb{C}_\infty).$$

The preimage of $\Gamma_1(\mathbb{C}_\infty)$ in $\mathrm{Aut}(\mathbb{C}_\infty)$ is, by the next Lemma, a closed subgroup, which contains $\mathrm{Autf}_\mathrm{L}(\mathbb{C}_\infty)$. The automorphisms which fix almost every $\mathbb{C}_N$ are Lascar strong and form a dense subset of $\mathrm{Aut}(\mathbb{C}_\infty)$. Thus the preimage of $\Gamma_1(\mathbb{C}_\infty)$ is the whole $\mathrm{Aut}(\mathbb{C})$ group. □

---

[12]We take also the disjoint union of the languages.

We conclude with a general lemma. Let M be a model of T and consider the topology of *point–wise convergence* on $\mathrm{Aut}(M)$, with basic open sets

$$U_{a,b} = \{f \mid f(a) = b\},$$

where a, b are finite tuples from M.

Lemma 29 *The natural map* $\mathrm{Aut}(M) \to \mathrm{Gal_L}(T)$ *is continuous.*

PROOF: Let Ω be a neighborhood of the image of f in $\mathrm{Gal_L}(T)$. The preimage of Ω under[13] $\nu : \mathrm{S}_M(N) \to \mathrm{Gal_L}(T)$ contains a basic neighborhood

$$O = \{p \mid \varphi(x) \in p\}$$

of $\mathrm{tp}(f(M)/N)$. Let a be the tuple of elements of M which are enumerated by the free variables of φ. Then

$$O = \{\mathrm{tp}(g(M)/N) \mid \mathbb{C} \models \varphi(g(a))\} \subset \{\mathrm{tp}(g(M)/N) \mid g(a) = f(a)\}.$$

Whence $U_{a,f(a)}$ is a neighborhood of f which is mapped into Ω. □

References

[1] Steve Buechler, Anand Pillay, and Frank O. Wagner. Supersimple theories. *J. Amer. Math. Soc.*, 14:109–124, 2001.

[2] H. Casanovas, D. Lascar, A. Pillay, and M. Ziegler. Galois groups of first order theories. Preprint, 2000.

[3] Wilfried Hodges. *Model Theory*. Encyclopedia of Mathematics and its Applications. Cambridge University Press, 1993.

[4] Ehud Hrushovski. Simplicity and the lascar group. Preprint, 1997.

[5] Byunghan Kim. A note on lascar strong types in simple theories. *J. Symbolic Logic*, 63:926–936, 1998.

[6] Daniel Lascar. The category of models of a complete theory. *J. Symbolic Logic*, 47:249–266, 1982.

[7] Daniel Lascar and Anand Pillay. Hyperimaginaries and automorphism groups. *J. Symbolic Logic*, 66(1):127–143, 2001.

---

[13] N can be any small model.

[8] Anand Pillay. On groups and fields definable in o–minimal structures. *J. Pure Appl. Algebra*, 53:239–255, 1988.

[9] Bruno Poizat. Une théorie de galois imaginaire. *J. Symbolic Logic*, 48(4):1151–1170, December 1983.

[10] Lou van den Dries. A generalization of the Tarski–Seidenberg Theorem and some non–definability results. *Bull. Amer. Math. Soc. (N.S.)*, 15(2):189–193, October 1986.

[11] Frank O. Wagner. *Simple Theories*, volume 503 of *Mathematics and Its Applications*. Kluwer Academic Publishers, Dordrecht, NL, 2000.

[12] André Weil. *L'integration dans les groupes topologiques*. Actualités scientifiques et industrielles. Hermann, Paris, 2. edition, 1951.